Birkhäuser Advanced Texts Basler Lehrbücher

Series Editors
Steven G. Krantz, Washington University, St. Louis, USA
Shrawan Kumar, University of North Carolina at Chapel Hill, Chapel Hill, USA

This series presents, at an advanced level, introductions to some of the fields of current interest in mathematics. Starting with basic concepts, fundamental results and techniques are covered, and important applications and new developments discussed. The textbooks are suitable as an introduction for students and non–specialists, and they can also be used as background material for advanced courses and seminars.

Liviu C. Florescu

Selected Topics in Mathematical Analysis

Real Number System – Recurrences – Asymptotic Analysis – Integration in Finite Terms

Liviu C. Florescu
Faculty of Mathematics
"Al.I.Cuza" University
Iași, Romania

ISSN 1019-6242 ISSN 2296-4894 (electronic)
Birkhäuser Advanced Texts Basler Lehrbücher
ISBN 978-3-031-67783-0 ISBN 978-3-031-67784-7 (eBook)
https://doi.org/10.1007/978-3-031-67784-7

© The Editor(s) (if applicable) and The Author(s), under exclusive license to Springer Nature Switzerland AG 2024

This work is subject to copyright. All rights are solely and exclusively licensed by the Publisher, whether the whole or part of the material is concerned, specifically the rights of translation, reprinting, reuse of illustrations, recitation, broadcasting, reproduction on microfilms or in any other physical way, and transmission or information storage and retrieval, electronic adaptation, computer software, or by similar or dissimilar methodology now known or hereafter developed.
The use of general descriptive names, registered names, trademarks, service marks, etc. in this publication does not imply, even in the absence of a specific statement, that such names are exempt from the relevant protective laws and regulations and therefore free for general use.
The publisher, the authors and the editors are safe to assume that the advice and information in this book are believed to be true and accurate at the date of publication. Neither the publisher nor the authors or the editors give a warranty, expressed or implied, with respect to the material contained herein or for any errors or omissions that may have been made. The publisher remains neutral with regard to jurisdictional claims in published maps and institutional affiliations.

This book is published under the imprint Birkhäuser, www.birkhauser-science.com by the registered company Springer Nature Switzerland AG
The registered company address is: Gewerbestrasse 11, 6330 Cham, Switzerland

If disposing of this product, please recycle the paper.

Preface

The purpose of this book is to provide students and instructors with topics that are not generally covered in undergraduate programs. This endeavor aims to complement and deepen their knowledge and familiarize them with a variety of techniques and methods useful in their teaching activities. The content of the book was the subject of a course, "Special Chapters in Mathematical Analysis," taught by the author from 1998 to 2012 at "Al. I. Cuza" University of Iași. Four topics were selected to form the basis of the four chapters in the book. The choice of these topics was more subjective on the part of the author and constrained by the limited time available for the course, with many other topics that could have been included.

The first chapter aims to provide a detailed discussion of real numbers, the foundation of any mathematical construction. The first paragraph is dedicated to an axiomatic presentation of the set of real numbers, denoted as $\mathbb{R}$, defined as a Dedekind complete ordered field. Along with this axiomatic presentation, attention is drawn to certain meta-theoretical issues of this axiomatic theory, such as consistency and categoricity. Theorem 1.1.9 shows that any two Dedekind complete ordered fields are isomorphic. Consequently, the axiomatic system presented is categorical. From a syntactic perspective, an axiomatic system is consistent if it does not lead to contradictions. Semantically, this means that there exist models of the axiomatic system. Such a model of the real numbers is presented in the second paragraph—the decimal model. This model, although highly intuitive, is less explored in mathematical books due to the difficulties that arise in defining operations with infinite decimals. The second paragraph thoroughly addresses these difficulties by providing a comprehensive presentation of the decimal model. The third paragraph is dedicated to the study of normed fields and their completions. It sets the stage for the presentation of another model of real numbers, Cantor's model. This model is based on the process of completing the set of rational numbers with respect to an archimedean norm. The final subparagraph deals with the classification of norms on the set of rational numbers, $\mathbb{Q}$ (Ostrowski's theorem). It demonstrates that on $\mathbb{Q}$, apart from the trivial norm, there essentially exists a unique archimedean norm and the p-adic norms. The completion of $\mathbb{Q}$ with respect to the p-adic norms is presented in the fourth paragraph.

All models of the set of real numbers start with the assumption that we have somehow constructed the set of rational numbers. The construction of the latter set is followed in subsections two and three of the Appendix, starting from a model of the set of natural numbers, $\mathbb{N}$. However, does $\mathbb{N}$ have models? In the first subsection of the Appendix, the Peano axioms system is presented, along with several considerations regarding its consistency.

Chapter 2 of the book is dedicated to the study of sequences defined by recurrence relations. Linear recurrences have a comprehensive theory, the presentation of which is the subject of the first paragraph of the chapter. The general theory of homogeneous recurrences is presented, and as a remarkable particular case, the Fibonacci sequence is discussed. The last subsection is dedicated to non-homogeneous linear recurrences. Non-linear recurrences form a vast class for which there are no general methods. The second paragraph only deals with some non-linear recurrences. Contraction mappings, Newton's method for approximating the roots of a function, and Heron's method for square root extraction are discussed here. The last three subsections are dedicated to double non-linear recurrences. Special attention is given to the arithmetic-geometric mean algorithm and its role in the study of elliptic functions. Gauss and Legendre's proofs for determining the limit of this algorithm are presented, along with applications for the approximate calculation of the lemniscate and ellipse perimeters.

Chapter 3 explores certain problems in asymptotic analysis. It introduces Bachmann-Landau symbols and asymptotic series. The second paragraph of the chapter is dedicated to Laplace's method regarding the asymptotic behavior of generalized integrals. A remarkable particular application of Laplace's method is the Stirling formula. The final paragraph of the chapter is focused on Poincaré's theorem concerning the asymptotic behavior of solutions to certain linear recurrence relations, as well as its consequences.

The final chapter of the book discusses mathematical results related to "Integration in Finite Terms." The problem addressed here is the expression of the primitives of functions in terms of elementary functions. The central result of the chapter is Liouville's principle, the theorem that forms the basis of the Risch algorithm, which in turn is the foundation of the algorithms implemented in computer algebra systems; the most popular of such systems are "Mathematica" and "Maple." Field extensions are presented, with special attention given to Liouville fields. This allows for a precise definition of elementary functions. The second paragraph demonstrates Liouville's principle. The final paragraph utilizes this principle to derive integration algorithms in the specific case of polynomials with transcendental functions as variables.

The first chapter of the book is purely theoretical: the construction of the real number—the decimal model, a model that is not fully developed anywhere. This chapter is not suitable for exercises. In the other three chapters, there are several exercises presented under the heading of "Examples" illustrating the various situations that arise in the text. These can be successfully used as oral or written exam questions. In addition to these, at the end of Chaps. 2, 3, and 4, several exercises are added with brief solutions.

Each chapter of the book is accompanied by its respective bibliography.

The book is intended for readers with a level of maturity typically attained after completing a Bachelor of Science degree in mathematics or applied mathematics. To go through the four chapters, one needs mathematical knowledge acquired during undergraduate courses at any mathematics faculty. The specific courses considered include "Differential and Integral Calculus," "Linear Algebra," and "Complex Functions."

The book may be useful to a broader audience interested in the topics covered in the four chapters mentioned above.

Iaşi, Romania   Liviu C. Florescu
June 19, 2024

Contents

1 **Real Number System** 1
   1.1 An Axiomatic Definition of the Real Numbers 2
   1.2 Decimal Model 10
       1.2.1 Finite Decimals 11
       1.2.2 Infinite Decimals 13
   1.3 Absolute Values 24
       1.3.1 Non-Archimedean Norms 27
       1.3.2 Completing a Normed Field 29
       1.3.3 Classification of Norms on $\mathbb{Q}$ 32
   1.4 p-adic Numbers 35
       1.4.1 p-adic Integers 35
   1.5 Appendix 45
       1.5.1 Peano's Axioms 45
       1.5.2 The Set of Integers 53
       1.5.3 The Set of Rational Numbers 57
   References 60

2 **Recurrences** 63
   2.1 Linear Recurrences 64
       2.1.1 Difference Calculus 65
       2.1.2 Homogeneous Linear Recurrences 67
       2.1.3 Equations with Constant Coefficients 72
       2.1.4 Fibonacci Sequence 79
       2.1.5 Non-homogeneous Linear Equations 81
   2.2 Nonlinear Recurrences 88
       2.2.1 Linearizable Nonlinear Recurrences 88
       2.2.2 Contractions 90
       2.2.3 Newton's Method 91
       2.2.4 Heron's Method 96
       2.2.5 Double Algorithms Generated by Means 98

		2.2.6	The Arithmetic-Geometric Mean	104
		2.2.7	The Perimeter of the Lemniscate and the Ellipse	114
	2.3	Exercises		124
	References			126

3 Elements of Asymptotic Analysis ... 127
- 3.1 Asymptotic Sequences and Series ... 128
  - 3.1.1 The Bachmann–Landau Symbols ... 128
  - 3.1.2 Asymptotic Series ... 131
  - 3.1.3 Operations with Asymptotic Series ... 137
- 3.2 Laplace's Method ... 142
  - 3.2.1 Integrals of the Type $\int_\alpha^\beta e^{th(x)} dx$ ... 147
  - 3.2.2 Integrals of the Type $\int_0^{+\infty} e^{-tx} x^\lambda g(x) dx$ ... 152
  - 3.2.3 Integrals of the Type $\int_\alpha^\beta e^{-tx^2} h(x) dx$ ... 156
  - 3.2.4 Stirling's Formula ... 158
- 3.3 Poincaré's Theorem ... 162
- 3.4 Exercises ... 171
- References ... 172

4 Integration in Finite Terms ... 173
- 4.1 Field Extensions ... 176
  - 4.1.1 Liouville Fields ... 179
  - 4.1.2 Exponential and Logarithmic Functions ... 183
  - 4.1.3 Elementary Functions ... 185
- 4.2 Liouville's Principle ... 187
- 4.3 Particular Cases ... 198
  - 4.3.1 Integrating Expressions of the Form $A_0 \theta + A_1$ ... 198
  - 4.3.2 Integrating a Polynomial in θ ... 205
- 4.4 Exercises ... 212
- References ... 214

Index ... 215

Chapter 1
Real Number System

The system of real numbers is one of the important pillars supporting the edifice of mathematics. The determining role it plays justifies the attention it receives from those concerned with the foundations of mathematics.

The real number notion has a historical character; this notion has been clarified throughout the millennial cultural evolution of mankind. The first attempt at abstract presentation appears as early as Eudoxus of Cnidos (c. 408–c. 355 BC) for whom the number is the ratio of two quantities. However, it took more than 2200 years to arrive at the exact description of what we call today the set of real numbers. Around 1870 the first rigorous constructions of real numbers appear. They belong to the German mathematicians Karl Weierstrass (1815–1897), Georg Cantor (1845–1918), and Richard Dedekind (1831–1916). All these models start from the assumption that the set of rational numbers is well-constructed. The construction of the set of rational numbers uses the set of integers and it, in turn, is based on a model of natural numbers. But is there a model of the natural numbers? In the appendix, we will sketch constructions of models for the sets of integers and rational numbers starting from the assumption that there is a model of the set of natural numbers. The issue of consistency will also be discussed there. It is clear that the theory of real numbers will be consistent once that of natural numbers is. A brief discussion of the consistency problem will also be treated in the appendix.

In the first section of this chapter, we will present the axiomatic theory of real numbers. Almost everywhere in the literature, two models of this axiomatic theory appear: the geometric construction of Dedekind based on so-called cuts in the set of rational numbers and the construction of Cantor based on Cauchy sequences. Although much more intuitive, the decimal model is less present in the specialized literature. This is probably due to technical difficulties that arise in defining the operations of addition and multiplication. We will try in the second section to overcome these difficulties. It is well known that any two models of the axiomatic theory of real numbers are isomorphic (the axiomatic system is categorical) and therefore it does not matter in which of them we work. In Sect. 1.3 we will deal with the absolute value on a field. We will show that any field can be completed with respect to an absolute value. A specific example of this is Cantor's model of completing the set of rational numbers $\mathbb{Q}$ with respect to the usual absolute value. We will demonstrate Ostrowski's theorem on the classification of absolute values on $\mathbb{Q}$: every non-trivial absolute value on the rational numbers $\mathbb{Q}$ is equivalent to either the usual real absolute value or a p-adic absolute value. In the last section of the chapter, we will construct the completion of $\mathbb{Q}$ with respect to a p-adic norm; this is a non-Archimedean model of $\mathbb{R}$.

1.1 An Axiomatic Definition of the Real Numbers

The first axiomatic presentation of the set of real numbers was made by David Hilbert in 1899. In what follows, we will consider the set of real numbers to be a Dedekind complete ordered field. We will further elaborate on this concise presentation in the following definition.

Definition 1.1.1 Let K be a set with at least two elements equipped with two binary operations $+, \cdot : K \times K \to K$ (addition and multiplication) and a relation $\leqslant \subseteq K \times K$ (relation of ordering) that satisfies the following axioms:

(A1) $x + (y + z) = (x + y) + z$, for every $x, y, z \in K$.
(A2) $x + y = y + x$, for every $x, y \in K$.
(A3) There exists $0 \in \mathbb{R}$ such that $x + 0 = x$, for every $x \in K$.
(A4) For every $x \in K$, there exists $-x \in K$ such that $x + (-x) = 0$.
(A5) $x \cdot (y \cdot z) = (x \cdot y) \cdot z$, for every $x, y, z \in K$.
(A6) $x \cdot y = y \cdot x$, for every $x, y \in K$.
(A7) There exists $1 \in K$ such that $x \cdot 1 = x$, for every $x \in K$.
(A8) For every $x \in K \setminus \{0\}$, there exists $x^{-1} \in K$ such that $x \cdot x^{-1} = 1$.
(A9) $x \cdot (y + z) = x \cdot y + x \cdot z$, for every $x, y, z \in K$.
(A10) $x \leqslant x$, for every $x \in K$.
(A11) $x \leqslant y$ and $y \leqslant x \Rightarrow x = y$.
(A12) $x \leqslant y$ and $y \leqslant z \Rightarrow x \leqslant z$.
(A13) For every $x, y \in \mathbb{R}$ we have $x \leqslant y$ or $y \leqslant x$.
(A14) $x \leqslant y \Rightarrow x + z \leqslant y + z$, for every $z \in K$.
(A15) $x \leqslant y \Rightarrow x \cdot z \leqslant y \cdot z$, for every $z \in K, 0 \leqslant z$.
(A16) For every $\emptyset \neq A \subseteq K$, A bounded above, there exists $\sup A \in K$.

Remarks 1.1.2

(i) The first four axioms state that $(K, +)$ is a **commutative group**. The first nine axioms state that $(K, +, \cdot)$ is a **field**. The first fifteen axioms state that $(K, +, \cdot, \leqslant)$ is an **ordered field**. Finally, the sixteenth axiom states that $(K, \leqslant)$ is **Dedekind complete**.

1.1 An Axiomatic Definition of the Real Numbers

(ii) 0 and 1 are the neutral elements for addition and multiplication respectively. $-x$ is called the opposite of x, and x^{-1} is called the inverse of x, with respect to the operations of addition and multiplication defined on K.

(iii) Axiom $(A1)$ and $(A5)$ state that the operations of addition and multiplication are associative. Axiom $(A2)$ and $(A6)$ state that the operations of addition and multiplication are commutative. Axiom $(A9)$ states that multiplication is distributive over addition. Finally, axioms $(A10)$–$(A12)$ state that the relation $\leqslant$ is an order relation, and $(A13)$ state that it is a total order on $\mathbb{R}$. Axiom $(A14)$ and $(A15)$ state that the relation of ordering is compatible with the operations of addition and multiplication, respectively.

(iv) A set $A \subseteq (K, \leqslant)$ is said to be bounded above (upper-bounded) if there exists $M \in K$ such that for all elements x in A, $x \leqslant M$. Axiom $(A16)$ states that for any non-empty set A of K that is upper-bounded, there exists a least upper bound in K, meaning there exists a real number $L = \sup A \in K$ such that for all elements x in A, $x \leqslant L$ and for any real number $y \in K$ such that $x \leqslant y$, for all $x \in A$, it follows that $L \leqslant y$. In other words, L is the smallest real number that is an upper bound for A.

(v) In relation to the above axioms, two natural questions arise: Are there any sets that satisfy the axioms $(A1)$–$(A16)$? Therefore, is the axiomatic theory consistent? And is it categorical? In other words, are any two models isomorphic?

The answer to the first question is yes, there are sets that satisfy the axioms $(A1)$–$(A16)$; some of the most famous models of complete ordered fields can be defined in terms of Cauchy sequences or Dedekind cuts (see [14]). In the following paragraph, we will construct the decimal model of real numbers. However, we must make the observation that all known models start from the assumption that there is a model of the set of rational numbers. This fact leads to the idea that the axiomatic system for K is consistent as long as there are models of the set of rational numbers. A detailed discussion of consistency will be made in Appendix.

The answer to the second question is that this axiomatic theory is categorical, meaning that every two different models are isomorphic. This result will be proved in Theorem 1.1.9 in this paragraph.

Definition 1.1.3 From the above it follows that there is a "single" set that satisfies the axioms $(A1)$–$(A16)$ ("single" means up to an isomorphism). We will call this set the **set of real numbers** and, in what follows, we will denote it with $K = \mathbb{R}$.

We define the strict order relation by $x < y$ if $x \leqslant y$ and $x \neq y$.

We will denote $\mathbb{R}_+ = \{x \in \mathbb{R} : 0 \leqslant x\}$ as the set of all positive numbers, $\mathbb{R}_- = \{x \in \mathbb{R} : x \leqslant 0\}$ as the set of all negative numbers, $\mathbb{R}^* = \mathbb{R} \setminus \{0\}$, and $\mathbb{R}_+^* = \{x \in \mathbb{R} : 0 < x\}$ as the set of strictly positive numbers. We will also use the notations $x \geqslant y$ or $x > y$ instead of $y \leqslant x$ or $y < x$ respectively.

Also we will use the simplified notation $x - y$ instead of $x + (-y)$ and, when the context is clear, xy instead of $x \cdot y$.

Consequences 1.1.4 In the following, we present some immediate consequences of the axioms $(A1)$–$(A16)$. Although we present these consequences for the set of real numbers, they hold for any Dedekind complete ordered field K. Moreover, as we will note in Theorem 1.1.9, any Dedekind complete ordered field is isomorphic to $\mathbb{R}$.

(a) For every $x \in \mathbb{R}$, the opposite $-x$ is unique and, for every $x \in \mathbb{R}^*$, the inverse x^{-1} is unique. Indeed, if we suppose that $x + y = 0 = x + z$, then $y = y + 0 = y + (x + z) = (y + x) + z = (x + y) + z = 0 + z = z + 0 = z$. A similar demonstration for the inverse.

(b) $-(-x) = x$, for every $x \in \mathbb{R}$, and, for every $x \in \mathbb{R}^*$, $\left(x^{-1}\right)^{-1} = x$. Indeed, $x = x + 0 = x + [-x + (-(-x))] = [x + (-x)] + (-(-x)) = 0 + (-(-x)) = -(-x)$. A similar demonstration for the inverse.

(c) $-(x + y) = (-x) + (-y)$, for every $x, y \in \mathbb{R}$, and, for every $x, y \in \mathbb{R}^*$, $(xy)^{-1} = x^{-1} y^{-1}$. Indeed $(x+y) + [(-x) + (-y)] = [x + (-x)] + [y + (-y)] = 0 + 0 = 0$. Because the opposite of $x + y$ is unique, $(-x) + (-y) = -(x + y)$. A similar proof for the inverse.

(d) $x \cdot 0 = 0$, for every $x \in \mathbb{R}$. Indeed, $x \cdot 0 = x \cdot (0 + 0) = x \cdot 0 + x \cdot 0$. If we add $-(x \cdot 0)$ in both sides of the last equation we obtain $0 = x \cdot 0$.

(e) $-(xy) = (-x)y = x(-y)$, and $(-x)(-y) = xy$, for every $x, y \in \mathbb{R}$. Indeed, $xy + (-x)y = [x + (-x)]y = 0 \cdot y = 0$ from where $(-x)y = -(xy)$. Then $x(-y) = (-y)x = -(yx) = -(xy)$ and $(-x)(-y) = -[x(-y)] = -[-(xy)] = xy$.

(f) $0 < 1$. To demonstrate this inequality, we first observe that $0 \neq 1$, because if we were to assume that $0 = 1$, then we would have: $x = x \cdot 1 = x \cdot 0 = 0$, for all $x \in \mathbb{R}$, and this contradicts the assumption that $\mathbb{R}$ has at least two elements. From $(A13)$ we have $0 < 1$ or $1 < 0$. If we suppose that $1 < 0$ then $0 < -1$. Multiplying both sides of the last inequality by -1 and using axiom $(A15)$ we obtain $0 < (-1)(-1) = 1$, and this is absurd. An immediate consequence of the above inequality is: $-1 < 0$.

(g) We say that a subset of $I \subseteq \mathbb{R}$ is **inductive** if it satisfies the conditions:
$$\begin{cases} (1)\ 0 \in I \\ (2)\ x \in I \Rightarrow x + 1 \in I \end{cases}.$$ We observe that there exist inductive subsets of $\mathbb{R}$: $\mathbb{R}$ itself being inductive, $\mathbb{R}_+$ is also inductive. Let $\mathcal{I}$ be the class of inductive subsets of $\mathbb{R}$, and let $\mathbb{N} = \bigcap_{I \in \mathcal{I}} I$. It can be easily verified that $\mathbb{N} \in \mathcal{I}$, so $\mathbb{N}$ is the smallest inductive subset of $\mathbb{R}$; particularly $\mathbb{N} \subseteq \mathbb{R}_+$. We call $\mathbb{N}$ the set of **natural numbers**. We remark that $0 \in \mathbb{N}$ and then $1 \in \mathbb{N}$, $2 = 1 + 1 \in \mathbb{N}$, and so on. Any natural number n has a unique successor $n' = n + 1 \in \mathbb{N}$. 0 is not the successor of any natural number (if we suppose that there is $x \in \mathbb{N}$ such that $0 = x' = x + 1$, then $x = -1 \notin \mathbb{R}_+ \supseteq \mathbb{N}$). If $n' = m'$, then $n = m$. If $M \subseteq \mathbb{N}$ is such that $0 \in M$ and, for all $n \in M$, $n' = n + 1 \in M$, then $M \in \mathcal{I}$ and thus $M = \mathbb{N}$. These remarks confirm that $\mathbb{N}$ satisfies Peano's axioms (see Appendix) and it is reasonable to call it the set of natural numbers. We also observe that a natural number and its successor are consecutive numbers (there is no other natural number between them). Indeed, for any natural number

1.1 An Axiomatic Definition of the Real Numbers

$n \in \mathbb{N}$, let $A = \{0, 1, \cdots n\} \cup \{x \in \mathbb{R} : x \geq n+1\}$. Then $A \in \mathcal{I}$ and thus $\mathbb{N} \subseteq A$. It follows that, for any $k \in \mathbb{N}, k \leq n$ or $k \geq n+1$.

Let's denote by $\mathbb{Z} = \{k \in \mathbb{R} : k \in \mathbb{N} \text{ or } -k \in \mathbb{N}\}$ the set of **integers**; we remark that, for every $k, l \in \mathbb{Z}, k < l+1 \Rightarrow k \leq l$. Let $\mathbb{Q} = \{p \cdot q^{-1} \in \mathbb{R} : p \in \mathbb{Z}, q \in \mathbb{N}^*\}$, where $\mathbb{N}^* = \mathbb{N} \setminus \{0\}$; $\mathbb{Q}$ is the set of **rational numbers** of $\mathbb{R}$. Obviously $\mathbb{N} \subseteq \mathbb{Z} \subseteq \mathbb{Q} \subseteq \mathbb{R}$.

(h) **Archimedean property**: For every $x, y \in \mathbb{R}, x > 0$, there exists $n \in \mathbb{N}$ such that $nx > y$. Indeed, if we suppose that $n \leq x^{-1}y$, for all $n \in \mathbb{N}$, then $\mathbb{N}$ is a non-empty subset of $\mathbb{R}$ that is bounded above. From $(A16)$ there exists $x_0 = \sup \mathbb{N}$. Let now $n_0 \in \mathbb{N}$ such that $x_0 - 1 < n_0$ or $x_0 < n_0 + 1$. Since $n_0 + 1$ is a natural number, the last relationship contradicts the quality of x_0 as an upper bound of $\mathbb{N}$. In particular, this property states that the set of natural numbers is not bounded above. Clearly, the set of integers is not bounded either above or below.

(i) $\mathbb{R}$ is an **integral domain**: $x \cdot y = 0 \Longrightarrow x = 0$ or $y = 0$.
If we suppose that $x \neq 0$, then there is $x^{-1} \in \mathbb{R}$ and so $x^{-1} \cdot x \cdot y = x^{-1} \cdot 0 = 0$, implying that $y = 0$.

(j) For every $A \subseteq \mathbb{R}, A \neq \emptyset$, A bounded below, there exists $\inf A \in \mathbb{R}$. If A is bounded below, then the set $-A = \{-x : x \in A\}$ is a non-empty, bounded above set. Let $x_0 = \sup(-A) \in \mathbb{R}$. It is easily verified that $-x_0 \in \mathbb{R}$ is the greatest lower bound of A.

(k) For every $x \in \mathbb{R}$, there exists $k_0 \in \mathbb{Z}$ such that $k_0 \leq x < k_0 + 1$; k_0 is the **integer part** of x (the largest integer not greater than x), and is denoted by $\lfloor x \rfloor$.
Let $A = \{k \in \mathbb{Z} : k \leq x\}$. Then A is a non-empty, bounded above set. Let $x_0 = \sup A$ and let $k_0 \in A$ such that $x_0 - 1 < k_0 \leq x$. For every $k \in A, k \leq x_0 < k_0 + 1$ and so $k \leq k_0$; therefore k_0 is the maximum of A and so $k_0 = x_0$. Obviously $k_0 \leq x < k_0 + 1$.

(l) **Well-ordering principle**: For every $M \subseteq \mathbb{N}, M \neq \emptyset$, there exists $\min M \in M$. Because M is non-empty and bounded below, there exists $x_0 = \inf M$. Let $x \in M$ such that $x < x_0 + 1$; then $x - 1 < x_0 \leq x$. For every $y \in M, x - 1 < x_0 \leq y$, so that $x \leq y$. Therefore x is the minimum of M and so $x_0 = x$.

(m) There is $x_0 \in \mathbb{R}_+$ such that $x_0^2 = 2$; $x_0 = \sqrt{2}$ is the square root of 2. We consider the set $A = \{x \in \mathbb{R}_+ : x^2 < 2\}$. A is a non-empty set bounded above: for every $x \in A, x^2 < 2 < 4 \Longrightarrow (x-2)(x+2) < 0 \Longrightarrow x < 2$. Let $x_0 = \sup A$; because $1 \in A, 1 \leq x_0$.
If we suppose that $x_0^2 < 2$ then
(*) there exists h such that $0 < h < 1$, and $(x_0 + h)^2 < 2$.
Indeed $(x_0 + h)^2 = x_0^2 + 2x_0 h + h^2 < x_0^2 + (2x_0 + 1)h < 2$, inequality that is true if we choose $0 < h < \min\left(1, \frac{2-x_0^2}{2x_0+1}\right)$. But from (*) it turns out that $x_0 + h \in A$ and therefore x_0 is not an upper bound for A.
If we suppose that $x_0^2 > 2$ then
(**) there exists h such that $0 < h < 1$, and $(x_0 - h)^2 > 2$.

Indeed $(x_0 - h)^2 = x_0^2 - 2x_0 h + h^2 > x_0^2 - 2x_0 h > 2$, inequality that is true if we choose $0 < h < \min\left(1, \frac{x_0^2 - 2}{2x_0}\right)$. But from $(**)$ it turns out that $x_0 - h$ is an upper bound for A and therefore x_0 is not the least upper bound for A. Therefore $x_0^2 = 2$.

Before continuing with the presentation of the consequences of axioms $(A1)$–$(A16)$, we will introduce the concepts of convergent sequence and Cauchy sequence.

The sequence $(x_n)_{n \in \mathbb{N}} \subseteq \mathbb{R}$ **converges to** $x \in \mathbb{R}$ if for every $\varepsilon \in \mathbb{R}_+^*$ there exists $n_0 \in \mathbb{N}$ such that $x - \varepsilon < x_n < x + \varepsilon$, for every $n \geqslant n_0$. We denote this fact by $x_n \to x$. x is called the **limit** of the sequence $(x_n)_{n \in \mathbb{N}}$. Let $(k_n)_{n \in \mathbb{N}}$ be a strictly increasing sequence of natural numbers ($k_n < k_{n+1}, \forall n \in \mathbb{N}$); the sequence $(x_{k_n})_{n \in \mathbb{N}}$ is called a **subsequence** of the sequence $(x_n)_{n \in \mathbb{N}}$. If $(x_n)_n$ converges to x, then any subsequence of it converges to x. A sequence $(x_n)_{n \in \mathbb{N}}$ is **convergent** if there exists $x \in \mathbb{R}$ such that $(x_n)_{n \in \mathbb{N}}$ converges to x. It can be easily shown that if a sequence is convergent, its limit is unique. For example, using the Archimedean property, it immediately follows that the sequence $(\frac{1}{n})_{n \in \mathbb{N}^*}$ converges to 0. $(x_n)_{n \in \mathbb{N}} \subseteq \mathbb{R}$ is a **Cauchy sequence** if for every $\varepsilon \in \mathbb{R}_+^*$, there exists $n_0 \in \mathbb{N}$ such that $-\varepsilon < x_m - x_n < \varepsilon$, for every $m, n \geqslant n_0$. It can be easily shown that any convergent sequence is a Cauchy sequence. At point **(p)** below, we will show that the converse of this statement is also true.

(n) $\mathbb{Q}$ is dense in $\mathbb{R}$.

(1) For every $x, y \in \mathbb{R}$ with $x < y$, there is $r \in \mathbb{Q}$ such that $x < r < y$.
From **(h)**, there exists $n \in \mathbb{N}^*$ such that $n > (y - x)^{-1}$ or $ny - nx > 1$. Let $m = \lfloor nx \rfloor + 1 \in \mathbb{Z}$ (see **(k)**). Then $nx < m \leqslant nx + 1 < ny$ and so $x < m \cdot n^{-1} < y$.

(2) For every $x \in \mathbb{R}$, $x = \sup\{r \in \mathbb{Q} : r < x\}$.
Let $A = \{r \in \mathbb{Q} : r \leqslant x\}$ From **(h)**, there exists $n \in \mathbb{N}$ such that $n > -x$ or $-n < x$ so that $-n \in A$. Hence A is a non-empty set. Because A is bounded above (x is an upper bound), there exists $y = \sup A$; obviously $y \leqslant x$. If we suppose that $y < x$, then, from (1), there exists $r \in \mathbb{Q}$ such that $y < r < x$. Then $r \in A$ and this contradicts the fact that y is an upper bound of A. It follows that $y = x$.

(3) For every $x \in \mathbb{R}$, $x = \inf\{r \in \mathbb{Q} : x \leqslant r\}$.

(4) For every $x \in \mathbb{R}$, there exists $(r_n)_{n \in \mathbb{N}} \subseteq \mathbb{Q}$ such that $r_n \to x$.
Indeed, for every $n \in \mathbb{N}^*$, there is $r_n \in \mathbb{Q}$ such that $x - \frac{1}{n} < r_n < x + \frac{1}{n}$ (see above point 1). Obviously $r_n \to x$.

(o) Any bounded sequence has convergent subsequences.
Let $(x_n)_{n \in \mathbb{N}} \subseteq \mathbb{R}$ be a bounded sequence and let then $a, b \in \mathbb{R}$ such that $a \leqslant x_n \leqslant b, \forall n \in \mathbb{N}$, or $(x_n)_{n \in \mathbb{N}} \subseteq [a, b] = \{x \in \mathbb{R} : a \leqslant x \leqslant b\}$. At least one of the two semi-intervals of $[a, b]$ will contain an infinity of terms of the sequence $(x_n)_{n \in \mathbb{N}}$. Thus, there exists an infinite set of natural numbers $N_1 \subseteq \mathbb{N}$

1.1 An Axiomatic Definition of the Real Numbers

such that (1) $a \leqslant x_n \leqslant \frac{a+b}{2}$, for every $n \in N_1$ or (2) $\frac{a+b}{2} \leqslant x_n \leqslant b$, for every $n \in N_1$. In case (1) we note $a_1 = a, b_1 = \frac{a+b}{2}$ and in case (2) we note $a_1 = \frac{a+b}{2}, b_1 = b$. Thus we obtain two real numbers a_1, b_1 such that $a \leqslant a_1 \leqslant x_n \leqslant b_1 \leqslant b, \forall n \in N_1 \subseteq \mathbb{N}$ and $b_1 - a_1 = \frac{b-a}{2}$. Reasoning inductively, at the step p we obtain an infinite set $N_p \subseteq N_{p-1} \subseteq \cdots \subseteq N_1 \subseteq \mathbb{N}$ and two numbers a_p, b_p such that $a \leqslant a_1 \leqslant \cdots \leqslant a_p \leqslant x_n \leqslant b_p \leqslant \cdots \leqslant b_1 \leqslant b, \forall n \in N_p$ and $b_p - a_p = \frac{b-a}{2^p}$ and so on. We choose now an arbitrary element of $(x_n)_{n \in \mathbb{N}}$, $x_{k_0} \in [a, b]$; there is a $k_1 \in N_1, k_1 > k_0$, such that $x_{k_1} \in [a_1, b_1]$ (N_1 is infinite). At the step p, we choose $k_p \in N_p, k_p > k_{p-1}$, such that $x_{k_p} \in [a_p, b_p]$ (N_p is an infinite) and so on. We remark that $\{a_p : p \in \mathbb{N}^*\}$ has b as upper bound and $\{b_p : p \in \mathbb{N}^*\}$ has a as lower bound. Therefore there are $\alpha = \sup_p a_p, \beta = \inf_p b_p, a_p \leqslant \alpha \leqslant \beta \leqslant b_p$, and $\beta - \alpha \leqslant b_p - a_p = \frac{b-a}{2^p}$, for every $p \in \mathbb{N}^*$. Since $b_p - a_p \to 0$, $\alpha = \beta$ and then $a_p \to \alpha$ and $\beta_p \to \alpha$. As $a_p \leqslant x_{k_p} \leqslant b_p$, for every $p \in \mathbb{N}^*$, $x_{k_p} \to \alpha$ and then $(x_{k_p})_{p \in \mathbb{N}}$ is a convergent subsequence of $(x_n)_{n \in \mathbb{N}}$.

(p) Every Cauchy sequence is convergent in $\mathbb{R}$.
Let $(x_n)_{n \in \mathbb{N}} \subseteq \mathbb{R}$ be a Cauchy sequence. Then there exists $n_0 \in \mathbb{N}$ such that $-1 < x_m - x_n < 1$, for every $m, n \geqslant n_0$, so that $x_{n_0} - 1 < x_m < x_{n_0} + 1$, for every $m \geqslant n_0$. Let $a = \min\{x_0, \cdots, x_{n_0-1}, x_{n_0} - 1\}$ and $b = \max\{x_0, \cdots, x_{n_0-1}, x_{n_0} + 1\}$; then $a \leqslant x_n \leqslant b$, for every $n \in \mathbb{N}$ so that $(x_n)_{n \in \mathbb{N}}$ is a bounded sequence. Using the consequence (o), $(x_n)_{n \in \mathbb{N}}$ has a subsequence, $(x_{k_n})_{n \in \mathbb{N}}$ convergent to a number x. Then, for every $\varepsilon > 0$ there exist $n_1 \in \mathbb{N}$ such that, for every $n \geqslant n_1$, $-\frac{\varepsilon}{2} < x_n - x_{k_n} < \frac{\varepsilon}{2}$, and $x - \frac{\varepsilon}{2} < x_{k_n} < x + \frac{\varepsilon}{2}$. Adding the last two double inequalities, we obtain $x - \varepsilon < x_n < x + \varepsilon$, for every $n \geqslant n_1$, which means that $x_n \to x$.

Remark 1.1.5 The last consequence ensures that $\mathbb{R}$ is Cauchy complete. Thus any Dedekind complete ordered field is Cauchy complete. It can be shown that, within the context of ordered fields, Dedekind completeness is equivalent to Cauchy completeness plus the Archimedean property. There are Cauchy complete ordered fields that are not Dedekind complete. The field of rational functions in one variable is a classic example of such of ordered field. It is Cauchy complete but not Dedekind complete because it does not satisfy the Archimedean property (see Example 7, page 17 in [6]).

We will next define the prime field of a field and mention the role it plays in the case of Dedekind complete ordered fields.

Definition 1.1.6 Let $(K, +, \cdot)$ be a field with at least two elements and let $\underline{0}$ be the additive identity and $\underline{1} \neq \underline{0}$ be the multiplicative identity. If there is no $n \in \mathbb{N}^*$ such that $n \cdot \underline{1} = \underbrace{\underline{1} + \underline{1} + \cdots + \underline{1}}_{n \text{ times}} = \underline{0}$, then K is said to have the **characteristic** 0. Otherwise, the **characteristic** of K is the smallest positive integer $n \in \mathbb{N}^*$ such that $n \cdot \underline{1} = \underline{0}$; obviously $n > 1$. We denote by $Ch(K)$ the characteristic of K. For example the characteristic of $\mathbb{Q}$, the field of rational numbers, is 0; indeed, for every

$n \in \mathbb{N}^*$, $n \cdot 1 > 0$. If p is a prime number and $\mathbb{Z}$ is the ring of integers then the field of integers modulo p ($\mathbb{F}_p = \mathbb{Z}/p\mathbb{Z} = \{\bar{0}, \bar{1}, \cdots, \overline{p-1}\}$) has the characteristic p.

Any intersection of subfields of K is still a subfield. Then $K_0 = \bigcap \{C : C \text{ is a subfield of } K\}$ is the smallest subfield of K. K_0 has no proper subfields; it is called the **prime** field of K.

The prime field of a field K can be characterized using the characteristic of K.

Theorem 1.1.7 *For every field K, $Ch(K) = 0$ or $Ch(K)$ is a prime number. If $Ch(K) = 0$, then the prime field of K is isomorphic to the field of rational numbers. If $Ch(K) = p$, then the prime field is isomorphic to the field of integers modulo p.*

Proof Let $(K, +, \cdot)$ be a field and let $\underline{0}$ be the additive identity and $\underline{1}$ be the multiplicative identity. If $Ch(K) \neq 0$, there exists $n \in \mathbb{N}^*$ such that $n \cdot \underline{1} = \underline{0}$. Let $n = p_1^{a_1} \cdots p_r^{a_r}$ be the prime factorization of n. Since K is an integral domain, there exists $p \in \{p_1, \cdots, p_r\}$ such that $p \cdot \underline{1} = \underline{0}$. For any other prime number q, there exist $k, l \in \mathbb{Z}$ such that $kp + lq = 1$.[1] Then $lq \cdot \underline{1} = kp \cdot \underline{1} + lq \cdot \underline{1} = (kp + lq) \cdot \underline{1} = 1 \cdot \underline{1} = \underline{1}$ and so, $q \cdot \underline{1} \neq \underline{0}$.

Therefore $Ch(K) = p$.

Let now K_0 be the prime field of K; $\underline{0}, \underline{1} \in K_0$. We define $0 \cdot \underline{1} = \underline{0}$ and, $\forall k \in \mathbb{Z} \setminus \mathbb{N}, k \cdot \underline{1} = -((-k) \cdot \underline{1})$ (the opposite of $(-k) \cdot \underline{1}$ in K). We remark that $R_K = \{k \cdot \underline{1} : k \in \mathbb{Z}\}$ is a subring of K_0, and so it is an integral domain. Let $F_K \subseteq K_0$ be the field of fractions of R_K, and since K_0 is the smallest subfield of K, $F_K = K_0$.

If $Ch(K) = 0$, then R_K is isomorphic to the ring of integers $\mathbb{Z}$, and then $K_0 = F_K$ is isomorphic with $\mathbb{Q}$. If $Ch(K) = p$ is a prime number, then R_K is isomorphic with $\mathbb{Z}/p\mathbb{Z}$ (the ring of integers modulo p) which is a field. Therefore $F_K = R_K = K_0$, and then K_0 is isomorphic with the field of integers modulo p, $\mathbb{Z}/p\mathbb{Z} = \{\bar{0}, \bar{1}, \cdots, \overline{p-1}\}$. □

Remark 1.1.8 An important example of a field of characteristic 0 is provided by ordered fields (see (i) of Remark 1.1.2). Indeed, in such a field K, $\underline{0} < \underline{1}$ (for a proof see **(f)** in Remark 1.1.4) and then $\underline{0} < n \cdot \underline{1}$, $\forall n \in \mathbb{N}^*$. It follows that the characteristic of K is 0 and then the prime field of K is isomorphic to $\mathbb{Q}$. We will now show that, up to an isomorphism, there exists only one Dedekind complete ordered field.

Theorem 1.1.9 *Any two Dedekind complete ordered fields are isomorphic.*

Proof A Dedekind complete ordered field K satisfies the axioms $(A1) - (A16)$; as such it has all the properties listed in Remark 1.1.4. Its prime field is isomorphic to $\mathbb{Q}$ and is dense in K (see **(n)** of Remark 1.1.4).

---

[1] Let $M = \{xp + yq : x, y \in \mathbb{Z}, xp + yq > 0\}$; $M \subseteq \mathbb{N}$ and $M \neq \emptyset$ ($p = 1 \cdot p + 0 \cdot q \in M$). According with well-ordering principle (see **(l)** of Remark 1.1.4), there exists $d = kp + lq = \min M$. The Euclidean division of p and q by d give us $p = dc_1 + r_1, q = dc_2 + r_2$, with $0 \leqslant r_1, r_2 < d$. If $r_1 > 0$, then $r_1 = p - dc_1 = (1 - kc_1)p - lc_1q \in M$ and then $r_1 \geqslant d$, which is absurd. Similarly, if we assume that $r_2 > 0$. Thus, $r_1 = r_2 = 0$, which implies that d is a common divisor for p and q. Since these are prime, it follows that $d = 1$, and therefore $kp + lq = 1$.

1.1 An Axiomatic Definition of the Real Numbers

Let now $(K', +, \cdot, \leqslant)$ and $(K'', +, \cdot, \leqslant)$ be two Dedekind complete ordered fields with $0'$ and $0''$ neutral elements for addition and $1'$ and $1''$ neutral elements for multiplication. Let Q' and Q'' be the prime fields of K' and respectively K''; then Q' and Q'' are dense in K' and K'' respectively and, being each of them isomorphic to the set of rational numbers, they are isomorphic to each other. Let $\varphi : Q' \to Q''$ be a field isomorphism. From **(f)** of Remark 1.1.4, $0' < 1'$ and $0'' < 1''$ and then $n \cdot 1' < (n+1) \cdot 1', n \cdot 1'' < (n+1) \cdot 1''$, for every $n \in \mathbb{N}$. Since $\varphi(0') = 0''$ and $\varphi(1') = 1''$, it follows that $\varphi(n \cdot 1') = n \cdot 1'' < (n+1) \cdot 1'' = \varphi((n+1) \cdot 1')$, for every $n \in \mathbb{N}$. Thus, φ is strictly increasing on $N' = \{n \cdot 1' : n \in \mathbb{N}\}$. For every $n \in \mathbb{N}$, we define $(-n) \cdot 1' = -(n \cdot 1')$. It is easily shown that φ is strictly increasing on the set of integers of K', $Z' = \{k \cdot 1' : k \in \mathbb{Z}\}$. For every $r \in Q'$, there exist $k \in \mathbb{Z}$ and $n \in \mathbb{N}^*$ such that $r = (k \cdot 1') \cdot (n \cdot 1')^{-1} \equiv (kn^{-1}) \cdot 1'$. Therefore $Q' = \{q \cdot \underline{1} : q \in \mathbb{Q}\}$. Let $r_1, r_2 \in Q'$ with $r_1 < r_2$; if $r_i = k_i n_i^{-1}$, where $k_i \in \mathbb{Z}, n_i \in \mathbb{N}^*, i = 1, 2$, then $k_1 n_2 < k_2 n_1$ and so $\varphi((k_1 n_2) \cdot 1') < \varphi((k_2 n_1) \cdot 1')$, from where $\varphi(k_1 \cdot 1') \cdot \varphi(n_2 \cdot 1') < \varphi(k_2 \cdot 1') \cdot \varphi(n_1 \cdot 1')$ or $\varphi(r_1) = \varphi((k_1 n_1^{-1}) \cdot 1') < \varphi((k_2 n_2^{-1}) \cdot 1') = \varphi(r_2)$. Therefore φ is an isomorphism of ordered fields between Q' and Q''.

We now extend this isomorphism to K'. For every $x \in K'$, $x = \sup\{r \in Q' : r \leqslant x\}$ (see **(n)** of Remark 1.1.4). Then, for every $x \in K'$, we consider the set $A = \{\varphi(r) : r \in Q', r \leqslant x\} \subseteq K''$. Because every Dedekind complete ordered field is Archimedean (see **(h)** of Remark 1.1.4), for every $x \in K'$, there exists $n \in \mathbb{N}$ such that $-n \cdot 1' < x < n \cdot 1'$. So, $\varphi(-n \cdot 1') \in A$ and then $A \neq \emptyset$. Furthermore, for every $r \in Q', r \leqslant x$ it follows that $r < n \cdot 1'$ and then $\varphi(r) < \varphi(n \cdot 1') = n \cdot 1''$. Therefore $n \cdot 1''$ is an upper bound for A. Because K'' is Dedekind complete, there exists $\sup A \in K''$. We define

$$\Phi(x) = \sup A = \{\varphi(r) : r \in Q', r \leqslant x\}.$$

Thus, we have defined a mapping $\Phi : K' \to K''$, $\Phi(x) = \sup\{\varphi(r) : r \in Q', r \leqslant x\}$. Obviously, $\Phi|_{Q'} = \varphi$.

Let us show that Φ is strictly increasing on K'. For every $x, y \in K', x < y$, there exist $r_1, r_2 \in Q'$ such that $x < r_1 < r_2 < y$. For every $r \in Q', r < x$ it follows that $r < r_1$ and then $\varphi(r) < \varphi(r_1)$. Therefore, $\Phi(x) \leqslant \varphi(r_1) < \varphi(r_2) \leqslant \Phi(y)$.

As Φ is strictly increasing, it follows in particular that it is injective on K'.

We will now show that Φ preserves the addition operation. For every $x, y \in K'$ and for every $r, q \in Q', r < x, q < y$ it follows that $r + q < x + y$ and then $\Phi(x+y) \geqslant \varphi(r+q) = \varphi(r) + \varphi(q)$, for every $r, q \in Q', r < x, q < y$. Then $\Phi(x+y) \geqslant \Phi(x) + \Phi(y)$.

Now, for every $r \in Q', r < x+y, r-x < y$, and then there exists $q \in Q'$ such that $r - x < q < y$. It follows that $r - q < x$, and then $\varphi(r) - \varphi(q) = \varphi(r-q) \leqslant \Phi(x)$. Therefore $\varphi(r) \leqslant \varphi(q) + \Phi(x) \leqslant \Phi(y) + \Phi(x)$, for every $r \in Q', r < x+y$. Then $\Phi(x+y) \leqslant \Phi(x) + \Phi(y)$. The two inequalities lead us to $\Phi(x+y) = \Phi(x) + \Phi(y)$, for every $x, y \in K'$.

Let's now show that Φ preserves the multiplication operation. Let $x, y \in K', x, y > 0'$; for every $r, q \in Q', 0' < r < x, 0' < q < y$ it follows that $r \cdot q < x \cdot y$ and so $\varphi(r) \cdot \varphi(q) = \varphi(r \cdot q) \leqslant \Phi(x \cdot y)$, from where $\Phi(x \cdot y) \geqslant \Phi(x) \cdot \Phi(y)$.

For every $r \in Q', r < x \cdot y$ it follows that $r \cdot x^{-1} < y$ and then, there exists $q \in Q'$ such that $r \cdot x^{-1} < q < y$, from where $r \cdot q^{-1} < x$. Therefore $\varphi(r) \cdot \varphi(q)^{-1} = \varphi(r \cdot q^{-1}) \leqslant \Phi(x)$, from where $\varphi(r) \leqslant \Phi(x) \cdot \varphi(q) \leqslant \Phi(x) \cdot \Phi(y)$, for every $r \in Q', r < x \cdot y$, so that $\Phi(x \cdot y) \leqslant \Phi(x) \cdot \Phi(y)$. The two inequalities lead us to $\Phi(x \cdot y) = \Phi(x) \cdot \Phi(y)$, for every $x, y \in K', x, y > 0'$.

If $x = 0', y > 0'$, then $\Phi(x \cdot y) = \varphi(0') = 0'' = 0'' \cdot \Phi(y) = \Phi(x) \cdot \Phi(y)$. If $x < 0' < y$, then $\Phi(x \cdot y) = \Phi(-(-x) \cdot y) = -\Phi((-x) \cdot y) = -\Phi(-x) \cdot \Phi(y) = \Phi(x) \cdot \Phi(y)$. Similar reasoning applies in the other possible cases.

To complete the proof, we need to show that Φ is surjective. Let $y \in K''$ be an arbitrary point, and let $A = \{r \in Q' : \varphi(r) < y\} \subseteq K'$. Because K'' is Archimedean, there exists $n \in \mathbb{N}$ such that $-n \cdot 1'' < y < n \cdot 1''$. From the first inequality, we obtain that $-n \cdot 1' \in A$, and so $A \neq \emptyset$; from the second inequality, we obtain that, for every $r \in A, \varphi(r) < y < n \cdot 1'' = \varphi(n \cdot 1')$ and so, $r < n \cdot 1'$. Therefore A is bounded above and then there exists $x = \sup A = \sup\{r \in Q' : \varphi(r) < y\} \in K'$. For every $r \in Q', r < x$ there exists $r_1 \in Q'$ such that $r < r_1, \varphi(r_1) < y$. Then $\varphi(r) < y$ and so $\Phi(x) \leqslant y$.

If we suppose that $\Phi(x) < y$, there exists $q \in Q''$, such that $\Phi(x) < q < y$. Let $r \in Q'$ such that $\varphi(r) = q$; then $r \in A$ and so $r \leqslant x$. Therefore $\varphi(r) \leqslant \Phi(x) < q = \varphi(r)$, which is absurd. So $\Phi(x) = y$, which shows that Φ is surjective. □

Remark 1.1.10 Since any two models that satisfy $(A1)$–$(A16)$ are isomorphic, this system of axioms is categorical. From this point of view, the set of real numbers is the unique Dedekind complete ordered field. This answers the second question posed in (v) in Remark 1.1.2.

1.2 Decimal Model

As mentioned, the rigorous constructions of real numbers have been exposed by Karl Weierstrass, Georg Cantor, and Richard Dedekind. Each of these constructions starts from a certain interpretation given to the real number. Thus, for Cantor, the number is a "limit" of a Cauchy sequence of rational numbers, for Dedekind, it is a cut, which is the set of rational numbers located to the left of a "real number", and for Weierstrass, it is the "sum" of a series of rational numbers. A much more intuitive interpretation can be found in Simon Stevin (1548–1620). He published in 1585 in Flemish "De Thiende" (an English version is [13]). In this book, Stevin views the real number as an infinite decimal, making it closer to rational numbers, which are finite or infinite repeating decimals. Stevin did not produce a rigorous construction of the real number. Furthermore, at the time, it was not clear what was meant by a real number. Although the decimal construction is much more intuitive, it presents a series of technical difficulties. These have caused it to be somewhat avoided. However, recently, several authors have advocated for this approach, although they have only presented sketches of this construction (see [7, 10, 14]).

1.2 Decimal Model

1.2.1 Finite Decimals

We assume that the set of rational numbers $\mathbb{Q}$ is known and that $(\mathbb{Q}, +, \cdot, \leqslant)$ is an Archimedean-ordered field. In appendix, we will present constructions of the sets of natural, integer, and rational numbers and we make comments about the consistency of the axiomatic theories that underlie these constructions.

Definition 1.2.1 A **decimal fraction** is a number in the form of $\dfrac{a}{10^n}$, where $a \in \mathbb{Z}$, $n \in \mathbb{N}$ and 10 does not divide a.

Let us denote by $D = \left\{ \dfrac{a}{10^n} : a \in \mathbb{Z}, n \in \mathbb{N}, 10 \nmid a \right\}$ the set of decimal fractions.

Proposition 1.2.2 *D is the set of all finite decimals.*

Proof Let $x = \frac{a}{10^n} \in D$. We denote by $a_0 = \lfloor \frac{a}{10^n} \rfloor \in \mathbb{Z}$ the integer part of $\frac{a}{10^n}$ and by $b = a - a_0 \cdot 10^n = a - \lfloor \frac{a}{10^n} \rfloor \cdot 10^n \in \mathbb{Z}$. Then $\frac{b}{10^n} = \frac{a}{10^n} - \lfloor \frac{a}{10^n} \rfloor \in [0, 1[$ and therefore $b \in [0, 10^n[\cap \mathbb{N}$. Obviously $x = a_0 + \frac{b}{10^n}$. Because b is a natural number between 0 and $10^n - 1$ we can write $b = a_1 \cdot 10^{n-1} + a_2 \cdot 10^{n-2} + \cdots a_{n-1} \cdot 10 + a_n$, where, for every $k = 1, \cdots n$, $a_k \in \{0, 1, \cdots 9\}$ are the digits of b. Then $x = a_0 + \frac{b}{10^n} = a_0 + 0.a_1 a_2 \cdots a_n$, where $a_0 \in \mathbb{Z}$ and $a_1, \cdots, a_n \in \{0, \cdots 9\}$ represents writing of the number x as a finite decimal.

Conversely, any finite decimal

$$x = a_0 + 0.a_1 a_2 \cdots a_n = \frac{a_0 \cdot 10^n + a_1 \cdot 10^{n-1} + a_2 \cdot 10^{n-2} + \cdots a_{n-1} \cdot 10 + a_n}{10^n}$$

is a decimal fraction. □

Proposition 1.2.3 $\mathbb{N} \subseteq \mathbb{Z} \subseteq D \subseteq \mathbb{Q}$, *and* $(D, +, \cdot, \leqslant)$ *is an Archimedean ordered subring of* $(\mathbb{Q}, +, \cdot, \leqslant)$.

Proof Obviously, $\mathbb{N} \subseteq \mathbb{Z} \subseteq D \subseteq \mathbb{Q}$, therefore on D we consider the operations $+, \cdot$ and order relation $\leqslant$ induced by those on $\mathbb{Q}$. It is easy to see that $(D, +)$ is an additive subgroup of $(\mathbb{Q}, +)$. Multiplication is a law of internal composition on D and $1 = \frac{1}{10^0}$ is the neutral element for multiplication. Because $\mathbb{Q}$ is Archimedean, D is Archimedean also (for every $x, y \in D$, $x > 0$, there is $n \in \mathbb{N}$ such that $nx > y$). □

It can be easily shown that, for every $x, y \in D$, $x = \frac{a}{10^n} = a_0 + 0.a_1 \cdots a_n$, $y = \frac{b}{10^m} = b_0 + 0.b_1 \cdots b_m$, we have $x < y$ if and only if there is $i \in \mathbb{N}$, $i \leqslant \max\{m, n\}$, such that $a_i < b_i$ and $a_k = b_k$, for every $k < i$.

Proposition 1.2.4 *Any rational number is a finite or an infinite repeating decimal.*

Proof Let $\dfrac{p}{q} \in \mathbb{Q}$, with $p \in \mathbb{Z}$ and $q \in \mathbb{N}^*$, p and q relatively prime numbers. Let us suppose that there exist $i, j \in \mathbb{N}$ such that $q = 2^i \cdot 5^j \cdot q_1$, where q_1 does not have 2 and 5 as divisors. Let $l = \max\{i, j\}$ and let $x = \dfrac{10^l \cdot p}{q} = \dfrac{p_1}{q_1}$. We denote by $k = \left\lfloor \dfrac{p_1}{q_1} \right\rfloor \in \mathbb{Z}$ be the integer part of $\dfrac{p_1}{q_1}$ and by $y = \dfrac{p_1}{q_1} - \left\lfloor \dfrac{p_1}{q_1} \right\rfloor \in [0, 1[$.

We have two cases to analyze $q_1 = 1$ and $q_1 > 1$.

If $q_1 = 1$, then $x = p_1 \in \mathbb{Z}$ and $\dfrac{p}{q} = \dfrac{p_1}{10^l} \in D$. As we have observed in Proposition 1.2.2, the numbers in set D are finite decimals.

Let us suppose now that $q_1 > 1$.

For every $m \in \mathbb{N}$, $10^m y - \lfloor 10^m y \rfloor = 10^m \cdot \dfrac{p_1}{q_1} - 10^m \left\lfloor \dfrac{p_1}{q_1} \right\rfloor - \lfloor 10^m y \rfloor \in [0, 1[$ is a rational number with denominator q_1, so that

$$10^m y - \lfloor 10^m y \rfloor \in \left\{0, \frac{1}{q_1}, \frac{2}{q_1}, \cdots, \frac{q_1 - 1}{q_1}\right\}.$$

Then there exist $m, n \in \mathbb{N}^*$ such that

$$10^{m+n} y - \lfloor 10^{m+n} y \rfloor = 10^m y - \lfloor 10^m y \rfloor.$$

Let $a_1 = \lfloor 10^{m+n} y \rfloor - \lfloor 10^m y \rfloor = 10^m (10^n - 1) y \in \mathbb{N}$; then

$$10^m y = \frac{a_1}{10^n - 1} = a_2 + \frac{a_3}{10^n - 1} = a_2 + \sum_{k=1}^{\infty} \frac{a_3}{10^{nk}} = a_2 + \frac{a_3}{10^n} + \frac{a_3}{10^{2n}} + \cdots,$$

where $a_2 = \left\lfloor \dfrac{a_1}{10^n - 1} \right\rfloor \in \mathbb{N}$ and $a_3 \in [0, 10^n - 1[\cap \mathbb{N}$. Therefore a_3 can be written in base 10: $a_3 = b_1 \cdot 10^{n-1} + b_2 \cdot 10^{n-2} + \cdots + b_n$, where, for every $k = 1, \cdots n$, $b_k \in \{0, 1, \cdots 9\}$, and then $\sum_{k=1}^{\infty} \dfrac{a_3}{10^{nk}} = 0.b_1 b_2 \cdots b_n b_1 b_2 \cdots b_n \cdots = 0.\overline{b_1 b_2 \cdots b_n}$ is a periodic decimal fraction. Now we obtain that

$$10^m y = 10^m \left(\frac{p_1}{q_1} - \left\lfloor \frac{p_1}{q_1} \right\rfloor\right) = 10^m \left(10^l \frac{p}{q} - k\right) = a_2 + 0.\overline{b_1 b_2 \cdots b_n}$$

and thus

$$\frac{p}{q} = \frac{k}{10^l} + \frac{a_2}{10^{m+l}} + 0.\underbrace{0 \cdots 0}_{m+l} \overline{b_1 b_2 \cdots b_n}$$

is an infinite repeating decimal. □

Remark 1.2.5 As we observed in Proposition 1.2.2, the set of decimal fractions D is formed by finite decimals, and, from the previous theorem, it follows that $\mathbb{Q} \setminus D$ is formed by infinite repeating decimals.

Definition 1.2.6 A sequence $(x_n)_{n \in \mathbb{N}} \subseteq \mathbb{Q}$ **converges** to $x \in \mathbb{Q}$ if, for every $u, v \in \mathbb{Q}$ with $u < x < v$, there exists $n_0 \in \mathbb{N}$ such that, for every $n \geq n_0$, $u < x_n < v$;

1.2 Decimal Model

we denote this by $x_n \xrightarrow{\mathbb{Q}} x$. Using this definition, we can easily show that $\frac{1}{10^n} \xrightarrow{\mathbb{Q}} 0$. Indeed, for every $u, v \in \mathbb{Q}$ with $u < 0 < v$, there is $n_0 \in \mathbb{N}, n_0 > \frac{1}{v} = v^{-1}$ ($\mathbb{Q}$ is an Archimedean ordered field), and then, for every $n \geqslant n_0$, $10^n \geqslant n > \frac{1}{v}$ or $u < 0 < \frac{1}{10^n} < v$.

1.2.2 Infinite Decimals

Definition 1.2.7 A **real number** is a function $f : \mathbb{N} \to D$ with the following three properties:

(1) $f(0) \in \mathbb{Z}$.
(2) $10^n \cdot [f(n) - f(n-1)] \in \{0, 1, \cdots, 9\}$, for all $n \in \mathbb{N}^*$.
(3) For every $n \in \mathbb{N}$, there is $p > n$ such that $10^p \cdot [f(p) - f(p-1)] < 9$.

We will denote by $\mathbb{R}$ the set of all real number.

Remarks 1.2.8

(i) Let f be a real number; for every $n \in \mathbb{N}^*$, we denote by $a_n = 10^n \cdot [f(n) - f(n-1)]$ and let $a_0 = f(0)$. For every $n \in \mathbb{N}$, we get:

$$f(0) = a_0,$$
$$f(1) - f(0) = \frac{a_1}{10},$$
$$\cdots\cdots$$
$$f(n) - f(n-1) = \frac{a_n}{10^n}.$$

Summing up the above relationships, it follows that, for every $n \in \mathbb{N}$,

$$f(n) = a_0 + \frac{a_1}{10} + \cdots + \frac{a_n}{10^n} = a_0 + 0.a_1 a_2 \cdots a_n.$$

Thus, f can be viewed as an infinite decimal:

$$f = \sum_{n=0}^{\infty} \frac{a_n}{10^n} = a_0 + 0.a_1 a_2 \cdots a_n \cdots.$$

For each $n \in \mathbb{N}$, $f(n)$ is the nth-order truncation of f ($f(n)$ gives the first n exact decimal places of f).

(ii) Condition (1) says that $f(0) \in \mathbb{Z}$; condition (2) shows that the decimals a_n of f are digits from 0 to 9, and condition (3) prevents the situation where $f = a_0 + 0.a_1 \cdots a_n \overline{9} = a_0 + 0.a_1 \cdots a_n 999 \cdots$ ($a_n < 9$), in which case $f = a_0 + 0.a_1 \cdots (a_n + 1)00 \cdots$.

We will show that $\mathbb{R}$ can be structured as a Dedekind complete ordered field, meaning that the axioms $(A1)$–$(A16)$ in Definition 1.1.1 are satisfied. At each step of the construction we will mention which of these axioms is verified.

Proposition 1.2.9 *Let* $f = a_0 + 0.a_1 \cdots a_n \cdots, g = b_0 + 0.b_1 \cdots b_n \cdots \in \mathbb{R}$; *then* $f = g$ *if and only if* $a_n = b_n$, *for every* $n \in \mathbb{N}$.

Proof If $f = g$ then $f(n) = g(n)$, for every $n \in \mathbb{N}$. Then $a_0 = f(0) = g(0) = b_0$ and $a_n = 10^n \cdot [f(n) - f(n-1)] = 10^n \cdot [g(n) - g(n-1)] = b_n$, for every $n \in \mathbb{N}^*$.

Reciprocally, let us suppose that, for every $n \in \mathbb{N}$, $10^n \cdot [f(n) - f(n-1)] = a_n = b_n = 10^n \cdot [g(n) - g(n-1)]$. It follows that $f(0) = a_0 = b_0 = g(0)$ and $f(n) - f(n-1) = g(n) - g(n-1)$, for every $n \in \mathbb{N}^*$. Then, for every $n \in \mathbb{N}^*$,

$$f(0) = g(0)$$
$$f(1) - f(0) = g(1) - g(0)$$
$$\cdots$$
$$f(n) - f(n-1) = g(n) - g(n-1).$$

Adding these $n + 1$ relations, we obtain $f(n) = g(n)$, for every $n \in \mathbb{N}^*$; therefore $f = g$. □

Remark 1.2.10 We have noticed in Proposition 1.2.4 that any rational number is either a finite decimal or an infinite repeating decimal one, i.e. $x = p \cdot q^{-1} = a_0 + 0.a_1 \cdots a_m$ or $x = a_0 + 0.a_1 \cdots a_m \overline{b_1 \cdots b_n} = a_0 + 0.a_1 \cdots a_m b_1 \cdots b_n b_1 \cdots b_n \cdots$, where $a_0 \in \mathbb{Z}$ and $a_i, b_j \in \{0, 1 \cdots, 9\}$, for all $i \in \{1, \cdots, m\}, j \in \{1, \cdots n\}$. This representation is unique except in the case where $a_m < 9$ and $b_1 = \cdots = b_n = 9$, in which case for the rational number the form $a_0 + 0.a_1 \cdots a_{m-1}(a_m + 1)$ is chosen. Let us define a mapping $\varphi : \mathbb{Q} \to \mathbb{R}$ by

$$\varphi(x) = \begin{cases} a_0 + 0.a_1 a_2 \cdots a_m \bar{0} \cdots, & x = a_0 + 0.a_1 a_2 \cdots a_m \in D \\ a_0 + 0.a_1 \cdots a_m \overline{b_1 \cdots b_n}, & x = a_0 + 0.a_1 \cdots a_m \overline{b_1 \cdots b_n} \in \mathbb{Q} \setminus D. \end{cases}$$

φ is an injection of $\mathbb{Q}$ into $\mathbb{R}$, $\varphi(0) = \underline{0}$ and $\varphi(1) = \underline{1}(\underline{0}, \underline{1} : \mathbb{N} \to D, \underline{0}(n) = 0, \underline{1}(n) = 1$, for every $n \in \mathbb{N}$).

In particular, it follows from here that $\mathbb{R}$ is a set with at least two elements.

We will start by defining an order relation on $\mathbb{R}$.

Definition 1.2.11 Let $f = a_0 + 0.a_1 \cdots a_n \cdots, g = b_0 + 0.b_1 \cdots b_n \cdots \in \mathbb{R}$. We say that f is **strictly less than** g, and denote this by $f < g$, if there exists $n \in \mathbb{N}$ such that $a_n < b_n$ and, for every $k < n, a_k = b_k$.

$f \leqslant g$ if $f < g$ or $f = g$.

Remarks 1.2.12

(i) The above-defined function φ preserves the relation $\leqslant$: for every $x, y \in \mathbb{Q}, x \leqslant y \iff \varphi(x) \leqslant \varphi(y)$. Next, we will identify $x \in \mathbb{Q}$ with $\varphi(x)$ and think $\mathbb{Q}$ as a subset of $\mathbb{R}$.

1.2 Decimal Model

(ii) Let $f = a_0 + 0.a_1 \cdots a_n \cdots$, $g = b_0 + 0.b_1 \cdots b_n \cdots \in \mathbb{R}$. If, for all $n \in \mathbb{N}$, $a_n \leqslant b_n$, then $f \leqslant g$. The proof is similar to the "if" part in Proposition 1.2.9, but in this case, the converse is not true. Indeed, let $x = 0,12$, $y = 0.21 \in D \subseteq \mathbb{R}$; $x < y$ but $a_2 > b_2$. Based on this observation, we can state that $a_0 \equiv \varphi(a_0) \leqslant f = a_0 + 0.a_1 \cdots a_n \cdots$.

(iii) If $a_0 < b_0$, then $f < g$; respectively $g \leqslant f \implies b_0 \leqslant a_0$. It follows that $f < \underline{0} \Leftrightarrow a_0 < 0$ or $f \geqslant \underline{0} \Leftrightarrow a_0 \geqslant 0$. This observation leads to inequality $f = a_0 + 0.a_1 \cdots a_n \cdots < \varphi(a_0 + 1) \equiv a_0 + 1$ which, combined with $a_0 \leqslant f = a_0 + 0.a_1 \cdots a_n \cdots$, obtained at (ii), shows us that $a_0 = \lfloor f \rfloor$ is the **integer part** of f. We observe also that $\{f\} = 0.a_1 a_2 \cdots a_n \cdots \in [0, 1[$; the number $\{f\} = f - \lfloor f \rfloor = 0.a_1 a_2 \cdots a_n \cdots$ is the **decimal part** of f.

(iv) Let $f = \lfloor f \rfloor + \{f\} = a_0 + 0.a_1 a_2 \cdots a_n \cdots \in \mathbb{R}$ and, for every $n \in \mathbb{N}$, let $x_n = a_0 + 0.a_1 a_2 \cdots a_n \in D \subseteq \mathbb{R}$ be the nth-order truncation of f; then

$$x_n \equiv \varphi(x_n) \leqslant f < \varphi\left(x_n + \frac{1}{10^n}\right) \equiv x_n + \frac{1}{10^n}, \text{ for every } n \in \mathbb{N}.$$

The inequality on the left results from (ii). For the inequality on the right consider an arbitrary n in $\mathbb{N}$. If $a_n < 9$, then $f = a_0 + 0.a_1 a_2 \cdots a_{n-1} a_n \cdots < a_0 + 0.a_1 a_2 \cdots a_{n-1}(a_n + 1)00 \cdots = \varphi\left(x_n + \frac{1}{10^n}\right)$. If $a_n = 9$ but $a_{n-1} < 9$, $f = a_0 + 0.a_1 a_2 \cdots a_{n-1} a_n \cdots < a_0 + 0.a_1 a_2 \cdots (a_{n-1} + 1)00 \cdots = \varphi\left(x_n + \frac{1}{10^n}\right)$, and so on. Finally, if $a_1 = a_2 = \cdots = a_n = 9$, then $f = a_0 + 0.99 \cdots 9 a_{n+1} \cdots < a_0 + 1 = \varphi\left(x_n + \frac{1}{10^n}\right)$.

Theorem 1.2.13 *The relation defined in Definition 1.2.11, $\leqslant$, is a total order on $\mathbb{R}$, and $\varphi(D)$ (and obviously $\varphi(\mathbb{Q})$) is dense in $\mathbb{R}$ with respect to this relation ((A10)–(A13)).*

Proof From Definition 1.2.11, we have that $f \leqslant f$, $\forall f \in \mathbb{R}$ ($\leqslant$ is reflexive). From the definition of $<$ it can be seen that $f < g$ excludes the possibility that $g \leqslant f$. Then $f \leqslant g$ and $g \leqslant f$ trains $f = g$ ($\leqslant$ is antisymmetric).

Let's prove the transitivity. We suppose that $f \leqslant g$ and $g \leqslant h$, where $f = a_0 + 0.a_1 a_2 \cdots a_l \cdots$, $g = b_0 + 0.b_1 b_2 \cdots b_m \cdots$, $h = c_0 + 0.c_1 c_2 \cdots c_n \cdots \in \mathbb{R}$. If ($f = g$ and $g \leqslant h$) or ($f \leqslant g$ and $g = h$), then $f \leqslant h$, obviously.

It remains to treat the case where $f < g$ and $g < h$. In this case, there exit $m, n \in \mathbb{N}$ such that $a_m < b_m$, $a_k = b_k$, $\forall k < m$, and $b_n < c_n$, $b_l = c_l$, $\forall l < n$. If $m < n$, then $a_k = c_k$, $\forall k < m$ and $a_m < b_m = c_m$. If $n < m$, then $a_k = c_k$, $\forall k < n$ and $a_n = b_n < c_n$. Finally, if $m = n$, then $a_k = c_k$, $\forall k < m$ and $a_m < b_m = b_n < c_n = c_m$. In all cases $f < h$ and thus $f \leqslant h$.

Let us show that the order $\leqslant$ is a total one on $\mathbb{R}$. Let $f = a_0 + 0.a_1 a_2 \cdots a_n \cdots$, $g = b_0 + 0.b_1 b_2 \cdots b_n \cdots \in \mathbb{R}$ with $f \neq g$. From Proposition 1.2.9, there is $n \in \mathbb{N}$ such that $a_n \neq b_n$. Let n_0 be the smallest natural number n for which $a_n \neq b_n$ (see (k) of Remark 1.1.4). Therefore, for every $k < n_0$, $a_k = b_k$ and $a_{n_0} \neq b_{n_0}$. Then $f < g$ if $a_{n_0} < b_{n_0}$ and $g < f$ if

$b_{n_0} < a_{n_0}$. Let's now prove that D is dense with respect to the order relation on $\mathbb{R}$. Let $f = a_0 + 0.a_1 a_2 \cdots a_n \cdots < b_0 + 0.b_1 b_2 \cdots b_n \cdots = g \in \mathbb{R}$ and let $i \in \mathbb{N}$ such that $a_k = b_k, \forall k < i$ and $a_i < b_i$. From (3) of Definition 1.2.7, there is $p > i$ such that $a_p < 9$. Let $x = a_0 + 0.a_1 a_2 \cdots a_i \cdots a_{p-1}(a_p + 1) \in D$. Then $f < \varphi(x) < g$. Therefore $D \equiv \varphi(D)$ is dense in $\mathbb{R}$. □

On every totally ordered set can be defined the **order topology**. As we will only use the convergence of sequences in this topology, we will only define this last notion.

Definition 1.2.14 A sequence $(f^n)_{n \in \mathbb{N}} \subseteq \mathbb{R}$ **converges** to $f \in \mathbb{R}$ if, for every $g, h \in \mathbb{R}$ with $g < f < h$, there exists $n_0 \in \mathbb{N}$ such that $g < f^n < h$, for every $n \geq n_0$; f is called the **limit** of (f_n) and we denote this fact by $f^n \xrightarrow[\mathbb{R}]{} f$ or $f^n \to f$. The sequence (f^n) is **convergent** if there is $f \in \mathbb{R}$ such that $f^n \xrightarrow[\mathbb{R}]{} f$.

Remarks 1.2.15

(i) Note that, in the previous definition, the intervals of the form $]g, h[= \{l \in \mathbb{R} : g < l < h\}$, where $g < f < h$, play the role of neighborhoods of f.

(ii) If a sequence (f^n) is convergent, its limit is unique. Indeed, if we suppose that $f^n \xrightarrow[\mathbb{R}]{} f$ and $f^n \xrightarrow[\mathbb{R}]{} g$ with $f \neq g$, then, because the order $\leq$ is total, we have $f < g$ or $g < f$. Let us suppose that $f < g$ and let $x \in D$ such that $f < \varphi(x) < g$. Then there exists $n_0 \in \mathbb{N}$ such that $f^n < \varphi(x)$ and $\varphi(x) < f^n$, for every $n \geq n_0$, which is absurd.

(iii) Due to $\varphi(D)$'s density in $\mathbb{R}$, $f^n \xrightarrow[\mathbb{R}]{} f$ if and only if for every $x, y \in D$ with $x \equiv \varphi(x) < f < \varphi(y) \equiv y$ there exists $n_0 \in \mathbb{N}$ such that $x \equiv \varphi(x) < f^n < \varphi(y) \equiv y$, for every $n \geq n_0$.

(iv) $x_n \xrightarrow[\mathbb{Q}]{} x$ if and only if $\varphi(x_n) \xrightarrow[\mathbb{R}]{} \varphi(x)$. Indeed, if we suppose that $x_n \xrightarrow[\mathbb{Q}]{} x$ (see Definition 1.2.6), then, for every $u, v \in \mathbb{Q}$ with $\varphi(u) < \varphi(x) < \varphi(v)$ it follows that $u < x < v$ (see the (i) of Remark 1.2.12). Let now $n_0 \in \mathbb{N}$ such that $u < x_n < v$, for every $n \geq n_0$. Then $\varphi(u) < \varphi(x_n) < \varphi(v), \forall n \geq n_0$, hence $\varphi(x_n) \xrightarrow[\mathbb{R}]{} \varphi(x)$. Reciprocally, if $\varphi(x_n) \xrightarrow[\mathbb{R}]{} \varphi(x)$, then, for every $u, v \in D$ with $u < x < v$, $\varphi(u) < \varphi(x) < \varphi(v)$. There is $n_0 \in \mathbb{N}$ such that $\varphi(u) < \varphi(x_n) < \varphi(v)$, and so $u < x_n < v, \forall n \geq n_0$, and then $x_n \xrightarrow[\mathbb{Q}]{} x$.

Therefore φ is a topological embedding of $\mathbb{Q}$ in $\mathbb{R}$.

(v) Using the previous remark, we can show that $\varphi(\frac{1}{10^n}) \xrightarrow[\mathbb{R}]{} \varphi(0) = \underline{0}$ (in Definition 1.2.6 we noted that $\frac{1}{10^n} \xrightarrow[\mathbb{Q}]{} 0$).

(vi) The introduced convergence is compatible with the order relation, i.e. $f^n \xrightarrow[\mathbb{R}]{} f$, $g^n \xrightarrow[\mathbb{R}]{} g$ and $f^n \leq g^n, \forall n \in \mathbb{N}$, implies that $f \leq g$. Indeed, if we suppose that $g < f$ and if we take $u, v, w \in D$ such that $u < g < v < f < w$, then there is $n_0 \in \mathbb{N}$ such that $u < g^n < w < f^n < w, \forall n \geq n_0$, which is absurd.

1.2 Decimal Model

(vii) Every convergent sequence in $\mathbb{R}$ is bounded. Indeed, let $(f^n) \subseteq \mathbb{R}$ be a convergent sequence and let f be its limit. Let u_0, v_0 fixed in $\mathbb{Q}$ such that $u_0 < f < v_0$, and let $n_0 \in \mathbb{N}$ such that $u_0 < f^n < v_0, \forall n \geqslant n_0$. There exist $g = \min\{f^0, f^1, \cdots, f^{n_0-1}, u_0\}$ and $h = \max\{f^0, f^1, \cdots, f^{n_0-1}, v_0\}$ such that $g \leqslant f^n \leqslant h, \forall n \in \mathbb{N}$. Therefore $(f^n)_{n \in \mathbb{N}}$ is bounded in $\mathbb{R}$.

(viii) If (f^n) is a constant sequence ($\forall n \in \mathbb{N}, f^n = f$), then $f^n \underset{\mathbb{R}}{\to} f$.

We will now present a useful result in the construction of the algebraic structure on $\mathbb{R}$.

Theorem 1.2.16 (The Convergence Theorem of Monotone Sequences) *Every bounded monotone sequence converges.*

Proof We will do the demonstration for the case of increasing sequences, and the case of decreasing ones being treated similarly. Let $(f^n)_{n \in \mathbb{N}} \subseteq \mathbb{R}$ be an increasing sequence bounded above by $l \in \mathbb{R}$. We suppose that, for every $n \in \mathbb{N}$, $f^n = a_0^n + 0.a_1^n \cdots a_m^n \cdots$ and $l = l_0 + 0.l_1 \cdots l_m \cdots$. Because $f^n \leqslant f^{n+1} \leqslant l$, for every $n \in \mathbb{N}$, it follows by (iii) of Remark 1.2.12 that $a_0^n \leqslant a_0^{n+1} \leqslant l_0$, for every $n \in \mathbb{N}$. Then $(a_0^n)_{n \in \mathbb{N}} \subseteq \mathbb{Z}$ is an increasing and bounded above sequence of integers. There exist $n_0 \in \mathbb{N}$ such that $a_0^n = a_0^{n_0} \in \mathbb{Z}, \forall n \geqslant n_0$.

Indeed, if, by absurd, we suppose that, for every $n \in \mathbb{N}$, there exists $m > n$ such that $a_0^m > a_0^n$ then $a_0^m \geqslant a_0^n + 1$. Then for $n = 1$ there exists $m_1 > 1$ such that $a_0^{m_1} \geqslant a_0^1 + 1$; for $n = m_1$ there exists $m_2 > m_1$ such that $a_0^{m_2} \geqslant a_0^{m_1} + 1$, and so on. At the step p, for $n = m_{p-1}$ there exists $m_p > m_{p-1}$ such that $a_0^{m_p} \geqslant a_0^{m_{p-1}} + 1$. If we add these p inequalities we get $l_0 \geqslant a_0^{m_p} \geqslant a_0^1 + p$, and this is for any $p \in \mathbb{N}$, which contradicts the hypothesis that $\mathbb{Q}$ is Archimedean (see Proposition 1.5.49).

Thus, we have shown that $N_0 = \{m \in \mathbb{N} : a_0^n = a_0^m, \forall n \geqslant m\} \neq \emptyset$ ($n_0 \in N_0$). Since $(\mathbb{N}, \leqslant)$ is well-ordered, it follows that n_0 can be chosen as the smallest element in N_0.

Therefore, for every $n < n_0$, $a_0^n < a_0^{n_0}$, so that $f^n < f^{n_0}$. For every $n \geqslant n_0$, $f^n = a_0^{n_0} + 0.a_1^n \cdots a_m^n \cdots \leqslant a_0^{n_0} + 0.a_1^{n+1} \cdots a_m^{n+1} \cdots = f^{n+1}$. It follows that $a_1^n \leqslant a_1^{n+1} \leqslant 9$, for every $n \geqslant n_0$. With similar reasoning, we deduce that there is $n_1 \geqslant n_0$ such that, for every $n \geqslant n_1$, $a_1^n = a_1^{n_1}$. It follows that the set $N_1 = \{m \in \mathbb{N} : m \geqslant n_0, a_1^n = a_1^m, \forall n \geqslant m\} \neq \emptyset$ ($n_1 \in N_1$). Again, using the well-ordering principle, n_1 can be chosen as the smallest element in the set N_1. It follows that, for every $n_0 \leqslant n < n_1, a_1^n < a_1^{n_1}$, from where $f^n < f^{n_1}$. For every $n \geqslant n_1$, $f^n = a_0^{n_0} + 0.a_1^{n_1} a_2^n \cdots$.

Continuing in the same manner, at step p, we will determine $n_p \geqslant n_{p-1}$ such that, for every $n \geqslant n_p, a_p^n = a_p^{n_p}$ (at step p, the p-th decimal of f^n stabilizes). Similarly, it is shown that, for every $n < n_p, f^n < f^{n_p}$ and, for every $n \geqslant n_p$,

$$f^n = a_0^{n_0} + 0.a_1^{n_1} \cdots a_p^{n_p} a_{p+1}^n \cdots,$$

and so on.

Let now $f = a_0^{n_0} + 0.a_1^{n_1}a_2^{n_2}\cdots a_p^{n_p}a_{p+1}^{n_{p+1}}\cdots$. We remark that $a_0^{n_0} \in \mathbb{Z}$, and $a_p^{n_p} \in \{0, 1, \cdots, 9\}$, for every $p \geqslant 1$. If we suppose that, for every $p \in \mathbb{N}^*, a_p^{n_p} = 9$, then $f = a_0^{n_0} + 1$, and we will re-grade $a_0^{n_0}$ to be $a_0^{n_0} + 1$ and, for every $p \geqslant 1$, $a_p^{n_p}$ to be 0. If there is $p_0 \in \mathbb{N}^*$ such that $a_{p_0}^{n_{p_0}} < 9$ and, for every $p > p_0, a_p^{n_p} = 9$, then $f = a_0^{n_0} + 0.a_1^{n_1}\cdots a_{p_0-1}^{n_{p_0-1}}(a_{p_0}^{n_{p_0}} + 1)$, and we will re-grade $a_{p_0}^{n_{p_0}}$ to be $a_{p_0}^{n_{p_0}} + 1$ and, for every $p \geqslant p_0 + 1$, $a_p^{n_p}$ to be 0. With these possible changes $f \in \mathbb{R}$.

Let's prove that, for every $n \in \mathbb{N}$, $f^n \leqslant f$.

For every $n \in \mathbb{N}$, $f^n = a_0^n + 0.a_1^n a_2^n \cdots$ and $a_0^n \leqslant a_0^{n_0}$.

If $a_0^n < a_0^{n_0}$, then $f^n < f$. If $a_0^n = a_0^{n_0}$, then $n \geqslant n_0$ and so $a_1^n \leqslant a_1^{n_1}$.

We distinguish two cases in this last situation:

If $a_1^n < a_1^{n_1}$, then $f^n = a_0^{n_0} + 0.a_1^n a_2^n \cdots < a_0^{n_0} + 0.a_1^{n_1}a_2^{n_2}\cdots = f$. If $a_1^n = a_1^{n_1}$, then $n \geqslant n_1$ and so $a_2^n \leqslant a_2^{n_2}$, and so on.

If there is p such that $a_0^n = a_0^{n_0}, a_1^n = a_1^{n_1}, \cdots a_p^n = a_p^{n_p}$, then $n \geqslant n_p$ and so $a_{p+1}^n \leqslant a_{p+1}^{n_{p+1}}$, and so on. Otherwise, if for every $p \in \mathbb{N}, a_p^n = a_p^{n_p}$, then $f^n = f$. Therefore, for every $n \in \mathbb{N}$, $f^n \leqslant f$.

This implies that $f^n \leqslant f$, for every $n \in \mathbb{N}$.

For every $g = b_0 + 0.b_1\cdots b_p\cdots, h \in \mathbb{R}$ with $g < f < h$, there is $i \in \mathbb{N}$ such that, for every $k < i$, $b_k = a_k^{n_k}$ and $b_i < a_i^{n_i}$. For every $n \geqslant n_i$ and for every $k < i$, $n_k \leqslant n_i \leqslant n$ and $b_k = a_k^{n_k} = a_k^n$ and $b_i < a_i^{n_i} = a_i^n$. It follows that, for every $n \geqslant n_i$, $g < f^n \leqslant f < h$, which shows that $f^n \underset{\mathbb{R}}{\to} f$. $\square$

Remark 1.2.17 From the above proof it follows that $f^n \leqslant f$, for every $n \in \mathbb{N}$ and that, for every $g < f$, there is $n_0 \in \mathbb{N}$ such that $g < f^n$, for every $n \geqslant n_0$. Therefore f is the least upper bound of the set $\{f^n : n \in \mathbb{N}\}$ or $f = \sup_n\{f^n : n \in \mathbb{N}\}$.

Similarly, if $(f^n)_{n \in \mathbb{N}}$ is a decreasing and bounded below sequence, then $f^n \to \inf_n\{f^n : n \in \mathbb{N}\}$

Definition 1.2.18 Let $f = a_0 + 0.a_1\cdots a_n\cdots \in \mathbb{R}$. For every $n \in \mathbb{N}$, we define $x_n = a_0 + 0.a_1\cdots a_n \in D \subseteq \mathbb{R}$ and $y_n = x_n + \frac{1}{10^n} \in D \subseteq \mathbb{R}$ with the remark that, if $a_n = 9$ then $y_n = a_0 + 0.a_1\cdots(a_{n-1} + 1)$ (if $a_n = a_{n-1} = 9$, then $y_n = a_0 + 0.a_1\cdots(a_{n-2} + 1)$ and so on; if $a_1 = a_2 = \cdots = a_n = 9$ then $y_n = a_0 + 1$).

(x_n) is called the sequence of **lower decimal approximations** of f and (y_n) is called the sequence of **upper decimal approximations** of f.

Proposition 1.2.19 Let $f = a_0 + 0.a_1\cdots a_n \cdots \in \mathbb{R}$ and let $(x_n)_{n \in \mathbb{N}}$ be the sequence of lower decimal approximations of f and $(y_n)_{n \in \mathbb{N}}$ be the sequence of upper decimal approximations of f; then (x_n) is an increasing sequence, $x_n \underset{\mathbb{R}}{\to} f$, and (y_n) is a decreasing sequence, $y_n \underset{\mathbb{R}}{\to} f$.

We will denote this by $x_n \uparrow f, y_n \downarrow f$.

1.2 Decimal Model

Proof Obviously, $x_n \leqslant x_{n+1} < y_{n+1}$, for every $n \in \mathbb{N}$. For every $n \in \mathbb{N}$, $10^{n+1}(y_{n+1} - y_n) = a_{n+1} - 9 \leqslant 0$; it follows that (y_n) is decreasing. The (iv) of Remark 1.2.12 implies that, for every $n \in \mathbb{N}$, $x_n \leqslant f < y_n$.

For every $g = b_0 + 0.b_1b_2\cdots b_n\cdots, h = c_0 + 0.c_1c_2\cdots c_n\cdots \in \mathbb{R}$ with $g < f < h$, there exist $i, j \in \mathbb{N}$ such that, for every $k < i$, $b_k = a_k$, $b_i < a_i$ and, for every $l < j$, $a_l = c_l$, $a_j < c_j$. Let $n_0 = \max\{i, j\}$; for every $n \geqslant n_0$, $g < x_n \leqslant f < y_n < h$. Therefore $x_n \underset{\mathbb{R}}{\to} f$, and $y_n \underset{\mathbb{R}}{\to} f$. □

Remarks 1.2.20

(i) In Theorem 1.2.13, we have shown that D is dense in $\mathbb{R}$ with respect to the order relation. It follows from this that D is also dense in $\mathbb{R}$ with respect to the topology defined by the order relation. The preceding proposition serves to confirm this fact once again.

(ii) The sequence $(y_n)_n$, defined by $y_n = \frac{1}{10^n} = 0.\underbrace{0\cdots 0}_{n-1}1$, for every $n \in \mathbb{N}$, is the sequence of upper decimal approximations of $\underline{0} \in \mathbb{R}$; hence $y_n \downarrow \underline{0}$.

Proposition 1.2.21 *Let $f = a_0 + 0.a_1\cdots a_n\cdots \in \mathbb{R}$ and let $(u_n), (v_n) \subseteq \mathbb{Q}$ such that $u_n \underset{\mathbb{R}}{\to} f$ and $v_n \underset{\mathbb{Q}}{\to} \underline{0}$. Then $u_n + v_n \equiv \varphi(u_n + v_n) \underset{\mathbb{R}}{\to} f$.*

Proof For every $u, v \in \mathbb{Q}$ with $u < f < v$ there exist $u', v' \in \mathbb{Q}$ such that $u < u' < f < v' < v$. Because $u_n \underset{\mathbb{R}}{\to} f$, there exists $n_0 \in \mathbb{N}$ such that $u' < u_n < v'$, for every $n \geqslant n_0$. It follows that $u' + v_n < u_n + v_n < v' + v_n, \forall n \geqslant n_0$. But $u' + v_n \underset{\mathbb{Q}}{\to} u'$ and $v' + v_n \underset{\mathbb{Q}}{\to} v'$, hence there exists $n_1 > n_0$ such that $u < u' + v_n < u_n + v_n < v' + v_n < v; \forall n \geqslant n_1$ and so $u_n + v_n \underset{\mathbb{R}}{\to} f$. □

Now we will introduce the **addition operation** on $\mathbb{R}$.

Definition 1.2.22 Let $f = a_0 + 0.a_1\cdots a_n\cdots$, $g = b_0 + 0.b_1\cdots b_n\cdots \in \mathbb{R}$ and let $(x_n)_n, (y_n)_n$ be the sequences of lower decimal approximations of f, respectively g: $x_n = a_0 + 0.a_1\cdots a_n$, $y_n = b_0 + 0.b_1\cdots b_n$, for every $n \in \mathbb{N}$. Then, for every $n \in \mathbb{N}$, $z_n = x_n + y_n \in D$ and $z_n \leqslant z_{n+1} < a_0 + b_0 + 1$. It follows that $(z_n)_n \subseteq \mathbb{R}$ is an increasing bounded sequence. According to Theorem 1.2.16, there exists $\lim_n z_n \in \mathbb{R}$.

Let us define $f + g = \lim_n (x_n + y_n) = \sup\{x_n + y_n : n \in \mathbb{N}\}$.

Apparently, the definition of addition depends on the sequences of lower decimal approximations of the two real numbers. In fact, this is not true.

Proposition 1.2.23 *For every $(u_n)_n, (v_n)_n \subseteq \mathbb{Q}$ with $u_n \underset{\mathbb{R}}{\to} f$, $v_n \underset{\mathbb{R}}{\to} g$, $u_n + v_n \underset{\mathbb{R}}{\to} f + g$.*

Proof Let $(x_n), (y_n)$ be the sequences of lower decimal approximations of f, respectively g; then, for every $n \in \mathbb{N}$, $u_n + v_n = x_n + y_n + z_n$, where $z_n = (u_n + v_n) - (x_n + y_n) = (u_n - x_n) + (v_n - y_n)$.

Let's prove that $u_n - x_n \underset{\mathbb{Q}}{\to} 0$ and $v_n - y_n \underset{\mathbb{Q}}{\to} 0$. For every $\varepsilon \in \mathbb{Q}, \varepsilon > 0$ and for every $a, b \in \mathbb{Q}$ such that $a < f < b$ and $b - a < \varepsilon$, there exists $n_0 \in \mathbb{N}$ such that, for every $n \geqslant n_0$, $b - \varepsilon < a < u_n < b < a + \varepsilon$ and $a - \varepsilon < x_n < b + \varepsilon$ or $-b - \varepsilon < -x_n < -a + \varepsilon$. Properly adding the inequalities, we get $-2\varepsilon < u_n - x_n < 2\varepsilon$, for every $n \geqslant n_0$ whence it follows that $u_n - x_n \underset{\mathbb{Q}}{\to} 0$. Similarly, it is shown that $v_n - y_n \underset{\mathbb{Q}}{\to} 0$. Therefore $z_n \underset{\mathbb{Q}}{\to} 0$. According to Proposition 1.2.21, $x_n + y_n + z_n \underset{\mathbb{R}}{\to} f + g$ and then $u_n + v_n \underset{\mathbb{R}}{\to} f + g$. □

Theorem 1.2.24 $(\mathbb{R}, +, \leqslant)$ *is a totally ordered commutative group* $((A1)$–$(A4)$, $(A14))$.

Proof For every $f, g, h \in \mathbb{R}$, let $(x_n), (y_n)$, and (z_n) be the sequences of lower decimal approximations of f, g respectively h. Then $x_n + y_n \underset{\mathbb{R}}{\to} f + g$ and $z_n \underset{\mathbb{R}}{\to} h$; according to Proposition 1.2.23, $(x_n + y_n) + z_n \underset{\mathbb{R}}{\to} (f + g) + h$. On the other hand $(x_n + y_n) + z_n = x_n + (y_n + z_n), \forall n \in \mathbb{N}$. With similar reasoning it is shown that $x_n + (y_n + z_n) \underset{\mathbb{R}}{\to} f + (g + h)$ and thus $(f + g) + h = f + (g + h)$. Also, from $x_n + y_n = y_n + x_n, x_n = x_n + 0, \forall n \in \mathbb{N}$, we deduce that $f + g = g + f$ and that $f = f + \underline{0}$.

We remark that the sequence $(-x_n)_{n \in \mathbb{N}}$ is a decreasing one. Furthermore, because $x_n \leqslant f = a_0 + 0.a_1 \cdots a_n \cdots < a_0 + 1$, then $-(a_0 + 1) < -x_n, \forall n \in \mathbb{N}$. Therefore $(-x_n)_{n \in \mathbb{N}}$ is decreasing and bounded below, hence there exists $\lim_n (-x_n) = \inf\{-x_n : n \in \mathbb{N}\}$; we will denote this limit as $-f$ and we will call it the opposite of the real number f. Because $x_n \underset{\mathbb{R}}{\to} f$ and $-x_n \underset{\mathbb{R}}{\to} -f$ we deduce that $f + (-f) = \underline{0}$.

Let now $f = a_0 + 0.a_1 a_2 \cdots a_n \cdots$, $g = b_0 + 0.b_1 b_2 \cdots b_n \cdots$, $h = c_0 + 0.c_1 c_2 \cdots c_n \cdots \in \mathbb{R}$ with $f \leqslant g$. If $f = g$, then $f + h = g + h$ and then $f + h \leqslant g + h$. If $f < g$ there exists $i \in \mathbb{N}$ such that, for every $k < i$, $a_k = b_k$ and $a_i < b_i$. For every $n \in \mathbb{N}$, let $x_n = a_0 + 0.a_1 a_2 \cdots a_n$, $y_n = b_0 + 0.b_1 b_2 \cdots b_n$ and $z_n = c_0 + 0.c_1 c_2 \cdots c_n$. Then, for every $n \geqslant i$, $x_n < y_n$ and hence $x_n + z_n < y_n + z_n \leqslant g + h$ But $x_n + z_n \underset{\mathbb{R}}{\to} f + h$ and thus, according to (vi) of Remark 1.2.15, $f + h \leqslant g + h$. □

Remarks 1.2.25

(i) $f \leqslant \underline{0}$ if and only if $-f \geqslant \underline{0}$. Indeed, if $f \leqslant \underline{0}$, then $\underline{0} = f + (-f) \leqslant \underline{0} + (-f) = -f$. If $-f \geqslant \underline{0}$, then $\underline{0} = (-f) + f \geqslant \underline{0} + f = f$.

(ii) The convergence structure on $\mathbb{R}$ is compatible with the addition operation. This means that, if $f^n \underset{\mathbb{R}}{\to} f$ and $g^n \underset{\mathbb{R}}{\to} g$, then $f^n + g^n \underset{\mathbb{R}}{\to} f + g$. Indeed, for every $u, v \in D$ with $u < f + g < v$ it follows that $u - f < g < v - f$. According to Theorem 1.2.13, there exist $u_1, v_1 \in D$ such that $u - f < u_1 < g < v_1 < v - f$ or $u - u_1 < f < v - v_1$ and $u_1 < g < v_1$. There exists $n_0 \in \mathbb{N}$ such that $u - u_1 < f^n < v - v_1$ and $u_1 < g^n < v_1$, for every $n \geqslant n_0$. Therefore, for every $n \geqslant n_0$, $u < f^n + g^n < v$, and so $f^n + g^n \underset{\mathbb{R}}{\to} f + g$.

1.2 Decimal Model

(iii) If $f^n \underset{\mathbb{R}}{\to} f$, then $-f^n \underset{\mathbb{R}}{\to} -f$. Indeed, for every $u, v \in D$ with $u < -f < v$ we have $u + f < \underline{0} < v + f$ or $-v < f < -u$. Let now $n_0 \in \mathbb{N}$ such that, for every $n \geqslant n_0$, $-v < f^n < -u$ or $u < -f^n < v$, which implies that $-f^n \underset{\mathbb{R}}{\to} -f$.

We will now define the **multiplication operation**.

Definition 1.2.26 Let $f = a_0 + 0.a_1a_2\cdots a_n\cdots$, $g = b_0 + 0.b_1b_2\cdots b_n\cdots \in \mathbb{R}$, $f, g \geqslant \underline{0}$ and let, for every $n \in \mathbb{N}$, $x_n = a_0 + 0.a_1a_2\cdots a_n$, $y_n = b_0 + 0.b_1b_2\cdots b_n$ be the lower decimal approximations of order n for f, respectively g. Then the sequence $(x_n \cdot y_n)_{n \in \mathbb{N}}$ is increasing and, for every $n \in \mathbb{N}$, $x_n \cdot y_n \leqslant (a_0+1)(b_0+1)$. Then the sequence $(x_n \cdot y_n)_{n \in \mathbb{N}}$ is convergent (see Theorem 1.2.16). We define $f \cdot g = \lim_n x_n \cdot y_n = \sup\{x_n \cdot y_n : n \in \mathbb{N}\}$.

If $f < \underline{0}, g \geqslant \underline{0}$, then we define $f \cdot g = -[(-f) \cdot g]$. Similarly, if $f \geqslant \underline{0}, g < \underline{0}$, then $f \cdot g = -[f \cdot (-g)]$ and in the case where $f < \underline{0}, g < \underline{0}$, $f \cdot g = (-f) \cdot (-g)$.

We often write fg instead of $f \cdot g$.

Remarks 1.2.27

(i) Taking into account that, for every $n \in \mathbb{N}$, $x_n \cdot 0 = 0$, it is obvious that $f \cdot \underline{0} = \underline{0} \cdot f = \underline{0}$, for every $f \in \mathbb{R}$.

(ii) For every $f \in \mathbb{R}$ and every $p \in \mathbb{N}^*$, $\underbrace{f + \cdots + f}_{p} = p \cdot f$. Indeed, if (x_n) is the sequence of lower decimal approximations of f, then $p \cdot x_n = \underbrace{x_n + \cdots + x_n}_{p} \underset{\mathbb{R}}{\to} \underbrace{f + \cdots + f}_{p}$. The constant sequence (y_n), $y_n = p$, for every $n \in \mathbb{N}$, is the sequence of lower decimal approximations of $p \in \mathbb{N} \subseteq D \subseteq \mathbb{R}$ and then $p \cdot x_n \underset{\mathbb{R}}{\to} p \cdot f$.

As in the case of addition, we will show that the definition of multiplication does not depend on the sequences of lower decimal approximations.

Proposition 1.2.28 Let $f, g \in \mathbb{R}$ and $(u_n), (v_n) \subseteq \mathbb{Q}$; if $u_n \underset{\mathbb{R}}{\to} f$ and $v_n \underset{\mathbb{R}}{\to} g$, then $u_n \cdot v_n \underset{\mathbb{R}}{\to} f \cdot g$.

Proof Let us first assume that $f, g \geqslant \underline{0}$ and let $(x_n), (y_n) \subseteq D$ be the sequences of lower decimal approximations of f, respectively, g; then $f \cdot g = \lim_n x_n \cdot y_n$.

For every $n \in \mathbb{N}$, we denote by $z_n = u_n \cdot v_n - x_n \cdot y_n = u_n \cdot v_n - u_n \cdot y_n + u_n \cdot y_n - x_n \cdot y_n = u_n(v_n - y_n) + (u_n - x_n)y_n$. In the proof of Proposition 1.2.23 we have shown that $u_n - x_n \underset{\mathbb{Q}}{\to} 0$ and $v_n - y_n \underset{\mathbb{Q}}{\to} 0$. The sequences (u_n) and (y_n) are convergent, so they are bounded (see (vii) of Remark 1.2.15). In $\mathbb{Q}$, the product of a sequence convergent to 0 and a bounded sequence converges to 0. Therefore $z_n \underset{\mathbb{R}}{\to} \underline{0}$. According to Proposition 1.2.21, $u_n \cdot v_n = x_n \cdot y_n + z_n \underset{\mathbb{R}}{\to} f \cdot g$.

If we suppose that $f \geqslant \underline{0}$ and $g < \underline{0}$, then $-v_n \underset{\mathbb{R}}{\to} -g$ (see (iii) of Remark 1.2.25). Therefore, $-(u_n \cdot v_n) = u_n \cdot (-v_n) \underset{\mathbb{R}}{\to} f \cdot (-g) = -(f \cdot g)$, and then $u_n \cdot v_n \underset{\mathbb{R}}{\to} f \cdot g$.

Similar demonstrations can be made in the cases $f < \underline{0}, g \geqslant \underline{0}$, and $f < \underline{0}, g < \underline{0}$. □

Theorem 1.2.29 $(\mathbb{R}, +, \cdot, \leqslant)$ *is an Archimedean totally ordered field* $((A5)$–$(A9)$, $(A15))$.

Proof We have shown that $(\mathbb{R}, +, \leqslant)$ is a linearly ordered commutative group. Let us show that the multiplication operation is associative and commutative.

Let $f, g, h \in \mathbb{R}$ and let $(x_n), (y_n)$ and (z_n) be the sequences of lower decimal approximations of f, g respectively, h. Using the previous proposition, $(x_n \cdot y_n) \cdot z_n \underset{\mathbb{R}}{\to} (f \cdot g) \cdot h$ and $x_n \cdot (y_n \cdot z_n) \underset{\mathbb{R}}{\to} f \cdot (g \cdot h)$. But, for every $n \in \mathbb{N}$, $(x_n \cdot y_n) \cdot z_n = x_n \cdot (y_n \cdot z_n)$, and because the limit of a convergent sequence is unique (see (ii) of Remark 1.2.15), $(f \cdot g) \cdot h = f \cdot (g \cdot h)$.

Similar demonstrations lead us to $f \cdot g = g \cdot f$ and $f \cdot (g + h) = f \cdot g + f \cdot h$.

We define $\underline{1} = \varphi(1) = 1 + 0.00 \cdots \in \mathbb{R}$; then, the sequence of lower decimal approximations of $\underline{1}$ is a constant sequence (for every $n \in \mathbb{N}, u_n = 1$). For every $f \in \mathbb{R}$, let (x_n) be the sequence of lower decimal approximations of f. Then $u_n \cdot x_n = x_n \uparrow f$, so that $\underline{1} \cdot f = f$ and then $\underline{1}$ is the neutral element in multiplication.

Let us show that every non-zero element, $f \in \mathbb{R}, f \neq \underline{0}$, has an inverse element under multiplication. From Theorem 1.2.13, $f > \underline{0}$ or $f < \underline{0}$.

Firstly, let $f = a_0 + 0.a_1 \cdots a_n \cdots > \underline{0}$; there exists $n_0 \in \mathbb{N}$ such that $a_{n_0} > 0$, and, for every $k < n_0, a_k = 0$. Then, for every $n \geqslant n_0, x_n = a_0 + 0.a_1 \cdots a_n > 0$ and so, there exists $x_n^{-1} \in \mathbb{Q}$ such that $x_n \cdot x_n^{-1} = 1$. Because $x_n \leqslant x_{n+1}$ it follows that, for every $n \geqslant n_0, x_{n+1}^{-1} \leqslant x_n^{-1}$. Since $x_n < a_{n_0} + 1$, for every $n \in \mathbb{N}, (a_{n_0} + 1)^{-1} < x_n^{-1}$. Therefore $(x_n^{-1})_{n \geqslant n_0}$ is a bounded below decreasing sequence; according to Theorem 1.2.16, there exists $\lim_n x_n^{-1} \in \mathbb{R}$; we will denote this limit with f^{-1}.

Now, let $f = a_0 + 0.a_1 \cdots a_n \cdots < \underline{0}$; according with (iii) of Remark 1.2.12, $a_0 \leqslant -1$ and then, for every $n \in \mathbb{N}, x_n = a_0 + 0.a_1 \cdots a_n < 0$. It follows that $(x_n^{-1})_{n \in \mathbb{N}}$ is a decreasing sequence. There is $p \geqslant 1$ such that $a_p \leqslant 8$ and therefore $f < -1 + 0.\underbrace{9 \cdots 9}_{p} = -\frac{1}{10^p}$. Since, for every $n \in \mathbb{N}, x_n \leqslant f < -\frac{1}{10^p}$, it follows that $-10^p < x_n^{-1}$. Therefore $(x_n^{-1})_{n \in \mathbb{N}}$ is a bounded below decreasing sequence and then there exists $\lim_n x_n^{-1} \in \mathbb{R}$; we will denote $\lim_n x_n^{-1} = f^{-1}$.

Since $x_n \underset{\mathbb{R}}{\to} f$ and $x_n^{-1} \underset{\mathbb{R}}{\to} f^{-1}$ it follows by Proposition 1.2.28 that $x_n \cdot x_n^{-1} \underset{\mathbb{R}}{\to} f \cdot f^{-1}$. But, for every $n \geqslant n_0, x_n \cdot x_n^{-1} = 1$ and then $f \cdot f^{-1} = \underline{1}$.

We know that $\leqslant$ is a total order on $\mathbb{R}$ compatible with addition. Let us show that it is also compatible with multiplication. Let $f, g, h \in \mathbb{R}$ with $f \leqslant g$ and $h \geqslant \underline{0}$. If $f = g$ or $h = \underline{0}$, then $f \cdot h = g \cdot h$ (see (i) of Remark 1.2.27). In both cases $f \cdot h \leqslant g \cdot h$.

1.2 Decimal Model

Now, let us suppose that $f < g$ and $h > \underline{0}$ and let $(x_n), (y_n), (z_n)$ be the sequences of lower decimal approximations of f, g, and respectively h. There exists $n_0 \in \mathbb{N}$ such that, for every $n \geqslant n_0$, $x_n < y_n$ and, for every $n \in \mathbb{N}$ $z_n > 0$. Therefore, for every $n \geqslant n_0$, $x_n \cdot z_n < y_n \cdot z_n$ and $x_n \cdot z_n \underset{\mathbb{R}}{\to} f \cdot h$, $y_n \cdot z_n \underset{\mathbb{R}}{\to} g \cdot h$. If we suppose that $g \cdot h < f \cdot h$, then there exists $u \in D$ such that $g \cdot h < u < f \cdot h$ and then there would be an $n_1 > n_0$ such that, for every $n \geqslant n_1$, $y_n \cdot z_n < u < x_n \cdot z_n$, which is absurd. Therefore $f \cdot h \leqslant g \cdot h$.

Let us also show that $\mathbb{R}$ verifies the Archimedean property. Let $f, g \in \mathbb{R}$ with $f > \underline{0}$, let (x_n) be the sequence of lower decimal approximations of f and (y_n) be the sequence of upper decimal approximations of g; f being strictly positive, there exists $n_0 \in \mathbb{N}$ such that $x_{n_0} > 0$. Because D is Archimedean, there exists $p \in \mathbb{N}$ such that $p \cdot x_{n_0} > y_{n_0}$ and, since (x_n) is increasing, for every $n \geqslant n_0$, $p \cdot x_n \geqslant p \cdot x_{n_0} > y_{n_0}$. As $p \cdot x_n \underset{\mathbb{R}}{\to} p \cdot f$, the (vi) of Remark 1.2.15 assures us that $p \cdot f > y_{n_0} > g$. □

Remark 1.2.30 $\underline{0} < \underline{1}$. Indeed, $\underline{0} = \varphi(0) < \varphi(1) = \underline{1}$. We used here the fact that $0 < 1$ in D (in any totally ordered commutative ring with at least two elements this inequality occurs).

We will now prove that $\mathbb{R}$ is Dedekind complete.

Theorem 1.2.31 *For every non-empty upper-bounded set $A \subseteq \mathbb{R}$, there exists the least upper bound of A in $\mathbb{R}$ ((A16)).*

Proof Let M be the set of upper bounds of A; because $A, M \neq \emptyset$, there exist $f^0 \in A$ and $g^0 \in M$, $f^0 \leqslant g^0$. In the case where $f^0 \in M$ then f^0 is the least upper bound of A, and thus the demonstration ends. Then we suppose that $f^0 \in \mathbb{R} \setminus M$, and then $f^0 < g^0$. Let $h = 0.5 \cdot (f^0 + g^0)$; it can be easily shown that $f^0 \leqslant h \leqslant g^0$, at least one of the inequalities being strict.

If $h \in M$, then we denote $f^1 = f^0$ and $g^1 = h$; if $h \in \mathbb{R} \setminus M$, then $f^1 = h$ and $g^1 = g^0$. We remark that, in both situations, we have:

$$f^0 \leqslant f^1 < g^1 \leqslant g^0, \text{ and } g^1 - f^1 = 0.5 \cdot (g^0 - f^0).$$

Let now $h = 0.5(f^1 + g^1)$; then $f^1 \leqslant h \leqslant g^1$, at least one of the inequalities being strict.

If $h \in M$, then we denote $f^2 = f^1$ and $g^2 = h$; if $h \in \mathbb{R} \setminus M$, then $f^2 = h$ and $g^2 = g^1$. We remark that, in both situations, we have:

$$f^0 \leqslant f^1 \leqslant f^2 < g^2 \leqslant g^1 \leqslant g^0, \text{ and } g^2 - f^2 = (0.5)^2 \cdot (g^0 - f^0), \text{ and so on.}$$

At step n we determine two numbers $f^n \in \mathbb{R} \setminus M$, $g^n \in M$ such that

$$f^0 \leqslant \cdots \leqslant f^n < g^n \leqslant \cdots \leqslant g^0, \text{ and } g^n - f^n = (0.5)^n \cdot (g^0 - f^0).$$

Thus, we inductively construct an increasing sequence $(f^n) \subseteq \mathbb{R} \setminus M$ and a decreasing one $(g^n) \subseteq M$ such that, for every $n \in \mathbb{N}$, $g^n - f^n = (0.5)^n \cdot (g^0 - f^0)$. According to Theorem 1.2.16, there exist $\lim_n f^n = \sup\{f^n : n \in \mathbb{N}\} \leqslant \inf\{g^n : n \in \mathbb{N}\} = \lim_n g^n$. We will show that these limits coincide. Indeed, the sequence $((0.5)^n)_{n \in \mathbb{N}}$ is bounded below and decreasing and then it is convergent to $\inf\{(0.5)^n : n \in \mathbb{N}\} = \ell$. But $\ell = \inf\{(0.5)^{4n} : n \in \mathbb{N}\}$; because $0 < (0.5)^{4n} < (0.1)^n = \frac{1}{10^n}, \forall n \in \mathbb{N}$, and $\frac{1}{10^n} \underset{\mathbb{Q}}{\to} 0$ (see (v) of Remark 1.2.15), it follows that $\ell = 0 = \underline{0}$. Therefore $g^n - f^n \underset{\mathbb{R}}{\to} \underline{0}$, and then $\lim_n f^n = \lim_n g^n = l$.

Let us show that $l = \sup A$. For every $f \in A$, and every $n \in \mathbb{N}$, $f \leqslant g^n$ ($g^n \in M$), so that $f \leqslant \inf\{g^n : n \in \mathbb{N}\} = l$, and therefore l is an upper bound of A. For every $g \in M$, and every $n \in \mathbb{N}$, $f^n < g$ ($f^n \notin M \implies \exists f \in A$ such that $f^n < f \leqslant g$). It follows that $\sup\{f^n : n \in \mathbb{N}\} \leqslant g$ or $l \leqslant g$. So that l is least upper bound of A. □

Remarks 1.2.32

(i) From the proof of the preceding theorem it can easily be deduced that there are two sequences $(h^n) \subseteq A$ and $(g^n) \subseteq M$ such that $h^n \uparrow l = \sup A$, and $g^n \downarrow l$. Indeed, since, for every $n \in \mathbb{N}$, $f^n \notin M$, there exists $h^n \in A$ such that $f^n < h^n \leqslant g^n$. Then $h^n \underset{\mathbb{R}}{\to} l$ (for every $u, v \in D$ with $u < f < v$, there exists $n_0 \in \mathbb{N}$ such that $u < f^n < h^n \leqslant g^n < v, \forall n \geqslant n_0$). Since $h^n \leqslant l = \sup A, \forall n \in \mathbb{N}$, by changing the order of the terms (h^n) becomes increasing and so $h^n \uparrow l$.

(ii) Similarly it is shown that for every non-empty below-bounded set $A \subseteq \mathbb{R}$, there exists a greatest lower bound in $\mathbb{R}$.

It follows that $\mathbb{R}$, thus constructed, verifies the axioms $(A1)$–$(A16)$, therefore it is a model of the set of real numbers—the decimal model.

1.3 Absolute Values

In the first subsection of this section, we will present some general results about absolute values or norms on a field; we will characterize Archimedean and non-Archimedean absolute values and present remarkable examples of such norms. We show that any normed field admits a unique completion up to an isomorphism. We will present Ostrowski's theorem on the classification of absolute values in the field of rational numbers. Thus, the Archimedean norms on $\mathbb{Q}$ are equivalent to the usual absolute value on $\mathbb{Q}$; the completion of $\mathbb{Q}$ with respect to such a norm leads us to Cantor's model for real numbers. A non-Archimedean norm on $\mathbb{Q}$ is, or the trivial norm, or is equivalent to a p-adic norm. A source of inspiration for some of the results presented in this paragraph is Chap. 9 of Nathan Jacobson's book [9]; this book is also recommended for those who want to deepen their study of the theory of valued fields.

1.3 Absolute Values

Definition 1.3.1 Let $(K, +, \cdot)$ be a field and let $\underline{0}$ be the additive identity and $\underline{1}$ be the multiplicative identity. We denote by 0 the additive identity and 1 the multiplicative identity in the field of real numbers $\mathbb{R}$. An **absolute value** (or **norm**) on the field K is an application $|\cdot| : K \to \mathbb{R}_+$ that has the following properties:

(1) $|x| = 0$ if and only if $x = \underline{0}$.
(2) $|x + y| \leqslant |x| + |y|$, for every $x, y \in K$.
(3) $|x \cdot y| = |x||y|$, for every $x, y \in K$.

Example 1.3.2 The map $\text{Mod} : \mathbb{R} \to \mathbb{R}_+$ defined by

$$\text{Mod}(x) = \begin{cases} x, & x \geqslant 0, \\ -x, & x < 0, \end{cases}$$

is an absolute value on $\mathbb{R}$; it is called the **usual absolute value** on $\mathbb{R}$.

Remarks 1.3.3

(i) We present some immediate consequences of the definition above:

1. $|\underline{1}| = |-\underline{1}| = 1$. Indeed, $|\underline{1}| = |\underline{1}^2| = |\underline{1}|^2$; $|\underline{1}| \leqslant |\underline{1} - \underline{1}| + |-\underline{1}|$, from where $1 \leqslant |-\underline{1}|$. In the same way, the converse inequality is proved.
2. $|-x| = |x|$, for every $x \in K$. Indeed, for every $x \in K$, $|-x| = |(-\underline{1}) \cdot x| = |-\underline{1}||x| = |x|$.
3. $|x^{-1}| = \dfrac{1}{|x|}$, for every $x \in K \setminus \{\underline{0}\}$. Indeed, $1 = |\underline{1}| = |x \cdot x^{-1}| = |x||x^{-1}|$.
4. $|x^k| = |x|^k$, for every $x \in K$, and every $k \in \mathbb{Z}$.
5. $\text{Mod}(|x| - |y|) \leqslant |x - y|$, for every $x, y \in K$ ("Mod" is the usual absolute value on $\mathbb{R}$ - see Example 1.3.2). Indeed, for every $x, y \in K$, $|x| = |(x - y) + y| \leqslant |x - y| + |y|$ from where $|x| - |y| \leqslant |x - y|$; similarly, $|y| - |x| \leqslant |x - y|$.

(ii) The concept of a norm on a field should not be confused with that of a norm on a vector space. Properties (1) and (2) from the definition above are common, but property (3) clearly distinguishes the two concepts, as we will observe in what follows. Nevertheless, like norms on vector spaces, the norms of fields allow for the construction of metrics.

Definition 1.3.4 Let $|\cdot|$ be an absolute value on a field K; the mapping $d_{|\cdot|} : K \times K \to \mathbb{R}_+$ defined by $d_{|\cdot|}(x, y) = |x - y|$, for every $x, y \in K$, is a metric on K. It will be called the **metric induced** by the norm $|\cdot|$ and the topology induced by this metric, $\tau_{d_{|\cdot|}}$, will be denoted by $\tau_{|\cdot|}$ and will be referred to as the **topology induced** by the absolute value $|\cdot|$. By rapport at this metric, the **sphere** with its center at $x \in K$ and radius $r > 0$ is $S(x, r) = \{y \in K : |y - x| = r\}$ and the **ball** of radius $r > 0$ centered at $x \in K$ is $B(x, r) = \{y \in K : |y - x| < r\}$. A sequence $(x_n)_{n \in \mathbb{N}} \subseteq K$ is **convergent** with respect to $|\cdot|$ to $x \in K$ if and only if $|x_n - x| \to 0$.

Two absolute values on K, $|\cdot|_1$ and $|\cdot|_2$, are called **equivalent** if the topologies induced by them coincide: $\tau_{|\cdot|_1} = \tau_{|\cdot|_2}$; this situation will be denoted by $|\cdot|_1 \sim |\cdot|_2$. The topologies generated by two norms coincide if and only if they have the same convergent sequences. It follows that $|\cdot|_1 \sim |\cdot|_2$ if and only if $|x_n - x|_1 \to 0 \Leftrightarrow |x_n - x|_2 \to 0$, for every $(x_n) \subseteq K$, $x \in K$.

The first difference between norms on vector spaces and norms on fields can be noticed in the following theorem that characterizes equivalent norms on fields.

Theorem 1.3.5 *Let $|\cdot|_1, |\cdot|_2$ be two absolute values on a field K. The following statements are equivalent:*

(1) $|\cdot|_1 \sim |\cdot|_2$.
(2) $|x|_1 < 1 \Leftrightarrow |x|_2 < 1$.
(3) *There exists $\alpha > 0$ such that $|x|_1 = |x|_2^\alpha$.*

Proof (1) $\Rightarrow$ (2): We suppose that $|\cdot|_1 \sim |\cdot|_2$; then $\tau_1 \equiv \tau_{|\cdot|_1} = \tau_{|\cdot|_2} \equiv \tau_2$. For every $x \in K$ and every $n \in \mathbb{N}^*$, let $x^n = \underbrace{x \cdot \ldots \cdot x}_{n \text{ times}} \in K$; the sequence $(x^n)_{n \in \mathbb{N}} \subseteq K$ is τ_1-convergent to $\underline{0}$ if and only if $|x^n|_1 = |x|_1^n \to 0$ and this is equivalent with $|x|_1 < 1$. Therefore

$$\{x \in K : |x|_1 < 1\} = \left\{x : x^n \xrightarrow[K]{\tau_1} \underline{0}\right\} = \left\{x : x^n \xrightarrow[K]{\tau_2} \underline{0}\right\} = \{x \in K : |x|_2 < 1\},$$

which concludes the proof.

(2) $\Rightarrow$ (3): First, we note that, under assumption (2), $|x|_1 = 1 \Leftrightarrow |x|_2 = 1$. Indeed, if we suppose that there is $x \in K^*$ with $|x|_1 = 1$ and $|x|_2 \neq 1$, then either $|x|_2 < 1 \stackrel{2)}{\Rightarrow} |x|_1 < 1$, which is absurd, or $|x|_2 > 1 \Rightarrow |x^{-1}|_2 < 1 \stackrel{2)}{\Rightarrow} |x^{-1}|_1 < 1 \Rightarrow |x|_1 > 1$, which is absurd.

(a) If $|x|_1 = 1$, for every $x \in K^* = K \setminus \{0\}$, then it follows that $|x|_2 = 1$, for every $x \in K^*$, so that (3) is checked with any $\alpha > 0$.

(b) Let us suppose that there is $a \in K^*$ with $|a|_1 \neq 1$; we can even assume that $|a|_1 > 1$, otherwise we replace a with a^{-1}.

For any $x \in K^*$, let $\beta = \dfrac{\log |x|_1}{\log |a|_1}$; then $|x|_1 = |a|_1^\beta$ (log is the natural logarithm).

Then, for every $p \in \mathbb{Z}$, $q \in \mathbb{N}^*$ with $pq^{-1} > \beta$ it follows that $|x|_1 < |a|_1^{\frac{p}{q}}$, or equivalent $\left|\dfrac{x^q}{a^p}\right|_1 < 1$; then, from (2), it follows that $\left|\dfrac{x^q}{a^p}\right|_2 < 1$ or equivalent $|x|_2 < |a|_2^{\frac{p}{q}}$. Since pq^{-1} is an arbitrary rational number greater than β, it follows that $|x|_2 \leq |a|_2^\beta$. We reason similarly for rational numbers that are smaller than β and obtain the inverse inequality and therefore $|x|_2 = |a|_2^\beta$.

1.3 Absolute Values

Then $\beta = \dfrac{\log |x|_2}{\log |a|_2}$ and so $\dfrac{\log |x|_2}{\log |a|_2} = \dfrac{\log |x|_1}{\log |a|_1}$ from where $|x|_2 = |x|_1^\alpha$, where $\alpha = \dfrac{\log |a|_2}{\log |a|_1}$.

(3) $\Rightarrow$ (1): In hypothesis (3) it follows that, for every $(x_n) \subseteq K$ and every $x \in K$, $|x_n - x|_1 \to 0 \Leftrightarrow |x_n - x|_2 \to 0$ and so $|\cdot|_1 \sim |\cdot|_2$. □

Corollary 1.3.6 *If two absolute values $|\cdot|_1, |\cdot|_2$ on a field are equivalent then $|x|_1 = 1 \Leftrightarrow |x|_2 = 1$.*

Remark 1.3.7 We remark that, if two norms on a field are equivalent, the spheres and the balls with a center at $\underline{0}$ and radius 1 coincide. The norms of vector spaces do not have this geometric property. For example, in $\mathbb{R}^2$, the Euclidean norm $\|x\|_2 = \sqrt{x_1^2 + x_2^2}$ and the norm $\|x\|_1 = |x_1| + |x_2|$, for all $x = (x_1, x_2) \in \mathbb{R}^2$, are equivalent. So, while the unit sphere for the first one is a circle with its center at the origin, the unit sphere for the second one is a square with its center at the origin and sides parallel to the two bisectors.

1.3.1 Non-Archimedean Norms

Definition 1.3.8 Let $(K, +, \cdot)$ be a field with $\underline{0}, \underline{1}$ the additive, respectively multiplicative, identities. An application $|\cdot| : K \to \mathbb{R}_+$ is a **non-Archimedean absolute value** (or **non-Archimedean norm**) if it satisfies the following conditions:

(1) $|x| = 0$ if and only if $x = \underline{0}$.
(2) $|x + y| \leqslant \max\{|x|, |y|\}$, for every $x, y \in K$.
(3) $|x \cdot y| = |x||y|$, for every $x, y \in K$.

We note that condition (2) in this definition is stronger than condition (2) in Definition 1.3.1; thus, any non-Archimedean norm is a field norm.

A norm on the field K is called an **Archimedean norm** if it is not non-Archimedean. So $|\cdot|$ is an Archimedean norm if it satisfies conditions (1), (2), and (3) in Definition 1.3.1 and there exist at least two elements $x, y \in K$ such that: $|x + y| > \max\{|x|, |y|\}$.

Examples 1.3.9

(1) Let K be an arbitrary field; we define the **trivial** absolute value on K: $|\cdot|_0 : K \to \mathbb{R}_+$ by

$$|x|_0 = \begin{cases} 0, & x = \underline{0}, \\ 1, & x \in K \setminus \{\underline{0}\}, \end{cases} \text{ for all } x \in K.$$

$|\cdot|_0$ is a non-Archimedean absolute value.

(2) A more interesting class of non-Archimedean norms is that of p-adic norms on $\mathbb{Q}$. Let p be a prime number; any rational number $x \in \mathbb{Q}^*$ admits a unique

representation of the form $x = p^k \cdot \frac{a}{b}$, where $k \in \mathbb{Z}$, and $a \in \mathbb{Z}^*, b \in \mathbb{N}^*$ are two numbers prime with p. Then we define

$$|x|_p = p^{-k} \text{ and } |0|_p = 0.$$

The mapping $|\cdot|_p : \mathbb{Q} \to \mathbb{R}_+$ is a non-Archimedean absolute value on $\mathbb{Q}$. Indeed, condition (1) in Definition 1.3.8 is clearly satisfied. Let now $x = p^k \cdot \frac{a}{b}$, $y = p^l \cdot \frac{c}{d}$, where $k, l \in \mathbb{Z}, a, c \in \mathbb{Z}^*, b, d \in \mathbb{N}^*$ are prime with p. If $k < l$, then $x + y = p^k \cdot \dfrac{ad + p^{l-k}bc}{bd}$. Because $ad + p^{l-k}bc, bd$ are prime with p,

$$|x + y|_p = p^{-k} = \max\{|x|_p, |y|_p\}.$$

If $l < k$, the reasoning is the same. If $k = l$, then $x + y = p^k \cdot \dfrac{ad + bc}{bd}$; there exists $m, e \in \mathbb{Z}, m \geqslant k$ such that $x + y = p^m \cdot \dfrac{e}{bd}$, where e, bd are prime with p. It follows that

$$|x + y|_p = p^{-m} \leqslant p^{-k} = \max\{|x|_p, |y|_p\}.$$

Finally, let us verify condition (3) of Definition 1.3.8; $x \cdot y = p^{k+l} \dfrac{ac}{bd}$ and, since ac, bd are prime with p,

$$|x \cdot y|_p = p^{-k-l} = |x|_p |y|_p.$$

The norms $|\cdot|_p$ are called the *p*-**adic** norms or *p*-**adic** absolute values on $\mathbb{Q}$.

We remark that $|p^n|_p = |p|_p^n = p^{-n}$, for every $n \in \mathbb{N}$; so that $p^n \xrightarrow{|\cdot|_p} 0$.

(3) The modulus norm defined in Example 1.3.2 is an Archimedean norm on $\mathbb{R}$ and thus also on $\mathbb{Q}$. Indeed, $\text{Mod}(1 + 1) = 2 > 1 = \max\{\text{Mod}(1), \text{Mod}(1)\}$.

Remark 1.3.10 Let $|\cdot|$ be a non-Archimedean absolute value on the field K. Then

$$|k \cdot \underline{1}| \leqslant 1, \text{ for every } k \in \mathbb{Z}.$$

Indeed, if $k \in \mathbb{N}^*$, then $k \cdot \underline{1} = \underbrace{\underline{1} + \cdots + \underline{1}}_{k \text{ times}}$. From (1) of Remark 1.3.3, $|\underline{1}| = 1$, and then $|k \cdot \underline{1}| \leqslant \max\{|\underline{1}|, \cdots, |\underline{1}|\} = 1$. If $k = 0$, then $|k \cdot \underline{1}| = |\underline{0}| = 0$ (see Definition 1.3.1). Finally, if $k \in \mathbb{Z} \setminus \mathbb{N}$, then $|k \cdot \underline{1}| = |-((-k) \cdot \underline{1})| = |(-k) \cdot \underline{1}| \leqslant 1$ (see (2) of Remark 1.3.3).

In the following theorem, we will show that this property characterizes non-Archimedean norms on a field.

Theorem 1.3.11 *An absolute value $|\cdot|$ on a field K is non-Archimedean if and only if $|n \cdot \underline{1}| \leqslant 1$, for every $n \in \mathbb{N}$.*

1.3 Absolute Values

Proof We observed in the preceding remark that, if $|\cdot|$ is a non-Archimedean norm on K, then $|n \cdot \underline{1}| \leqslant 1$, for every $n \in \mathbb{N}$.

Now, let's assume that $|\cdot|$ is a norm on the field K that satisfies the condition $|n \cdot \underline{1}| \leqslant 1$, for every $n \in \mathbb{N}$. According to (4) of Remark 1.3.3, for every $n \in \mathbb{N}$, and every $x, y \in K$,

$$|x+y|^n = |(x+y)^n| = \left| x^n + \binom{n}{1} \cdot x^{n-1} \cdot y + \cdots + \binom{n}{n} \cdot y^n \right| \leqslant$$

$$\leqslant |x|^n + \left| \binom{n}{1} \cdot \underline{1} \right| |x|^{n-1}|y| + \cdots + \left| \binom{n}{n} \cdot \underline{1} \right| |y|^n \leqslant |x|^n + |x|^{n-1}|y| + \cdots + |y|^n \leqslant$$

$$\leqslant (n+1) \max\{|x|, |y|\}^n, \text{ from where}$$

$$|x+y| \leqslant (n+1)^{\frac{1}{n}} \max\{|x|, |y|\}, \forall n \in \mathbb{N}.$$

If we take the limit in the last inequality for n tending to infinity, we obtain

$$|x+y| \leqslant \max\{|x|, |y|\}, \text{ for every } x, y \in K,$$

which implies that the norm $|\cdot|$ is non-Archimedean. □

Corollary 1.3.12 *An absolute value $|\cdot|$ on the field K with $\underline{1}$ the multiplicative identity is an Archimedean norm if and only if there is $n_0 \in \mathbb{N}$ such that $|n_0 \cdot \underline{1}| > 1$.*

Remarks 1.3.13

(i) With the notations in the previous corollary, we observe that $|n_0^p \cdot \underline{1}| = |(n_0 \cdot \underline{1})^p| = |n_0 \cdot \underline{1}|^p \xrightarrow[p \to +\infty]{} +\infty$. It follows that a normed field is Archimedean if and only if the set $\mathbb{N} \cdot \underline{1} = \{n \cdot \underline{1} : n \in \mathbb{N}\}$ ("the natural numbers" of K) is unbounded above in norm.

(ii) Let K be a field; if K's prime field is a field of integers modulo the prime number p (see Theorem 1.1.7), then any norm on K is non-Archimedean. In fact, in this case, $\mathbb{N} \cdot \underline{1} = \{n \cdot \underline{1} : n \in \mathbb{N}\} = \{\underline{0}, \underline{1}, \cdots, (p-1) \cdot \underline{1}\}$ is a finite set and thus it cannot be unbounded in any norm $|\cdot|$ on K. The observation above ensures us that $|\cdot|$ is non-Archimedean.

1.3.2 Completing a Normed Field

As mentioned in Definition 1.3.4, a metric can be defined on any normed field $(K, |\cdot|)$ by $d_{|\cdot|}(x, y) = |x - y|$, for every $x, y \in K$. If the metric space $(K, d_{|\cdot|})$ is complete, we say that $(K, |\cdot|)$ is a complete normed field. If not, from a completion result for metric spaces (the Hausdorff theorem), the metric space $(K, d_{|\cdot|})$ has a unique completion up to an isometry; in this subsection, we will show that in the case of normed fields, the completion is also a normed field and the isometry above is also a field isomorphism. One common way to complete a normed field is by using Cauchy sequences. Given a normed field $(K, |\cdot|)$, a **Cauchy sequence** in K is a sequence $(x_n)_{n \in \mathbb{N}}$ such that for every $\varepsilon > 0$, there exists an $n_0 \in \mathbb{N}$ such that for all $n, m \geqslant n_0$, we have $|x_n - x_m| < \varepsilon$. It is easy to verify that any convergent sequence in $(K, |\cdot|)$ (see Definition 1.3.4) is a Cauchy sequence. The converse of this statement is not true in any normed field. A normed field in which any Cauchy sequence is convergent is called a **complete** normed field.

Theorem 1.3.14 *For any normed field $(K, |\cdot|)$, there exists a complete normed field $(\widehat{K}, |\cdot|\hat{\,})$ and a field homomorphism $i : K \to \widehat{K}$ such that i is an isometry between K and $i(K)$ and $i(K)$ is dense in $\widehat{K}$. The field $(\widehat{K}, |\cdot|\hat{\,})$ is unique up to an isometric field isomorphism.*

Proof We will present only a sketch of the proof of this theorem.

Let $\mathcal{K} = \{(x_n)_n \subseteq K : (x_n)_n$ is a Cauchy sequence in $K\}$. If K is complete, the $\mathcal{K}$ coincides with the set of convergent sequences on K; in this case, $\widehat{K} = K, |\cdot|\hat{\,} = |\cdot|$ and i is the identity.

Now, we assume that there exist non-converging Cauchy sequences in K. We define two operations on $\mathcal{K}$ by:

$$(x_n)_n + (y_n)_n = (x_n + y_n)_n \text{ and } (x_n)_n \cdot (y_n)_n = (x_n \cdot y_n)_n.$$

Then $(\mathcal{K}, +, \cdot)$ is a unitary commutative ring. The set $\mathcal{O} = \{(x_n)_n : x_n \to 0\}$ is a maximal ideal in $\mathcal{K}$. Then the quotient of $\mathcal{K}$ by $\mathcal{O}$, $\widehat{K} = \mathcal{K}/\mathcal{O}$, is a field; we recall that $\widehat{x} \in \widehat{K}$ is of form $\widehat{x} = [(x_n)] = \{(y_n)_n \in \mathcal{K} : x_n - y_n \to 0\}$.

Let $i : K \to \widehat{K}, i(x) = [\bar{x}]$, where $\bar{x}$ is the constant sequence with the general term x ($i(x)$ is the set of all sequences converging to x). The mapping i is a field homomorphism.

For every $\widehat{x} = [(x_n)] \in \widehat{K}$, $(|x_n|)_n$ is a Cauchy sequence in $\mathbb{R}$; the statement is a consequence of the inequality $\text{Mod}(|x_m| - |x_n|) \leqslant |x_m - x_n|$—see 5 of Remark 1.3.3. Therefore there exists $\lim_{n \to \infty} |x_n| \in \mathbb{R}$ (see **(p)** of Remark 1.1.4); this limit does not depend on $(x_n) \in \widehat{x}$.

Thus, we can consistently define $|\cdot|\hat{\,} : \widehat{K} \to \mathbb{R}_+$ by $|\widehat{x}|\hat{\,} = \lim_n |x_n|$.

It is easily shown that $|\cdot|\hat{\,}$ is an absolute value on the field $\widehat{K}$ and that i is an isometry between K and $i(K)$ (for every $x \in K$, $|i(x)|\hat{\,} = |x|$).

Let $\widehat{x} = [(x_n)] \in \widehat{K}$; for every $\varepsilon > 0$ there exists $n_0 \in \mathbb{N}$ such that, for every $m, n \geqslant n_0$, $|x_m - x_n| < \frac{\varepsilon}{2}$.

Then $\widehat{x_0} = [\bar{x}_{n_0}] \in i(K)$ and $|\widehat{x} - \widehat{x_0}|\hat{\,} = \lim_n |x_m - x_n| \leqslant \frac{\varepsilon}{2} < \varepsilon$.

It follows that $i(K)$ is dense in $\widehat{K}$.

Let us show that $(\widehat{K}, |\cdot|\hat{\,})$ is a complete normed field.

Let $(\widehat{x^p})_{p \in \mathbb{N}} \subseteq (\widehat{K}, |\cdot|\hat{\,})$ be a Cauchy sequence. Then, for every $\varepsilon > 0$, there exists $p_0 \in \mathbb{N}$ such that, for every $p, q \geqslant p_0$, $|\widehat{x^p} - \widehat{x^q}|\hat{\,} < \frac{\varepsilon}{3}$.

Since $i(K)$ is dense in $\widehat{K}$, for every $p \in \mathbb{N}^*$, there exists $x_p \in K$ such that $|i(x_p) - \widehat{x^p}|\hat{\,} < \frac{1}{p}$.

1.3 Absolute Values

Let $p_1 \geqslant p_0$ be such that $\frac{1}{p} < \frac{\varepsilon}{3}$, for every $p \geqslant p_1$. Then, for every $p, q \geqslant p_1$,

$$|x_p - x_q| = |i(x_p) - i(x_q)|\hat{\;} \leqslant |i(x_p) - \widehat{x^p}|\hat{\;} + |\widehat{x^p} - \widehat{x^q}|\hat{\;} + |\widehat{x^q} - i(x_q)|\hat{\;} <$$

$$< \frac{1}{p} + \frac{\varepsilon}{3} + \frac{1}{q} < \varepsilon.$$

It follows that $(x_p)_p \subseteq K$ is a Cauchy sequence and then $\hat{x} = [(x_p)] \in \widehat{K}$. Furthermore,

$$|\hat{x} - \widehat{x^p}|\hat{\;} \leqslant |\hat{x} - i(x_p)|\hat{\;} + |i(x_p) - \widehat{x^p}|\hat{\;} \leqslant \frac{1}{p} + \lim_q |x_p - x_q| \xrightarrow[p \to \infty]{} 0.$$

Therefore $(\widehat{K}, |\cdot|\hat{\;})$ is a complete normed field that contains a dense subfield isomorph and isometric with K.

Let now $(K_1, |\cdot|_1)$, $(K_2, |\cdot|_2)$ be two complete normed fields and let $i_j : K \to K_j$, $j = 1, 2$ be two field homomorphisms such that every i_j is an isometry between K and $i_j(K)$ and $i_j(K)$ is dense in K_j, $j = 1, 2$. Then $i_2 \circ i_1^{-1} : i_1(K_1) \to i_2(K_2)$ is an isometric isomorphism between the fields $i_1(K_1)$ and $i_2(K_2)$. We can extend it to an isometric isomorphism between K_1 and K_2. □

Definition 1.3.15 The field $(\widehat{K}, |\cdot|\hat{\;})$ constructed in the previous theorem is called the **completion** of the field K; as a result of the theorem, it is unique up to an isometric isomorphism of normed fields.

Remarks 1.3.16

(i) Cantor's construction of the real numbers is similar to the above construction: the real numbers are the completion of the rational numbers using the usual absolute value on $\mathbb{Q}$. Notice, however, that the construction from this theorem makes explicit use of the completeness of the real numbers, so completion of the rational numbers needs a slightly different treatment.

(ii) The completion $(\widehat{K}, |\cdot|\hat{\;})$ is non-Archimedean if and only if $(K, |\cdot|)$ is non-Archimedean. Indeed, if we suppose that $(K, |\cdot|)$ is non-Archimedean, then, for every $\hat{x}, \hat{y} \in \widehat{K}$ and for every $\varepsilon > 0$, there exist $x, y \in K$ such that $|i(x) - \hat{x}|\hat{\;} < \varepsilon$ and $|i(y) - \hat{y}|\hat{\;} < \varepsilon$. Then

$$|\hat{x} + \hat{y}|\hat{\;} \leqslant |\hat{x} - i(x) + i(x) + i(y) - i(y) + \hat{y}|\hat{\;} \leqslant |\hat{x} - i(x)|\hat{\;} + |x + y| + |\hat{y} - i(y)|\hat{\;} <$$

$$< 2\varepsilon + \max\{|x|, |y|\} \leqslant 2\varepsilon + \max\{\varepsilon + |\hat{x}|\hat{\;}, \varepsilon + |\hat{y}|\hat{\;}\} < 3\varepsilon + \max\{|\hat{x}|\hat{\;}, |\hat{y}|\hat{\;}\}.$$

In the above inequalities, we have used the fact that: $|x| = |i(x) - \hat{x} + \hat{x}|\hat{\;} \leqslant \varepsilon + |\hat{x}|\hat{\;}$ and similarly for y.

The converse statement is evident.

(iii) From (3) of Theorem 1.3.5, we note that, if $|\cdot|_1$ and $|\cdot|_2$ are two equivalent norms on the field K then they are uniformly equivalent (they have the same Cauchy sequences). From the proof of the previous theorem, it follows immediately that they have the same completed $\widehat{\mathcal{K}} = \mathcal{K}/\mathcal{O}$ and that the completed norms are equivalent: $|\cdot|\widehat{\,}_1 \sim |\cdot|\widehat{\,}_2$.

1.3.3 Classification of Norms on $\mathbb{Q}$

The set of rational numbers $\mathbb{Q}$ is a subfield of $\mathbb{R}$. Therefore the trace on $\mathbb{Q}$ of the usual norm Mod is an Archimedean absolute value on $\mathbb{Q}$ (see Example 1.3.2 and (3) of Example 1.3.9). Completion of $\mathbb{Q}$ with respect to this norm is $\mathbb{R}$. Two natural questions arise: are there other norm structures on $\mathbb{Q}$? If the answer is affirmative, what are the completions of $\mathbb{Q}$ with respect to these norms?

The following theorem will give a complete answer to the first question. Thus, we will show that any other Archimedean norm on $\mathbb{Q}$ is equivalent to the usual norm. Non-Archimedean norms are either the trivial norm or are equivalent to p-adic norms (see (2) of Example 1.3.9).

Theorem 1.3.17 (Ostrowski's Theorem)

(a) If $|\cdot|$ is an Archimedean norm on $\mathbb{Q}$, then there is $\alpha \in]0, 1]$ such that

$$|q| = (\mathrm{Mod}(q))^\alpha, \text{ for every } q \in \mathbb{Q}.$$

(b) If $|\cdot|$ is a non-Archimedean norm on $\mathbb{Q}$, then $|\cdot|$ is the trivial norm or there exist $\alpha > 0$ and $p \in \mathbb{N}$ a prime number such that

$$|q| = |q|_p^\alpha, \text{ for every } q \in \mathbb{Q}.$$

Proof

(a) Let $|\cdot|$ be an Archimedean norm on $\mathbb{Q}$. For every $n \in \mathbb{N}$,

$$|n \cdot \underline{1}| = |n \cdot 1| = |n| = |\underbrace{1 + \cdots + 1}_{n \text{ times}}| \leqslant \underbrace{|1| + \cdots + |1|}_{n \text{ times}} = n. \tag{1.1}$$

According to Corollary 1.3.12, there exists $a \in \mathbb{N}$ such that $|a| = |a \cdot \underline{1}| > 1$; from (1.1), $a \geqslant |a|$. It follows that $a \geqslant 2$.

We denote with $\alpha = \dfrac{\log |a|}{\log a} \in]0, 1]$; it follows that

$$|a| = a^\alpha. \tag{1.2}$$

For every $n \in \mathbb{N}^*$, there exists $k = \left\lfloor \dfrac{\log n}{\log a} + 1 \right\rfloor \in \mathbb{N}^*$ such that $a^{k-1} \leqslant n < a^k$; then n admits a writing in base a of the form: $n = x_0 + x_1 a + \cdots + x_{k-1} a^{k-1}$,

1.3 Absolute Values

where, for every $i = 0, \cdots, k-1$, $0 \leqslant x_i \leqslant a-1$ and $x_{k-1} \geqslant 1$. From (1.1) and (1.2) it follows that

$$|n| \leqslant |x_0| + |x_1| \cdot |a| + \cdots + |x_{k-1}| \cdot |a|^{k-1} \leqslant x_0 + x_1 a^{\alpha} + \cdots x_{k-1} a^{\alpha(k-1)} \leqslant$$

$$\leqslant (a-1) \cdot (1 + a^{\alpha} + \cdots a^{(k-1)\alpha}) = (a-1) \cdot \frac{a^{k\alpha} - 1}{a^{\alpha} - 1} <$$

$$< (a-1) \cdot \frac{a^{k\alpha}}{a^{\alpha} - 1} = \frac{(a-1) \cdot a^{\alpha}}{a^{\alpha} - 1} \cdot a^{(k-1)\alpha} \leqslant \frac{(a-1) \cdot a^{\alpha}}{a^{\alpha} - 1} \cdot n^{\alpha} = A \cdot n^{\alpha},$$

where $A = \frac{(a-1) \cdot a^{\alpha}}{a^{\alpha} - 1}$. Therefore

$$|n| \leqslant A \cdot n^{\alpha}, \text{ for every } n \in \mathbb{N}. \tag{1.3}$$

By replacing n with n^m in (1.3), we obtain $|n|^m = |n^m| \leqslant A \cdot n^{m\alpha}$, from where

$$|n| \leqslant \sqrt[m]{A} \cdot n^{\alpha}, \text{ for every } m \in \mathbb{N}. \tag{1.4}$$

If $m \to +\infty$ in (1.4), we obtain

$$|n| \leqslant n^{\alpha}, \text{ for every } n \in \mathbb{N}. \tag{1.5}$$

Now, again let $n \in \mathbb{N}$ be arbitrary, and $k = \left\lfloor \frac{\log n}{\log a} + 1 \right\rfloor \in \mathbb{N}^*$ such that $a^{k-1} \leqslant n < a^k$. If we denote by $b = a^k - n$, then $0 < b \leqslant a^k - a^{k-1}$. From (1.2)

$$|n| \geqslant |a^k| - |b| = a^{k\alpha} - |b|. \tag{1.6}$$

Since $b \in \mathbb{N}$, from (1.5) we obtain

$$|b| \leqslant b^{\alpha} \leqslant (a^k - a^{k-1})^{\alpha}. \tag{1.7}$$

Finally, (1.6) and (1.7) lead us to:

$$|n| \geqslant a^{k\alpha} - (a^k - a^{k-1})^{\alpha} = \left[1 - \left(1 - \frac{1}{a}\right)^{\alpha} \right] \cdot a^{k\alpha} = B \cdot a^{k\alpha} > B \cdot n^{\alpha},$$

where $B = \left[1 - \left(1 - \frac{1}{a}\right)^{\alpha} \right]$. Therefore

$$|n| > B \cdot n^{\alpha}, \text{ for every } n \in \mathbb{N}. \tag{1.8}$$

By replacing n with n^m in (1.8), we obtain $|n|^m = |n^m| > B \cdot n^{m\alpha}$, from where

$$|n| \geqslant \sqrt[m]{B} \cdot n^\alpha \text{ for every } n \in \mathbb{N}. \tag{1.9}$$

If in (1.9) $m \to +\infty$, we obtain

$$|n| \geqslant n^\alpha, \text{ for every } n \in \mathbb{N}. \tag{1.10}$$

From (1.5) and (1.10) we obtain

$$|n| = n^\alpha = (\text{Mod}(n))^\alpha, \text{ for every } n \in \mathbb{N}. \tag{1.11}$$

Let now $k \in \mathbb{Z} \setminus \mathbb{N}$ be arbitrary; then using (1.11),

$$|k| = |-k| = (\text{Mod}(-k))^\alpha = (\text{Mod}(k))^\alpha. \tag{1.12}$$

Finally, for every $q = \frac{k}{n} \in \mathbb{Q}$ ($k \in \mathbb{Z}, n \in \mathbb{N}^*$), we obtain using (1.11) and (1.12),

$$|q| = |k \cdot n^{-1}| = |k| \cdot \frac{1}{|n|} = \frac{(\text{Mod}(k))^\alpha}{(\text{Mod}(n))^\alpha} = \left(\text{Mod}\left(\frac{k}{n}\right)\right)^\alpha = (\text{Mod}(q))^\alpha.$$

(b) Let $|\cdot|$ be a non-Archimedean and non-trivial norm on $\mathbb{Q}$; from Theorem 1.3.11, $|n| \leqslant 1$, for all $n \in \mathbb{N}$. If we assume that $|n| = 1$, for all $n \in \mathbb{N}^*$, then for all $q \in \mathbb{Q}^*$, $|q| = 1$, which would mean that the norm $|\cdot|$ is a trivial norm.

Therefore, there exists $n_0 \in \mathbb{N}^*$ such that $0 < |n_0| < 1$. Using the prime factorization of $n_0 = p_1^{a_1} \cdots p_r^{a_r}$, we deduce that $0 < |p_1|^{a_1} \cdots |p_r|^{a_r} < 1$, and then there exists a prime number p such that $0 < |p| < 1$. Let $\alpha = -\frac{\log |p|}{\log n} > 0$; then

$$|p| = p^{-\alpha}. \tag{1.13}$$

For any $k \in \mathbb{Z}$, k not divisible by p, there exist $r, s \in \mathbb{Z}$ such that $rk + sp = 1$. If we assume that $|k| < 1$, since $|r| \leqslant 1$ and $|s| \leqslant 1$, it results that

$$|1| = |rk + sp| \leqslant \max\{|r||k|, |s||p|\} < 1,$$

which is absurd. Therefore, any $k \in \mathbb{Z}^*$ prime with p has the norm 1. Let $q = p^k \frac{m}{n} \in \mathbb{Q}$, where $m \in \mathbb{Z}$ and $n \in \mathbb{N}^*$ are prime with p; it follows from (1.13) that

$$|q| = |p|^k = p^{-k\alpha} = |q|_p^\alpha.$$

$\square$

1.4 *p*-adic Numbers

In this section, we will sketch the construction of non-Archimedean models of $\mathbb{R}$: the completions of $\mathbb{Q}$ with respect to *p*-adic norms.

We noticed in (iii) of Remark 1.3.16 that every two equivalent norms on a field lead to the same completed field. According to Ostrowski's theorem, any Archimedean norm on $\mathbb{Q}$ is equivalent to the usual norm Mod. It follows from the above observation that there is only one Archimedean completion of $\mathbb{Q}$, which is the set of real numbers.

Any non-Archimedean norm on $\mathbb{Q}$ is either the trivial norm or is equivalent to one of the *p*-adic norms. The trivial norm induces the discrete metric on $\mathbb{Q}$, with respect to which $\mathbb{Q}$ is a complete metric space and therefore a complete normed field.

In what follows, we will construct the completion $\mathbb{Q}_p$ of $\mathbb{Q}$ with respect to a *p*-adic norm. This completion is called the *p*-adic numbers field and, in ultrametric analysis, plays the same role that the set of real numbers plays in real analysis.

Let *p* be a prime number; following (2) of Example 1.3.9, for every $x \in \mathbb{Q}^*$, there is a unique $k \in \mathbb{Z}$ such that $x = p^k \cdot \frac{a}{b}$, where $a \in \mathbb{Z}^*, b \in \mathbb{N}^*$ are two numbers prime with *p*. Then $|x|_p = p^{-k}$ and $|0|_p = 0$.

The application $|\cdot|_p : \mathbb{Q} \to \mathbb{R}_+$ is a non-Archimedean norm on $\mathbb{Q}$. According to Theorem 1.3.14, the normed field $(\mathbb{Q}, |\cdot|_p)$ admits a unique completion $(\widehat{\mathbb{Q}}, \widehat{|\cdot|_p}) \equiv (\mathbb{Q}_p, |\cdot|_p)$ up to an isometric isomorphism, which is also a non-Archimedean field (see (ii) of Remark 1.3.16). In what follows, we aim to construct this completion effectively, starting from $\mathbb{Q}$.

1.4.1 *p*-adic Integers

We recall the congruence relation modulo q on $\mathbb{Z}$; thus, if $q \in \mathbb{N}, q > 1$, we define the following binary relation on $\mathbb{Z}$ called congruence modulo q: we say that $a, b \in \mathbb{Z}$ are **congruent modulo** q and write $a \equiv b(\bmod q)$ if q divides $a - b$ (that is $q|(a-b)$). The congruence modulo q is an equivalence relation that is compatible with the operations of addition, and multiplication.

Let *p* be a prime number and let

$$\mathcal{Z} = \{(x_n)_{n \in \mathbb{N}} \subseteq \mathbb{Z} : x_n \equiv x_{n-1}(\bmod p^n), \text{ for every } n \in \mathbb{N}^*\}.$$

Let's give examples of integer sequences found in $\mathcal{Z}$:

Examples 1.4.1

(1) Let $(x_n)_n \subseteq \mathbb{Z}, x_n = p^{n+1}$, for every $n \in \mathbb{N}$; obviously, for every $n \in \mathbb{N}^*, p^n|(x_n - x_{n-1})$ and so $(x_n)_n \in \mathcal{Z}$.
(2) Let $(x_n)_n \subseteq \mathbb{Z}, x_n = 1 + p + \cdots + p^n$, for every $n \in \mathbb{N}$; then $(x_n)_n \in \mathcal{Z}$.

(3) Let $x \in \mathbb{Z}$ and let $(x_n)_n \subseteq \mathbb{Z}$ be the constant sequence x ($x_n = x$, for every $n \in \mathbb{N}$); then $(x_n)_n \in \mathcal{Z}$.

On the set $\mathcal{Z}$, we define a relation by:

$$(x_n)_n \sim (y_n)_n \iff x_n \equiv y_n (\text{mod } p^{n+1}), \text{ for every } n \in \mathbb{N}.$$

It can be easily observed that $\sim$ is an equivalence relation on $\mathcal{Z}$.

Definition 1.4.2 The quotient set $\mathcal{Z}/\sim$ is called the set of *p*-**adic integers** and it is denoted by $\mathbb{Z}_p$.

Let $\widehat{x} = [(x_n)] \in \mathbb{Z}_p$ be the equivalence class of the sequence $(x_n)_n \subseteq \mathbb{Z}$, that is, the set of all sequences in $\mathcal{Z}$ equivalent to $(x_n)_n$. For all $n \in \mathbb{N}$, we denote by $\bar{x}_n$ the smallest non-negative integer with the property that:

$$x_n \equiv \bar{x}_n (\text{mod } p^{n+1}).$$

Obviously

$$0 \leqslant \bar{x}_n < p^{n+1}, \text{ for every } n \in \mathbb{N},$$

and

$$\bar{x}_n \equiv x_n \equiv x_{n-1} \equiv \bar{x}_{n-1} (\text{mod } p^n), \text{ for every } n \in \mathbb{N}^*.$$

Therefore $(\bar{x}_n)_n \in \mathcal{Z}$ and $(x_n)_n \sim (\bar{x}_n)_n$ so that $\widehat{x} = [(\bar{x}_n)_n]$.

The sequence $(\bar{x}_n)_n$ is called the **canonical representation** of the *p*-adic integer $\widehat{x}$.

Proposition 1.4.3

(1) Let $(\bar{x}_n)_n$ be the canonical representation of the p-adic integer $\widehat{x} \in \mathbb{Z}_p$; then there exists a sequence $(a_n)_n \subseteq \{0, 1, \cdots, p-1\}$ such that, for all $n \in \mathbb{N}$, $\bar{x}_n = \sum_{k=0}^{n} a_k p^k = a_0 + a_1 p + \cdots + a_n p^n$.

(2) Two p-adic integers coincide if and only if they have the same canonical representation.

Proof

(1) Let $(\bar{x}_n)_n$ be the canonical representation of the *p*-adic integer $\widehat{x} \in \mathbb{Z}_p$; then, for every $n \in \mathbb{N}^*$, $\bar{x}_n \equiv \bar{x}_{n-1} (\text{mod } p^n)$ and then there exists $a_n \in \mathbb{Z}$ such that $\bar{x}_n - \bar{x}_{n-1} = a_n p^n$. But $0 \leqslant \bar{x}_n < p^{n+1}$ and $0 \leqslant \bar{x}_{n-1} < p^n$, from where $-p^n < a_n p^n < p^{n+1}$ or $-1 < a_n < p$ and so $0 \leqslant a_n \leqslant p - 1$, for every $n \in \mathbb{N}^*$. We denote by $a_0 = \bar{x}_0$ and then we remark that $0 \leqslant a_0 \leqslant p - 1$. Thus, we have determined the sequence $(a_n)_n \subseteq \{0, 1, \cdots, p-1\}$ in such a way that $\bar{x}_n - \bar{x}_{n-1} = a_n p^n$, for every $n \in \mathbb{N}^*$. By summing up the previous relations from 1 to n, we obtain $\bar{x}_n = a_0 + a_1 p + \cdots + a_n p^n$.

1.4 p-adic Numbers

(2) Let $(\bar{x}_n)_n$ be the canonical representation of the p-adic integer $\widehat{x} \in \mathbb{Z}_p$ and $(\bar{y}_n)_n$ be the canonical representation of the p-adic integer $\widehat{y} \in \mathbb{Z}_p$. If $\widehat{x} = \widehat{y}$, then $(\bar{x}_n)_n \sim (\bar{y}_n)_n$ hence $\bar{x}_n \equiv \bar{y}_n \pmod{p^{n+1}}$, for every $n \in \mathbb{N}$. But, for every $n \in \mathbb{N}$, $0 \leqslant \bar{x}_n < p^{n+1}$ and $0 \leqslant \bar{y}_n < p^{n+1}$ and then $\bar{x}_n = \bar{y}_n$, for every $n \in \mathbb{N}$. □

Remark 1.4.4 From the previous proposition it follows that a p-adic integer is uniquely determined by its canonical representation. Then, $\mathbb{Z}_p$ is in bijective correspondence with the set of sequences $(\bar{x}_n)_n$ of the form $\bar{x}_n = a_0 + a_1 p + \cdots + a_n p^n$, where $a_n \in \{0, 1, \cdots, p-1\}$, for every $n \in \mathbb{N}$. It follows from this that $\mathbb{Z}_p$ has the same cardinality as the set of functions $f : \mathbb{N} \to \{0, \cdots, p-1\}$ which has the cardinality of the continuum: $p^{\aleph_0} = c$ (the cardinality of $\mathbb{Z}_p$ is the same as that of $\mathbb{R}$).

Examples 1.4.5

(1) Let $(x_n)_n \in \mathcal{Z}$, where, for every $n \in \mathbb{N}$, $x_n = p^{n+1}$ and let $\widehat{x} = [(x_n)] \in \mathbb{Z}_p$. Its canonical representation is $\bar{0}$ (the sequence constant zero). Therefore $\widehat{x} = [\bar{0}]$.
(2) Let $(x_n)_n \in \mathcal{Z}$, $x_n = 1 + p + \cdots + p^n$, for every $n \in \mathbb{N}$; according to (1) of previous proposition, $\bar{x}_n = x_n$, for every $n \in \mathbb{N}$.
(3) Let $x \in \mathbb{Z}$ and let $(x_n)_n = \bar{x} \in \mathcal{Z}$, $x_n = x$, for every $n \in \mathbb{N}$, the sequence constant x; then x admits a writing in base p of the form $x = a_0 + a_1 p + \cdots + a_q p^q$, where $a_i \in \{0, 1, \cdots, p-1\}$, for every $i = 0, 1, \cdots, q$. The canonical representation of $\widehat{x} = [(x_n)]$ is given by $(\bar{x}_n)_n$, where

$$\bar{x}_n = \begin{cases} a_0 + a_1 p + \cdots + a_n p^n, & n \leqslant q, \\ a_0 + a_1 p + \cdots + a_q p^q, & n > q. \end{cases}$$

The function $i : \mathbb{Z} \to \mathbb{Z}_p$ defined by $i(x) = [\bar{x}]$ is the **canonical injection** of the set of integers into the set of p-adic integers.

On $\mathbb{Z}_p$, we define two operations as follows: for every $\widehat{x} = [(x_n)], \widehat{y} = [(y_n)] \in \mathbb{Z}_p$:

$$\widehat{x} + \widehat{y} = [(x_n + y_n)], \quad \widehat{x} \cdot \widehat{y} = [(x_n \cdot y_n)].$$

It can be easily verified that the operations are consistently defined and that, with respect to these two operations, $\mathbb{Z}_p$ is a commutative ring with unity. The additive neutral element is $\widehat{0} = [\bar{0}]$ and the multiplicative neutral element is $\widehat{1} = [\bar{1}]$. The function i defined above becomes a ring homomorphism with respect to these two operations.

Let $U \subseteq \mathbb{Z}_p$ be the group of all invertible elements with respect to the multiplication operation, that is

$$U = \{\widehat{x} \in \mathbb{Z}_p : \text{there exists } \widehat{y} \in \mathbb{Z}_p \text{ such that } \widehat{x} \cdot \widehat{y} = \widehat{1}\}.$$

In the above definition of U, $\widehat{y} = \widehat{x}^{-1}$ is the inverse element of $\widehat{x}$ in the ring $\mathbb{Z}_p$ with respect to the multiplication operation. Obviously, $\widehat{x} \in U$ if and only if $\widehat{y} \in U$.

The following theorem gives a characterization of the invertible elements of the ring $\mathbb{Z}_p$ (hence of the elements of U).

Theorem 1.4.6 $\widehat{x} = [(x_n)] \in U$ if and only if $x_0 \not\equiv 0 \pmod{p}$ ($p \nmid x_0$).

Proof Let $\widehat{x} = [(x_n)] \in U$; there exists $\widehat{y} = [(y_n)] \in U$ such that $\widehat{x} \cdot \widehat{y} = \widehat{1}$. It follows that, for every $n \in \mathbb{N}$,

$$x_n \cdot y_n \equiv 1 \pmod{p^{n+1}}.$$

Particularly, $x_0 \cdot y_0 \equiv 1 \pmod{p}$, from where we deduce that $x_0 \not\equiv 0 \pmod{p}$.

Conversely, we assume that $x_0 \not\equiv 0 \pmod{p}$; because $x_1 \equiv x_0 \pmod{p}$, $x_2 \equiv x_1 \equiv x_0 \pmod{p}, \cdots, x_n \equiv \cdots \equiv x_0 \pmod{p}$, it follows that

$$x_n \not\equiv 0 \pmod{p}, \text{ for every } n \in \mathbb{N}. \tag{1.14}$$

In particular, for all $n \in \mathbb{N}$, x_n and p^{n+1} are prime to each other, and thus there exist two integers y_n and z_n such that

$$y_n \cdot x_n + z_n \cdot p^{n+1} = 1. \tag{1.15}$$

Because $x_n \equiv x_{n-1} \pmod{p^n}$, from the previous relationship it follows that $x_n \cdot y_n - x_{n-1} \cdot y_{n-1} = (x_n \cdot y_n - 1) + (1 - x_{n-1} \cdot y_{n-1})$ is divisible by p^n, for every $n \in \mathbb{N}$. Therefore

$$x_n \cdot y_n \equiv x_{n-1} \cdot y_{n-1} \pmod{p^n}. \tag{1.16}$$

Now, from (1.16), $x_n \cdot y_n - x_{n-1} \cdot y_{n-1} = x_n \cdot y_n - x_n \cdot y_{n-1} + x_n \cdot y_{n-1} - x_{n-1} \cdot y_{n-1} = x_n(y_n - y_{n-1}) + y_{n-1}(x_n - x_{n-1})$ is divisible by p^n; from (1.14), x_n is not divisible by p and, since $(x_n)_n \in \mathcal{Z}$, $x_n - x_{n-1}$ is divisible by p^n. It follows that $y_n - y_{n-1}$ is divisible by p^n, that is

$$y_n \equiv y_{n-1} \pmod{p^n}, \text{ for all } n \in \mathbb{N}.$$

this implies $(y_n)_n \in \mathcal{Z}$. Let $\widehat{y} = [(y_n)] \in \mathbb{Z}_p$; (1.15) says us that $x_n \cdot y_n \equiv 1 \pmod{p^{n+1}}$, for all $n \in \mathbb{N}$, what it means that $\widehat{x} \cdot \widehat{y} = \widehat{1}$. Therefore $\widehat{x} \in U$. □

Remarks 1.4.7

(i) Let $(\bar{x}_n)_n$ be the canonical representation of the p-adic integer $\widehat{x}$; then $\widehat{x} \in U$ if and only if $\bar{x}_0 \neq 0$.

(ii) If $x \in \mathbb{Z}$, then $i(x) = [\bar{x}] \in U$ if and only if $p \nmid x$ (p does not divide x). Indeed, according to example (3) from Example 1.4.5 and the previous observation, $i(x) \in U$ if and only if $a_0 \neq 0$, where a_0 is the remainder of dividing x by p. Therefore, $i(x) = [\bar{x}] \in U$ if and only if x is not divisible by p.

1.4 p-adic Numbers

If, for example, we choose $x = p-1$, then $i(x) = [\bar{x}] \in U$; in relationship (1.15) from the above theorem's proof, we can choose, for all $n \in \mathbb{N}$, $z_n = 1$, and then we find $y_n = \frac{1-p^{n+1}}{p-1} = -p^n - p^{n-1} - \cdots - p - 1$. It follows that $[\bar{x}]^{-1} = [(y_n)]$. The canonical representation of $\widehat{y} = [(y_n)]$ is $(\bar{y}_n)_n$, where, for all $n \in \mathbb{N}$, $\bar{y}_n = (p-1) + (p-2)p + \cdots + (p-2)p^n$.

Definition 1.4.8 We denote by $\mathbb{Q}' = \left\{ \frac{m}{n} : m \in \mathbb{Z}, n \in \mathbb{N}^* \text{ with } p \nmid n \right\}$; from the previous observation, we have that $\widehat{x} = i(n) \in U$, and thus there exists $\widehat{y} \in \mathbb{Z}_p$ such that $\widehat{x} \cdot \widehat{y} = \widehat{1}$. We will now extend the embedding i to $\mathbb{Q}'$ by $i\left(\frac{m}{n}\right) = i(m) \cdot \widehat{y}$. The extended embedding will still be denoted by i. Thus, we interpret $\widehat{y}$ as $i\left(\frac{1}{n}\right)$. The extended embedding is injective and also preserves the operations of addition and multiplication. We will then agree to identify $\mathbb{Q}'$ with $i(\mathbb{Q}')$ and therefore consider $\mathbb{Q}'$ as a subset of the p-adic integers, $\mathbb{Z}_p$.

The following theorem allows us to represent the elements of the ring $\mathbb{Z}_p$ using invertible elements.

Theorem 1.4.9 (The Representation of p-adic Integers) *For any $\widehat{x} \in \mathbb{Z}_p \setminus \{\widehat{0}\}$, there exist a unique number $m \in \mathbb{N}$ and a unique element $\widehat{x}_0 \in U$ such that:*

$$\widehat{x} = p^m \cdot \widehat{x}_0.$$

Proof If $\widehat{x} \in U$, then $m = 0$ and $\widehat{x}_0 = \widehat{x}$.

Now, let us suppose that $\widehat{x} = [(x_n)] \notin U$; then $x_0 \equiv 0 (\bmod\, p)$ (see Theorem 1.4.6). Since $\widehat{x} \neq \widehat{0}$, there exists $m \in \mathbb{N}$ such that $x_m \not\equiv 0(\bmod\, p^{m+1})$; let's consider that m is the smallest number with this property. Obviously, $m > 0$ and $x_{m-1} \equiv 0(\bmod\, p^m)$.

$$x_m \equiv x_{m-1}(\bmod\, p^m) \Longrightarrow x_m \equiv 0(\bmod\, p^m),$$

$$x_{m+1} \equiv x_m(\bmod\, p^{m+1}) \Longrightarrow x_{m+1} \equiv 0(\bmod\, p^m),$$

$$\ldots$$

$$x_{m+n} \equiv x_{m+n-1}(\bmod\, p^{m+n}) \Longrightarrow x_{m+n} \equiv 0(\bmod\, p^m), \text{ for all } n \in \mathbb{N}.$$

Therefore, for any $n \in \mathbb{N}$, $y_n = \frac{1}{p^m} \cdot x_{m+n} \in \mathbb{Z}$.

For any $n \in \mathbb{N}^*$, $x_{m+n} \equiv x_{m+n-1}(\bmod\, p^{m+n})$ and then $y_n \equiv y_{n-1}(\bmod\, p^n)$. Therefore $(y_n)_n \in \mathcal{Z}$. Furthermore, because $x_m \not\equiv 0(\bmod\, p^{m+1})$, $y_0 = \frac{1}{p^m} x_m \not\equiv 0(\bmod\, p)$. From Theorem 1.4.6, this means that $\widehat{x}_0 = [(y_n)] \in U$.

Now, for any $n \in \mathbb{N}$, $p^m y_n = x_{m+n} \equiv x_n (\bmod\, p^{n+1})$, hence $(p^m y_n)_n \sim (x_n)_n$, and therefore $p^m \cdot \widehat{x}_0 = \widehat{x}$.

Now, let's demonstrate that the representation of p-adic integers is unique.

Let us suppose that $\widehat{x} = p^m \cdot \widehat{y_0} = p^k \cdot \widehat{z_0}$, where $m, k \in \mathbb{N}$ and $\widehat{y_0} = [(y_n)], \widehat{z_0} = [(z_n)] \in U$. According to Theorem 1.4.6, $y_0 \not\equiv 0(\mod p)$, and e $z_0 \not\equiv 0(\mod p)$. Given the fact that $(y_n)_n, (z_n)_n \in \mathcal{Z}$, it follows that

$$y_n \not\equiv 0(\mod p), z_n \not\equiv 0(\mod p), \text{ for any } n \in \mathbb{N}. \tag{1.17}$$

On the other hand, because $(p^m y_n)_n \sim (p^k z_n)_n$,

$$p^m y_n \equiv p^k z_n (\mod p^{n+1}), \text{ for any } n \in \mathbb{N}. \tag{1.18}$$

In (1.18) we take $n = m - 1$ and obtain

$$p^m y_{m-1} \equiv p^k z_{m-1}(\mod p^m). \tag{1.19}$$

From (1.17) and (1.19) we deduce that $m \leqslant k$.

Now, we take in (1.18) $n = k - 1$, and obtain

$$p^m y_{k-1} \equiv p^k z_{k-1}(\mod p^k). \tag{1.20}$$

From (1.17) and (1.20), $m \geqslant k$, and then $m = k$.

Now, from (1.18), $p^m y_{m+n} \equiv p^m z_{m+n} (\mod p^{m+n+1})$, and so

$$y_{m+n} \equiv z_{m+n}(\mod p^{n+1}), \text{ for any } n \in \mathbb{N}. \tag{1.21}$$

Because $(y_n)_n, (z_n)_n \in \mathcal{Z}$,

$$y_{m+n} \equiv y_n(\mod p^{n+1}), z_{m+n} \equiv z_n(\mod p^{n+1}), \text{ for any } n \in \mathbb{N}. \tag{1.22}$$

From (1.21) and (1.22), for any $n \in \mathbb{N}$, $y_n \equiv z_n (\mod p^{n+1})$, and then $\widehat{y_0} = \widehat{z_0}$. $\square$

Corollary 1.4.10 $\mathbb{Z}_p$ *is an integral domain.*

Proof Assuming that there exist two non-zero elements of $\mathbb{Z}_p$, $\widehat{x}$ and $\widehat{y}$ such that $\widehat{x} \cdot \widehat{y} = \widehat{0}$, from the above representation theorem, we have $\widehat{x} = p^m \cdot \widehat{x_0}$ and $\widehat{y} = p^k \cdot \widehat{y_0}$, where $m, k \in \mathbb{N}$ and $\widehat{x_0}, \widehat{y_0} \in U$. Thus, we get $p^{m+k} \cdot \widehat{x_0} \cdot \widehat{y_0} = \widehat{0}$. Then $(p^{m+k} x_n y_n)_n \sim \bar{0}$, from where $p^{m+k} x_n y_n \equiv 0 (\mod p^{n+1})$, for any $n \in \mathbb{N}$. We recall the relationship (1.17) from the proof of the previous theorem: $y_n \not\equiv 0 (\mod p), z_n \not\equiv 0 (\mod p)$, for any $n \in \mathbb{N}$, which leads us to the obviously absurd conclusion $p^{n+1} \mid p^{m+k}$, for any $n \in \mathbb{N}$. $\square$

According to the previous corollary, $\mathbb{Z}_p$ is an integral domain; then we can construct its field of fractions, denoted by $\mathbb{Q}_p$. We will use Greek letters such as α, β, etc. to denote the elements of $\mathbb{Q}_p$, and we will use 0 and 1 to denote the neutral elements for addition and multiplication in $\mathbb{Q}_p$, respectively. Since $\mathbb{Z}_p$ contains $\mathbb{Z}$ as a subring, $\mathbb{Q}_p$ will contain $\mathbb{Q}$ as a subfield.

1.4 p-adic Numbers

In what follows, we will naturally identify $\mathbb{Z}_p$ with a subset of its field of fractions, $\mathbb{Q}_p$, and thus we can write: $\mathbb{Z} \subseteq \mathbb{Q}' \subseteq \mathbb{Z}_p \subseteq \mathbb{Q}_p$, and also $\mathbb{Q} \subseteq \mathbb{Q}_p$.

The following result is a theorem on the representation of elements in the field $\mathbb{Q}_p$.

Theorem 1.4.11 (The Representation of p-adic Numbers) *For any $\alpha \in \mathbb{Q}_p \setminus \{0\}$, there exist a unique number $m \in \mathbb{Z}$ and a unique element $\widehat{x}_0 \in U$ such that:*

$$\alpha = p^m \cdot \widehat{x}_0.$$

Proof Let's remember that, for the construction of the fraction field of the integral domain $\mathbb{Z}_p$, an equivalence relation is defined on $\mathbb{Z}_p \times \mathbb{Z}_p^*$ by

$$(\widehat{x}, \widehat{y}) \sim (\widehat{u}, \widehat{v}) \text{ if and only if } \widehat{x} \cdot \widehat{v} = \widehat{y} \cdot \widehat{u}.$$

$\mathbb{Q}_p$ is then the quotient space $(\mathbb{Z}_p \times \mathbb{Z}_p^*)/\sim$. The elements of $\mathbb{Q}_p$ are equivalence classes of the form: $\alpha = [(\widehat{x}, \widehat{y})] = \{(\widehat{u}, \widehat{v}) : (\widehat{x}, \widehat{y}) \sim (\widehat{u}, \widehat{v})\}$.

Let now $\alpha = [(\widehat{x}, \widehat{y})] \in \mathbb{Q}_p, \alpha \neq 0$; then $\widehat{z}, \widehat{y} \in \mathbb{Z}_p^*$, and so, from the representation theorem of p-adic integers (Theorem 1.4.9), there exist $m, k \in \mathbb{N}$, there exist $\widehat{x}_0, \widehat{y}_0 \in U$ such that $\widehat{x} = p^m \cdot \widehat{x}_0$, and $\widehat{y} = p^k \cdot \widehat{y}_0$.

Then $(\widehat{x}, \widehat{y}) = (p^m \cdot \widehat{x}_0, p^k \cdot \widehat{y}_0) \sim \begin{cases} (p^{m-k}\widehat{x}_0\widehat{y}_0^{-1}, 1), & \text{if } m \geq k \\ (\widehat{x}_0\widehat{y}_0^{-1}, p^{k-m}), & \text{if } m < k \end{cases}$, from where it follows that $\alpha = p^{m-k} \cdot \widehat{x}_0 \cdot \widehat{y}_0^{-1}$. The proof is complete if we observe that $\widehat{x}_0 \cdot \widehat{y}_0^{-1} \in U$.

Uniqueness follows from the uniqueness of the representation of p-adic integers and the construction of the field of fractions. □

The above representation theorem allows for the definition of a field norm on $\mathbb{Q}_p$.

Definition 1.4.12 Let $\alpha \in \mathbb{Q}_p$; if $\alpha \neq 0$, then there exist a unique $m \in \mathbb{Z}$ and a unique $\widehat{x} \in U$ such that $\alpha = p^m \cdot \widehat{x}$. The we define

$$|\alpha|_p = \begin{cases} p^{-m}, & \text{if } \alpha \neq 0, \\ 0, & \text{if } \alpha = 0. \end{cases}$$

Theorem 1.4.13 *The mapping $|\cdot|_p : \mathbb{Q}_p \to \mathbb{R}_+$ defined by $\alpha \mapsto |\alpha|_p$ is a non-Archimedean norm on $\mathbb{Q}_p$.*

Proof It is evident from the definition that $|\alpha|_p = 0$ if and only if $\alpha = 0$.

Let $\alpha = p^m \cdot \widehat{x}, \beta = p^k \cdot \widehat{y} \in \mathbb{Q}_p^*$, where $m, k \in \mathbb{Z}, \widehat{x}, \widehat{y} \in U$. Then $\alpha \cdot \beta = p^{m+k} \cdot \widehat{x} \cdot \widehat{y}$ and, since $\widehat{x} \cdot \widehat{y} \in U$, $|\alpha \cdot \beta|_p = p^{-(m+k)} = |\alpha|_p |\beta|_p$.

$\alpha + \beta = p^m \cdot \widehat{x} + p^k \cdot \widehat{y}$.

1. Let us assume first that $m > k$; then $\alpha + \beta = p^k \cdot (p^{m-k} \cdot \widehat{x} + \widehat{y})$. Let us assume that $\widehat{x} = [(x_n)], \widehat{y} = [(y_n)]$, where $(x_n)_n, (y_n)_n \in \mathcal{Z}$. From Theorem 1.4.6 it follows that $y_0 \not\equiv 0 \pmod{p}$ and hence $p^{m-k}x_0 + y_0 \not\equiv 0 \pmod{p}$. Therefore, by

the same Theorem 1.4.6, $\widehat{z} = p^{m-k} \cdot \widehat{x} + \widehat{y} \in U$ and $\alpha + \beta = p^k \cdot \widehat{z}$; by the definition of the norm, $|\alpha + \beta|_p = p^{-k} = \max\{p^{-m}, p^{-k}\} = \max\{|\alpha|_p, |\beta|_p\}$.

2. If $m < k$, the argument is analogous and one obtains $|\alpha + \beta|_p = p^{-m} = \max\{|\alpha|_p, |\beta|_p\}$.
3. If $m = k$, then $\alpha + \beta = p^m \cdot (\widehat{x} + \widehat{y})$. From the representation theorem for p-adic integers, $\widehat{x} + \widehat{y} = p^l \cdot \widehat{z}$, where $l \in \mathbb{N}$ and $\widehat{z} \in U$. Then $\alpha + \beta = p^{m+l} \cdot \widehat{z}$ and thus $|\alpha + \beta|_p = p^{-(m+l)} \leqslant p^{-m} = \max\{|\alpha|_p, |\beta|_p\}$.

$\square$

Remark 1.4.14 For every $x \in \mathbb{Q}^*$, there exists a unique $n \in \mathbb{Z}$ such that $x = p^n \cdot \frac{a}{b}$, where $a, b \in \mathbb{Z}^*$ are two numbers prime to p. Then $\frac{a}{b} \in \mathbb{Q}' \subseteq \mathbb{Z}_p$; moreover, since $p \nmid a$, $\frac{a}{b} \in U$. Then $|x|_p = p^{-n}$. Thus, the trace on $\mathbb{Q}$ of the non-Archimedean norm of $\mathbb{Q}_p$ is precisely the p-adic norm defined in (2) of Example 1.3.9.

The following theorem shows that $\mathbb{Q}$ is a dense subset of $\mathbb{Q}_p$.

Theorem 1.4.15

1. Let $\alpha = [(x_n)] \in \mathbb{Z}_p \subseteq \mathbb{Q}_p$; then $(x_n)_n \subseteq \mathbb{Z} \subseteq \mathbb{Q}_p$ and $x_n \xrightarrow[\mathbb{Q}_p]{|\cdot|_p} \alpha$.

2. For every $\alpha \in \mathbb{Q}_p$, there exists $(x_n)_n \subseteq \mathbb{Q}$ such that $x_n \xrightarrow[\mathbb{Q}_p]{|\cdot|_p} \alpha$.

Proof

1. Let $\alpha = [(x_n)] \in \mathbb{Z}_p \subseteq \mathbb{Q}_p$; the sequence $(x_n)_n \subseteq \mathbb{Z}$ can be thought of as a sequence in $\mathbb{Q}_p$ by identifying each term x_n with the equivalence class generated by the constant sequence $\bar{x}_n : x_n \equiv i(x_n) = [\bar{x}_n]$. Note that, for all $n \in \mathbb{N}$, $(x_m)_m \sim (x_{n+m})_m$; indeed, for all $m \in \mathbb{N}$, $x_{n+m} \equiv x_m \pmod{p^{m+1}}$. It follows that we can change the representative of $\alpha = [(x_{n+m})_m]$; thus, for all $n \in \mathbb{N}$,

$$\alpha - [\bar{x}_n] = [(x_{n+m} - x_n)_m].$$

But, for every $n \in \mathbb{N}$, $x_{n+m} - x_n$ is divisible by p^{n+1} and so

$$\alpha - [\bar{x}_n] = p^{n+1} \cdot [(x'_{n+m} - x'_n)_m],$$

from where $|\alpha - [\bar{x}_n]|_p \leqslant p^{-(n+1)}$, for any $n \in \mathbb{N}$, and then $x_n \equiv [\bar{x}_n] \xrightarrow[\mathbb{Q}_p]{|\cdot|_p} \alpha$.

2. If $\alpha \in \mathbb{Q}_p^*$, then, according to theorem of representation of p-adic numbers (1.4.11), $\alpha = p^m \cdot \widehat{x}$ with $m \in \mathbb{Z}$ and $\widehat{x} \in U \subseteq \mathbb{Z}_p$.

If $m \in \mathbb{N}$, then $\alpha \in \mathbb{Z}_p$ and then, using the previous point of this proof, α is the limit of a sequence of integers.

1.4 p-adic Numbers

If $m \in \mathbb{Z} \setminus \mathbb{N}$ and $\widehat{x} = [(x_n)]$, then, using again the previous point, $x_n \xrightarrow[\mathbb{Q}_p]{|\cdot|_p} \widehat{x}$.
Then $p^m \cdot x_n \xrightarrow[\mathbb{Q}_p]{|\cdot|_p} p^m \cdot \widehat{x} = \alpha$. The proof concludes if we further observe that $(p^m \cdot x_n)_n \subseteq \mathbb{Q}$.

□

The following result is of the same type as Bolzano–Weierstrass theorem from classical analysis on $\mathbb{R}$.

Theorem 1.4.16 *Any bounded sequence in the normed field $(\mathbb{Q}_p, |\cdot|_p)$ contains a convergent subsequence.*

Proof If the sequence $(\alpha_n) \subseteq \mathbb{Q}_p$ has a constant subsequence equal to 0, then this subsequence converges to 0. We can assume that $\alpha_n \in \mathbb{Q}_p^*$, for any $n \in \mathbb{N}$, and then, by the representation theorem of p-adic numbers, there exists a sequence of integers $(k_n)_n \subseteq \mathbb{Z}$ and a sequence $(\widehat{x}_n)_n \subseteq U \subseteq \mathbb{Z}_p$ such that

$$\alpha_n = p^{k_n} \cdot \widehat{x}_n, \text{ for any } n \in \mathbb{N}.$$

(1) First, assume that $(\alpha_n)_n \subseteq \mathbb{Z}_p$; then $(k_n)_n \subseteq \mathbb{N}$ and, since $|\alpha_n|_p = p^{-k_n} \leqslant 1$, it follows that any sequence in $\mathbb{Z}_p$ is bounded. If $(k_n)_n$ has an unbounded subsequence, then $(\alpha_n)_n$ has a subsequence that converges to 0, and the proof is complete.

Assume that the sequence $(k_n)_n$ is a bounded sequence of natural numbers; then it has a subsequence constant equal to some $k \in \mathbb{N}$. Thus, $(\alpha_n)_n$ has a subsequence, also denoted by $(\alpha_n)_n$, of the form $\alpha_n = p^k \cdot \widehat{x}_n$, for all $n \in \mathbb{N}$. For every $n \in \mathbb{N}$, let $\widehat{x}_n = [(x_m^n)_m]$, where $(x_m^n)_m$ is the canonical representation of the p-adic integer $\widehat{x}_n$. Then

$$0 \leqslant x_m^n < p^{m+1}, \text{ for all } m, n \in \mathbb{N}.$$

The sequence of integers $(x_0^n)_{n \in \mathbb{N}}$ takes values in the finite set $\{0, 1, \cdots, p-1\}$; thus, there exists an element $x_0 \in \{0, 1, \cdots, p-1\}$ and an infinite subset $\mathbb{N}_0 \subseteq \mathbb{N}$ such that $x_0^n = x_0$, for every $n \in \mathbb{N}_0$. The sequence of integers $(x_1^n)_{n \in \mathbb{N}_0}$ takes values in the finite set $\{0, 1, \cdots, p^2 - 1\}$; thus, there exists an element $x_1 \in \{0, 1, \cdots, p^2 - 1\}$ and an infinite subset $\mathbb{N}_1 \subseteq \mathbb{N}_0$ such that $x_1^n = x_1$, for every $n \in \mathbb{N}_1$. Inductively, at step $q \in \mathbb{N}$, we will find an element $x_q \in \{0, 1, \cdots, p^{q+1} - 1\}$ and an infinite subset $\mathbb{N}_q \subseteq \mathbb{N}_{q-1}$ such that $x_q^n = x_q$, for every $n \in \mathbb{N}_q$.

We choose $n_0 \in \mathbb{N}_0, n_1 \in \mathbb{N}_1$ with $n_1 > n_0, \cdots, n_q \in \mathbb{N}_q$ with $n_q > n_{q-1}, \cdots$.

Let now the sequence $(x_q)_q \subseteq \mathbb{Z}$; for any $q \in \mathbb{N}, n_q \in \mathbb{N}_q \subseteq \mathbb{N}_{q-1}$ and so $x_q^{n_q} = x_q$ and $x_{q-1}^{n_q} = x_{q-1}$; it follows that

$$x_q - x_{q-1} = x_q^{n_q} - x_{q-1}^{n_q} \equiv 0(\mod p^q).$$

Therefore $(x_q)_q \in \mathcal{Z}$. Let $\widehat{x} = [(x_q)] \in \mathbb{Z}_p$. We observe that for all $q \in \mathbb{N}$, $n_q \in \mathbb{N}_q \subseteq \mathbb{N}_{q-1} \subseteq \cdots \subseteq \mathbb{N}_1 \subseteq \mathbb{N}_0$, and then $x_i^{n_q} = x_i$, for all $i = 0, 1, , q$. Thus, for every $m \leqslant q$, $x_m^{n_q} - x_m = 0$, and then $x_m^{n_q} \equiv x_m \pmod{p^{q+1}}$, for every $m \in \mathbb{N}$. It follows that, for every $q \in \mathbb{N}$, $\widehat{x}^{n_q} - \widehat{x} = p^{q=1} \cdot \widehat{y}_q$, where $\widehat{y}_q \in \mathbb{Z}_p$. Therefore $|\widehat{x}^{n_q} - \widehat{x}|_p \leqslant p^{-(q+1)}$, for any $q \in \mathbb{N}$, and then $\widehat{x}^{n_q} \xrightarrow[\mathbb{Q}_p]{|\cdot|_p} \widehat{x}$.

So we have identified a subsequence of the sequence $(\alpha_n)_n$, $(\alpha_{n_q})_q$ that converges to $p^k \cdot \widehat{x}$.

(2) Assuming now that $(\alpha_n)_n \subseteq \mathbb{Q}_p \setminus \mathbb{Z}_p$, then $(k_n)_n \subseteq \mathbb{Z} \setminus \mathbb{N}$. Since the sequence $(\alpha_n)_n$ is bounded, there exists $M > 0$ such that $|\alpha_n|_p = p^{-k_n} \leqslant M$. It follows that the sequence $(k_n)_n \subseteq \mathbb{Z}$ is bounded and then it has a subsequence constant equal to some $k \in \mathbb{N}$. Thus, $(\alpha_n)_n$ has a subsequence, also denoted by $(\alpha_n)_n$, of the form $\alpha_n = p^k \cdot \widehat{x}_n$, for any $n \in \mathbb{N}$.

However, the sequence $(\widehat{x}_n)_n \subseteq \mathbb{Z}_p$ is bounded (any sequence in $\mathbb{Z}_p$ is bounded) and, according to the first part of the proof, it has a subsequence, also denoted by $(\widehat{x}_n)_n$, which converges to a $\widehat{x} \in \mathbb{Z}_p$. Therefore $\alpha_n \xrightarrow[\mathbb{Q}_p]{|\cdot|_p} p^k \cdot \widehat{x}$. □

The previous theorem allows us to prove the completeness of the normed field $(\mathbb{Q}_p, |\cdot|_p)$.

Theorem 1.4.17 *The normed field $(\mathbb{Q}_p, |\cdot|_p)$ is Cauchy complete.*

Proof Let $(\alpha_n)_{n \in \mathbb{N}} \subseteq \mathbb{Q}_p$ be a Cauchy sequence. Then there exists $n_0 \in \mathbb{N}$ such that $|\alpha_n - \alpha_{n_0}|_p < 1$, for any $n \geqslant n_0$, so that $|\alpha_n|_p \leqslant |\alpha_n - \alpha_{n_0}|_p + |\alpha_{n_0}|_p < 1 + |\alpha_{n_0}|_p$, for all $n \geqslant n_0$. Let $M = \max\{|\alpha_0|_p, \cdots, |\alpha_{n_0-1}|_p, 1 + |\alpha_{n_0}|_p\}$; then, for any $n \in \mathbb{N}$, $|\alpha_n|_p \leqslant M$, so that $(\alpha_n)_{n \in \mathbb{N}}$ is a bounded sequence. Using the previous theorem, $(\alpha_n)_{n \in \mathbb{N}}$ has a subsequence, $(\alpha_{k_n})_{n \in \mathbb{N}}$ convergent to $\alpha \in \mathbb{Q}_p$. Then, $|\alpha_n - \alpha|_p \leqslant |\alpha_n - \alpha_{k_n}|_p + |\alpha_{k_n} - \alpha|_p \to 0$, which means that $\alpha_n \xrightarrow[\mathbb{Q}_p]{|\cdot|_p} \alpha$. □

Remark 1.4.18 Since $(\mathbb{Q}_p, |\cdot|_p)$ is a complete field and contains $\mathbb{Q}$ as a dense subfield, it follows that $(\mathbb{Q}_p, |\cdot|_p)$ is a completion of $(\mathbb{Q}, |\cdot|_p)$. Because the completion of a normed field is unique up to an isometric isomorphism of fields, it follows that $(\mathbb{Q}_p, |\cdot|_p)$ is the completion of $(\mathbb{Q}, |\cdot|_p)$.

This chapter is dedicated to various constructions of real numbers. Ostrowski's theorem states that on the set of rational numbers, there are essentially two types of norms: the archimedean absolute value norm and the non-archimedean p-adic norms. The completion of $\mathbb{Q}$ with respect to the absolute value norm yields the set of real numbers. It seemed natural then to also construct completions of $\mathbb{Q}$ with respect to the p-adic norms. However, we do not intend here to develop the theory of p-adic numbers and non-archimedean analysis. There are many books dedicated to these topics for those interested. For those interested in delving deeper into this field, we mention a few of them: [1–4, 12].

1.5 Appendix

Natural numbers are among the first entities with which abstract human thinking has operated. Some authors even claim that: "We review evidence that a system for arithmetic reasoning with real numbers evolved before language evolved" (see the introduction in [5]).

As soon as we have natural numbers at our disposal, we can build the entire structure of real numbers. It is then evident the fascination that natural numbers have exerted on mathematicians. What are they? Where did they come from? In connection with the difficulty of answering these questions, the statement made in 1886 in a lecture for Berliner Naturforscher-Versammlung by Leopold Kronecker is often cited: "God created the natural numbers. All else is the work of men".

Attempts to reduce the definition of natural numbers to logical propositions were made at the end of the nineteenth century and the beginning of the twentieth century, primarily by Gottlob Frege and Bertrand Russell. Although seemingly promising, the logical project failed; one of the main reasons appears to be that the concept of actual infinity with which natural number arithmetic operates does not appear anywhere in logic.

The efforts to systematize knowledge about natural numbers have materialized in various axiomatizations of arithmetic. One of the established axiomatizations was that of Giuseppe Peano (see [11]).

1.5.1 Peano's Axioms

One of the first axiomatic versions of the natural number was presented in 1888 by Richard Dedekind. A year later, Giuseppe Peano simplified Dedekind's version; that's why some authors refer to this axiomatic system as Dedekind-Peano's axioms. Four out of the nine axioms of Peano's axioms relate to the properties of equality and are generally omitted as they are understood. We will also adopt the same simplifying approach and present here only the five axioms that govern Peano arithmetic (PA).

Definition 1.5.1 As primary concepts in PA, we consider the constant symbol 0 and the notion of "natural number" (a priori concepts that are not defined within the theory). We will denote the set of natural numbers by $\mathbb{N}$ and its elements by x, y, z, etc. We also consider a primary relation, the "successor relation" $s \subseteq \mathbb{N} \times \mathbb{N}$; we will usually denote $(x, y) \in s$ by $s(x) = y$. In such a situation, we will say that y is the **successor** of x.

These primary elements must satisfy the following axioms:

($PA1$) $0 \in \mathbb{N}$ (0 is a natural number).
($PA2$) For any $x \in \mathbb{N}$, there exists a unique $y \in \mathbb{N}$ such that $y = s(x)$ (every natural number has a unique successor).

(PA3) For every $x \in \mathbb{N}$, $s(x) \neq 0$ (0 is not the successor of any natural number).
(PA4) $s(x) = s(y) \Rightarrow x = y$ (two different natural numbers have different successors).
(PA5) For every subset $M \subseteq \mathbb{N}$ with the following two properties: $0 \in M$ and $x \in M \Rightarrow s(x) \in M$, it follows that $M = \mathbb{N}$.

The last axiom is called the induction axiom.

Remark 1.5.2 If we denote by $\mathbb{N}^* = \mathbb{N} \setminus \{0\}$, then we observe that the axioms (PA2), (PA3) and (PA4) say that s is an injective function defined on the set $\mathbb{N}$ with values in $\mathbb{N}^*$.

There are several meta-theoretical issues related to this system of axioms, the most important of which is: Is it consistent? This means that there is no proposition P in the theory developed based on the system of axioms such that both P and $\neg P$ are true ($\neg P$ denotes the negation of proposition P). Semantically, consistency reduces to showing that there exist models of the theory.

Another problem is completeness, which means that for any well-formed proposition P, it can be shown that either P is true or $\neg P$ is true.

In 1931, Kurt Gödel formulated his first incompleteness theorem, which states that PA is incomplete.

The consistency of the theory of natural numbers is a subject of ongoing research and debate in the field of mathematical logic. One of the most famous results in this area is Gödel's incompleteness theorems, which prove that any formal system that is powerful enough to express the basic concepts of number theory (such as addition, multiplication, and induction) cannot be both consistent and complete. However, it's important to note that Gödel's theorems only apply to formal systems and not to specific models of natural numbers, such as Peano's axioms.

The problem of the consistency of Peano arithmetic is ranked second among the 23 essential problems to be solved in the twentieth century, as stated by David Hilbert in 1900. In 1931, Kurt Gödel formulated his second incompleteness theorem, which implies that PA cannot prove its own consistency. We can construct a model of PA (and thus support the consistency of the theory) within the framework of ZFC set theory (the axiomatic theory of sets developed by Ernst Zermelo and Adolf Fraenkel, to which the axiom of choice is added). However, this does not solve the problem of consistency but rather shifts it to the consistency of ZFC theory.

This situation, along with the depersonalization of mathematical objects through abstraction, may have led Bertrand Russell to the paradoxical statement that: "Mathematics may be defined as the subject in which we never know what we are talking about, nor whether what we are saying is true" (see "The Study of Mathematics" in *Mysticism and Logic and Other Essays*, George Allen. Unwin LTD, London, 1917).

The reader interested in the meta-theoretical problems of Peano arithmetic can find additional information in [8].

1.5 Appendix

Next, we will show how, starting from the Peano axioms, we can introduce an algebraic structure on $\mathbb{N}$ that will allow us, in the following subsections, to construct the set of integers and the set of rational numbers.

We will present two immediate consequences of axioms $(PA1)$–$(PA5)$.

Remarks 1.5.3

(i) For every $x \in \mathbb{N}$, $s(x) \neq x$. Indeed, let $M = \{x \in \mathbb{N} : s(x) \neq x\}$. Since $s(0) \neq 0$ (see $(PA3)$), $0 \in M$. For every $x \in M$, $s(x) \neq x$, and then $s(s(x)) \neq s(x)$ (see $(PA4)$), so that $s(x) \in M$. It follows from $(PA5)$ that $M = \mathbb{N}$.

(ii) For every $y \in \mathbb{N}^*$, there exists $x \in \mathbb{N}$ such that $s(x) = y$. Indeed, let $M = \{0\} \cup \{y \in \mathbb{N}^* : \text{ there is } x \in \mathbb{N} \text{ such that } s(x) = y\}$. Obviously $0 \in M$. For every $y \in M$, either $y = 0$, in which case there is $0 \in \mathbb{N}$ such that $s(0) = s(y)$ and then $s(y) \in M$, or $y \neq 0$, in which case there is $x \in \mathbb{N}$ such that $s(x) = y$, from where, $s(s(x)) = s(y)$, and so $s(y) \in M$. From $(PA5)$ it follows that $M = \mathbb{N}$, so that $\mathbb{N}^* = \{y \in \mathbb{N}^* : \text{ there is } x \in \mathbb{N} \text{ such that } s(x) = y\}$. It follows from Remark 1.5.2 that $s : \mathbb{N} \to \mathbb{N}^*$ is a bijection.

Theorem 1.5.4 *For every $x \in \mathbb{N}$ there exists a unique function $f_x : \mathbb{N} \to \mathbb{N}$ such that $f_x(0) = x$ and $f_x(s(y)) = s(f_x(y))$, for every $y \in \mathbb{N}$.*

Proof

Existence Let M be the set

$$M = \{x \in \mathbb{N} : \exists f_x : \mathbb{N} \to \mathbb{N}, f_x(0) = x, f_x(s(y)) = s(f_x(y)), \forall y \in \mathbb{N}\}.$$

Let us remark that the function $f_0 : \mathbb{N} \to \mathbb{N}$, defined by $f_0(x) = x$, for every $x \in \mathbb{N}$, has the mentioned properties such that $0 \in M$.

Let now $x \in M$ and let $f_x : \mathbb{N} \to \mathbb{N}$ a function with the mentioned properties. We define $f_{s(x)} : \mathbb{N} \to \mathbb{N}$ by $f_{s(x)}(y) = s(f_x(y))$, for every $y \in \mathbb{N}$. Then $f_{s(x)}(0) = s(f_x(0)) = s(x)$ and, for every $y \in \mathbb{N}$, $f_{s(x)}(s(y)) = s(f_x(s(y))) = s(s(f_x(y))) = s(f_{s(x)}(y))$. It follows that $s(x) \in M$ and therefore $M = \mathbb{N}$.

Uniqueness Now, for every $x \in \mathbb{N}$, let $f_x, g_x : \mathbb{N} \to \mathbb{N}$ be two functions that satisfy the properties: $f_x(0) = g_x(0) = x$, and, for every $y \in \mathbb{N}$, $f_x(s(y)) = s(f_x(y))$, $g_x(s(y)) = s(g_x(y))$. Let $M = \{y \in \mathbb{N} : f_x(y) = g_x(y)\}$. Since $f_x(0) = x = g_x(0)$ it follows that $0 \in M$. For every $y \in M$, $f_x(y) = g_x(y)$ and then $f_x(s(y)) = s(f_x(y)) = s(g_x(y)) = g_x(s(y))$; therefore $s(y) \in M$. It follows that $M = \mathbb{N}$ and so $f_x = g_x$.

□

Proposition 1.5.5 *For every $x \in \mathbb{N}$, let f_x be a function whose existence and uniqueness have been demonstrated in the previous theorem. Then*

(a) $f_0(x) = x$, for every $x \in \mathbb{N}$, and
(b) $f_{s(x)}(y) = f_x(s(y))$, for every $x, y \in \mathbb{N}$.

Proof We recall that $f_x : \mathbb{N} \to \mathbb{N}$, $f_x(0) = x$ and $f_x(s(y)) = s(f_x(y))$, for every $y \in \mathbb{N}$.

(a) We take the set $M = \{x \in \mathbb{N} : f_0(x) = x\}$. $f_0(0) = 0$ and so $0 \in M$. For every $x \in M$, $f_0(s(x)) = s(f_0(x)) = s(x)$ and then $s(x) \in M$. Therefore $M = \mathbb{N}$.
(b) For an arbitrary $x \in \mathbb{N}$, we take the set $M = \{y \in \mathbb{N} : f_{s(x)}(y) = f_x(s(y))\}$. Then $f_{s(x)}(0) = s(x) = s(f_x(0)) = f_x(s(0))$ from where $0 \in M$. Now, for every $y \in M$, $f_{s(x)}(s(y)) = s(f_{s(x)}(y)) = s(f_x(s(y))) = f_x(s(s(y)))$ from where $s(y) \in M$. It follows that $s(y) \in M$, and then $M = \mathbb{N}$.
□

Definition 1.5.6 We define the addition operation $+ : \mathbb{N} \times \mathbb{N} \to \mathbb{N}$ by $+(x, y) \equiv x + y = f_x(y)$, for every $x, y \in \mathbb{N}$ (f_x is the function whose existence has been proven in Theorem 1.5.4).

We define also $s(0) = 1$

Proposition 1.5.7

(1) For every $x, y \in \mathbb{N}$, $s(x + y) = x + s(y) = s(x) + y$.
(2) For every $x \in \mathbb{N}$, $x + 0 = 0 + x = x$ (0 is the identity element of addition).
(3) For every $x \in \mathbb{N}$, $s(x) = x + 1 = 1 + x$.
(4) If $x + y = 0$, then $x = 0 = y$.

Proof

(1) According with Theorem 1.5.4 and Proposition 1.5.5, for every $x, y \in \mathbb{N}$, $f_x(s(y)) = s(f_x(y)) = f_{s(x)}(y)$ and then $s(x + y) = s(f_x(y)) = f_x(s(y)) = x + s(y)$ and $x + s(y) = f_x(s(y)) = f_{s(x)}(y) = s(x) + y$.
(3) From Proposition 1.5.5, for every $x \in \mathbb{N}$, $f_0(x) = x = f_x(0)$; then $x + 0 = f_x(0) = x = f_0(x) = 0 + x$.
(3) For every $x \in \mathbb{N}$, $x + 1 = x + s(0) = s(x + 0) = s(x) = f_0(s(x)) = f_{s(0)}(x) = s(0) + x = 1 + x$.
(4) If $x + y = 0$ and if we suppose that $x \neq 0$, then there exists $z \in \mathbb{N}$ such that $x = s(z)$ (see (ii) of Remark 1.5.3). Then $0 = x + y = s(z) + y = s(z + y)$, which is absurd (contradicts $(PA3)$). Therefore $x = 0$ and then $y = 0 + y = x + y = 0$.
□

Theorem 1.5.8 ($\mathbb{N}, +$) *is a commutative monoid with cancellation property.*

Proof We have already shown in point (2) of the previous proposition that 0 is a neutral element with respect to the addition operation. We still need to prove that the addition operation is associative and commutative.

We fix two arbitrary points $x, y \in \mathbb{N}$, and let $M = \{z \in \mathbb{N} : (x + y) + z = x + (y + z)\}$. Since $(x + y) + 0 = x + y = x + (y + 0)$ it follows that $0 \in M$. Now, for every $z \in M$, $(x + y) + s(z) = s((x + y) + z) = s(x + (y + z)) = x + s(y + z) = x + (y + s(z))$, from where $s(z) \in M$. According to the induction axiom, $M = \mathbb{N}$.

Now, we fix an arbitrary point $x \in \mathbb{N}$ and denote by $M = \{y \in \mathbb{N} : x + y = y + x\}$. From the previous proposition, $x + 0 = 0 + x$ and then $0 \in M$. For every

1.5 Appendix

$y \in M$, $x + s(y) = s(x + y) = s(y + x) = s(y) + x$ and so $s(y) \in M$. Thus we proved that $M = \mathbb{N}$.

To show that $(\mathbb{N}, +)$ has the cancellation property we must prove that $x + y = x + z \implies y = z$. For this we fix two arbitrary points $y, z \in \mathbb{N}$ such that $y \neq z$ and let $M = \{x \in \mathbb{N} : x + y \neq x + z\}$. From $y \neq z$ it follows that $0 \in M$. Let now $x \in M$; then $x + y \neq x + z$ and hence, from $(PA4)$, $s(x) + y = s(x + y) \neq s(x + z) = s(x) + z$. Therefore $s(x) \in M$ and then $(PA5)$ implies that $M = \mathbb{N}$. □

Theorem 1.5.9 *For every $x, y \in \mathbb{N}$, one and only one of the following relationships is satisfied:*

(1) $x = y$.
(2) There exists a unique $u \in \mathbb{N}^*$ such that $x = y + u$.
(3) There exists a unique $u \in \mathbb{N}^*$ such that $x + u = y$.

Proof Let us first show that any two of the three situations mentioned in the theorem statement cannot occur simultaneously.

(1) and (2) are incompatible: If $x = y$ and $x = y + u$ it follows that $y = y + u$ and, from the cancellation property, $u = 0 \notin \mathbb{N}^*$.

Similarly, it can be shown that (1) and (3) are incompatible.

(2) and (3) are incompatible: If $x = y + u$ and $x + v = y$, then $x = x + v + u$. Using again the cancellation property we obtain $v + u = 0$. From (4) of Proposition 1.5.7, $u = v = 0 \notin \mathbb{N}^*$.

Now let's show that every $x, y \in \mathbb{N}$ satisfy one of the conditions (1), (2), or (3). We fix an arbitrary $x \in \mathbb{N}$; let

$$M = \{y \in \mathbb{N} : x, y \text{ satisfy one of the conditions } (1), (2), (3)\}.$$

If $y = 0$, then, either $x = 0$, and so x, y satisfy (1), or $x \neq 0$ and then $x = 0 + x = y + x$ and so x, y satisfy (2) with $u = x \in \mathbb{N}^*$. For $y = 0$, condition (3) cannot occur. Therefore $0 \in M$.

Let $y \in M$; then x, y satisfy one of the conditions (1), (2), or (3)

In case (1), $x = y$, and then $s(y) = s(x) = x + 1$; it follows that $x, s(y)$ satisfy (3) and then $s(y) \in M$.

If x, and y satisfy (2), then there exists $u \in \mathbb{N}^*$ such that $x = y + u$. Let $v \in \mathbb{N}$ such that $s(v) = u$; then $x = y + s(v) = s(y) + v$. In this situation, we have two alternatives: either $v = 0$, in which case $x, s(y)$ satisfy (1), or $v \neq 0$, where $x, s(y)$ satisfy (3). In both cases, $s(y) \in M$.

If x, y satisfy (3), there exists $u \in \mathbb{N}^*$ such that $x + u = y$; it follows that $x + s(y) = s(x + u) = s(y)$, and then x and $s(y)$ satisfy (3) with $u = s(y) \in \mathbb{N}^*$.

In all cases $s(y) \in M$, for every $y \in M$. Therefore $M = \mathbb{N}$, and thus, for any $x, y \in \mathbb{N}$, one and only one of the conditions (1), (2), or (3) is satisfied. □

Now we will define a partial order relation on $\mathbb{N}$.

Definition 1.5.10 Let $x, y \in \mathbb{N}$; we say that x is **strictly smaller** than y if there exists $u \in \mathbb{N}^*$ such that $x + u = y$; we denote this by $x < y$. x is **less than or equal**

than y if there exists $u \in \mathbb{N}$ such that $x+u = y$, or, equivalently, if $x = y$ or $x < y$; in this case we write $x \leqslant y$.

Remarks 1.5.11

(i) According to Theorem 1.5.9, for every $x, y \in \mathbb{N}$, $x < y$, or $x = y$, or $y < x$ (law of trichotomy).
(ii) If $x < y$, then $s(x) = x + 1 \leqslant y$. Indeed, if $x < y$, there is $u \in \mathbb{N}^*$ such that $x + u = y$. Let $v \in \mathbb{N}$ such that $s(v) = u$; then $(x + 1) + v = s(x) + v = x + s(v) = y$, from where $s(x) = x + 1 \leqslant y$.

Theorem 1.5.12 $(\mathbb{N}, +, \leqslant)$ *is a well-ordered monoid.*

Proof From the definition $\leqslant$ is reflexive. If $x \leqslant y$ and $y \leqslant x$ then there exist $u, v \in \mathbb{N}$ such that $x + u = y$ and $y + v = x$. Then $y + (v + u) = (y + v) + u = y$. From the cancellation property, $v + u = 0$, and then, according to Proposition 1.5.7, $u = v = 0$. It follows that $x = y$ and so $\leqslant$ is antisymmetric. Let's now demonstrate transitivity. If $x \leqslant y$ and $y \leqslant z$, then there exist $u, v \in \mathbb{N}$ such that $x + u = y$ and $y + v = z$. Then $x + (u + v) = z$ and so $x \leqslant z$.

It follows from (i) of Remark 1.5.11 that $\leqslant$ is a total order on $\mathbb{N}$.

Let's observe that 0 is the smallest element in $\mathbb{N}$; indeed, for every $x \in \mathbb{N}$, $x = 0 + x$ and then $0 \leqslant x$. Let now $A \subseteq \mathbb{N}$ be a non-empty subset and let $P = \{x \in \mathbb{N} : x \leqslant y,$ for every $y \in A\}$. From the above remark, $0 \in P$. Since $A \neq \emptyset$ there is $y \in A$; then $y < y + 1$ and so $y + 1 \notin P$, so that $P \neq \mathbb{N}$. From $(PA5)$, it follows that there is $p \in P$ such that $s(p) = p + 1 \notin P$. If we suppose that $p \notin A$, then, for every $x \in A$, $p < x$; from (ii) of Remark 1.5.11, $p + 1 \leqslant x$, for every $x \in A$, which would lead to the absurd conclusion that $s(p) = p + 1 \in P$. Hence $p \in P \cap A$, and therefore p is the smallest element in A.

We have shown that $\leqslant$ is a well-ordering on $\mathbb{N}$; let us now demonstrate that $\leqslant$ is compatible with addition. For every $x, y \in \mathbb{N}$ with $x \leqslant y$ and for every $z \in \mathbb{N}$, there exists $u \in \mathbb{N}$ such that $x + u = y$ and so $(x + z) + u = (x + u) + z = y + z$, from where $x + z \leqslant y + z$. □

Remarks 1.5.13

(i) For every $x, y, z \in \mathbb{N}$, $x + z \leqslant y + z$ implies $x \leqslant y$. Indeed, from $x + z \leqslant y + z$ it follows that there is $u \in \mathbb{N}$ such that $x + z + u = y + z$. The cancellation property allows us to deduce that $x + u = y$ and so that $x \leqslant y$.
(ii) For every $x, y, z, w \in \mathbb{N}$, $x \leqslant y$ and $z \leqslant w$ imply $x + z \leqslant y + w$.
Indeed, from $x \leqslant y$ and $z \leqslant w$ it follows that there are $u, v \in \mathbb{N}$ such that $x + u = y$ and $z + v = w$. Then $(x + z) + (u + v) = y + w$ and hence $x + z \leqslant y + w$.

Let us now define the multiplication on $\mathbb{N}$.

Theorem 1.5.14 *For every $x \in \mathbb{N}$, there exists a unique function $g_x : \mathbb{N} \to \mathbb{N}$ such that $g_x(0) = 0$ and $g_x(s(y)) = g_x(y) + x$, for every $y \in \mathbb{N}$.*

1.5 Appendix

Proof

Existence We define a subset $M \subseteq \mathbb{N}$ by

$$M = \{x \in \mathbb{N} : \exists g_x : \mathbb{N} \to \mathbb{N}, g_x(0) = 0 \text{ and } g_x(s(y)) = g_x(y) + x, \forall y \in \mathbb{N}\}.$$

The function $g_0 : \mathbb{N} \to \mathbb{N}$, defined by $g_0(y) = 0$, for every $y \in \mathbb{N}$, fulfills the required properties such that $0 \in M$. Let now $x \in M$; we define $g_{s(x)} : \mathbb{N} \to \mathbb{N}$ by $g_{s(x)}(y) = g_x(y) + y$, for every $y \in \mathbb{N}$. Then $g_{s(x)}(0) = 0$ and, for every $y \in \mathbb{N}$, $g_{s(x)}(s(y)) = g_x(s(y)) + s(y) = g_x(y) + x + s(y) = g_x(y) + s(x) + y = g_{s(x)}(y) + s(x)$ and so $s(x) \in M$. According to $(PA5)$, $M = \mathbb{N}$.

Uniqueness Now, for every $x \in \mathbb{N}$, let $g_x, h_x : \mathbb{N} \to \mathbb{N}$ be two functions that satisfy the properties: $g_x(0) = h_x(0) = 0$ and $g_x(s(y)) = g_x(y)+x, h_x(s(y)) = h_x(y)+x$, for every $y \in \mathbb{N}$. Let $M = \{y \in \mathbb{N} : g_x(y) = h_x(y)\}$. Since $g_x(0) = 0 = h_x(0)$ it follows that $0 \in M$. For every $y \in M$, $g_x(s(y)) = g_x(y) + x = h_x(y) + x = h_x(s(y))$; therefore $s(y) \in M$. It follows that $M = \mathbb{N}$ and so $g_x = h_x$.
□

Proposition 1.5.15 *For every $x \in \mathbb{N}$, let g_x be a function whose existence and uniqueness have been demonstrated in the previous theorem. Then*

(a) $g_0(x) = 0$, for every $x \in \mathbb{N}$, and
(b) $g_{s(x)}(y) = g_x(y) + y$, for every $x, y \in \mathbb{N}$.

Proof We recall that $g_x : \mathbb{N} \to \mathbb{N}$, $g_x(0) = 0$ and $g_x(s(y)) = g_x(y) + x$, for every $y \in \mathbb{N}$.

(a) We take the set $M = \{x \in \mathbb{N} : g_0(x) = 0\}$. Obviously, $0 \in M$. If $x \in M$, then $g_0(s(x)) = g_0(x) + 0 = 0$; therefore $s(x) \in M$ and then $M = \mathbb{N}$.
(b) For an arbitrary $x \in \mathbb{N}$, we take the set $M = \{y \in \mathbb{N} : g_{s(x)}(y) = g_x(y) + y\}$. $g_{s(x)}(0) = g_x(0) + 0 = 0$ and so $0 \in M$. For every $y \in M$, $g_{s(x)}(s(y)) = g_{s(x)}(y) + s(x) = g_x(y) + y + s(x) = g_x(y) + s(y) + x = g_x(s(y)) + s(y)$, and then $s(y) \in M$. According to $(PA5)$, $M = \mathbb{N}$.
□

We will now define the multiplication operation.

Definition 1.5.16 For any $x \in \mathbb{N}$, let g_x be the unique function whose existence was demonstrated in Theorem 1.5.14. Then, for every $y \in \mathbb{N}$, let us define $x \cdot y = g_x(y)$. Thus, we have defined a function $\cdot : \mathbb{N} \times \mathbb{N} \to \mathbb{N}$ by $\cdot(x, y) = x \cdot y$, for every $x, y \in \mathbb{N}$. Usually, the symbol "·" is omitted, so we will write xy instead of $x \cdot y$. We also introduce an order of operation: first multiplication, then addition. So $x + y \cdot z = x + (y \cdot z)$.

Theorem 1.5.17 $(\mathbb{N}, +, \cdot, \leqslant)$ *is a well-ordered commutative semiring.*

Proof According to Proposition 1.5.15, the function g, whose existence and uniqueness have been demonstrated in Theorem 1.5.14, satisfies, for every $x, y \in \mathbb{N}$, the following relationships: $g_x(0) = 0 = g_0(x)$, $g_x(s(y)) = g_x(y) + x$ and $g_{s(y)}(x) = g_y(x) + x$.

Using the notations $x \cdot y = g_x(y)$, these relationships can be rewritten as:

$$x \cdot 0 = 0 = 0 \cdot x, \text{ for every } x \in \mathbb{N}. \tag{1.23}$$

$$x \cdot (y+1) = x \cdot y + x = (y+1) \cdot x, \text{ for every } x, y \in \mathbb{N}. \tag{1.24}$$

Let's show that multiplication is distributive over addition:

$$x \cdot (y+z) = x \cdot y + x \cdot z, \text{ for every } x, y, z \in \mathbb{N}. \tag{1.25}$$

We fix $x, y \in \mathbb{N}$ and denote by $M = \{z \in \mathbb{N} : x \cdot (y+z) = x \cdot y + x \cdot z\}$. Using the above relationship (1.23), $x(y+0) = xy = xy + 0 = xy + x \cdot 0$, and so $0 \in M$. For every $z \in M$, $x \cdot (y + s(z)) = x \cdot s(y+z) = x \cdot ((y+z)+1) \stackrel{(1.24)}{=} x \cdot (y+z) + x = (x \cdot y + x \cdot z) + x = x \cdot y + (x \cdot z + x) \stackrel{(1.24)}{=} x \cdot y + x \cdot (z+1) = x \cdot y + x \cdot s(z)$, from where $s(z) \in M$ and then $M = \mathbb{N}$.

Now we prove that multiplication is commutative:

$$x \cdot y = y \cdot x, \text{ for every } x, y \in \mathbb{N}. \tag{1.26}$$

We fix an arbitrary point $x \in \mathbb{N}$ and let $M = \{y \in \mathbb{N} : x \cdot y = y \cdot x\}$. Using the relationship (1.23), we obtain $x \cdot 0 = 0 = 0 \cdot x$, which implies that $0 \in M$. For every $y \in M$, $x \cdot s(y) = x \cdot (y+1) \stackrel{(1.24)}{=} x \cdot y + x = y \cdot x + x \stackrel{(1.24)}{=} (y+1) \cdot x = s(y) \cdot x$. Therefore $s(y) \in M$ and then $M = \mathbb{N}$.

Let us prove that 1 is the identity for multiplication. For every $x \in \mathbb{N}$, $x \cdot 1 = g_x(1) = g_x(s(0)) = g_x(0) + x = x$ so that

$$x \cdot 1 = 1 \cdot x = x, \text{ for every } x \in \mathbb{N}. \tag{1.27}$$

The multiplication is associative:

$$(x \cdot y) \cdot z = x \cdot (y \cdot z), \text{ for every } x, y, z \in \mathbb{N}. \tag{1.28}$$

We fix $x, y \in \mathbb{N}$ and we denote by $M = \{z \in \mathbb{N} : (x \cdot y) \cdot z = x \cdot (y \cdot z)\}$. $(x \cdot y) \cdot 0 \stackrel{(1.23)}{=} 0 = x \cdot (y \cdot 0)$, and so $0 \in M$. For every $z \in M$, $x \cdot (y \cdot s(z)) = x \cdot (y \cdot (z+1)) \stackrel{(1.24)}{=} x \cdot (y \cdot z + y) \stackrel{(1.25)}{=} x \cdot (y \cdot z) + x \cdot y = (x \cdot y) \cdot z + x \cdot y \stackrel{(1.25)}{=} (x \cdot y)(z+1) = (x \cdot y) \cdot s(z)$. Therefore $s(z) \in M$, and then $M = \mathbb{N}$.

In the end, let's demonstrate the compatibility between the order relation and multiplication:

$$\text{For every } x, y, z \in \mathbb{N} \text{ with } x \leqslant y \text{ it follows that } x \cdot z \leqslant y \cdot z. \tag{1.29}$$

1.5 Appendix   53

We fix $x, y \in \mathbb{N}$ with $x \leqslant y$ and we denote by $M = \{z \in \mathbb{N} : x \cdot z \leqslant y \cdot z\}$. From (1.23) it immediately follows that $0 \in M$. Let now $z \in M$; then $x \cdot s(z) = x \cdot (z+1) \stackrel{(1.24)}{=} x \cdot z + x \leqslant y \cdot z + y \stackrel{(1.24)}{=} y \cdot s(z)$. Therefore $s(z) \in M$, and then $M = \mathbb{N}$.

I have proven in Theorem 1.5.12 that $(\mathbb{N}, +, \leqslant)$ is a well-ordered commutative monoid. If we add properties (1.25)–(1.29) to this, it follows that $(\mathbb{N}, +, \cdot, \leqslant)$ is a well-ordered commutative semiring with 1 as identity. □

Remarks 1.5.18

(i) If $x \cdot y = 0$, then $x = 0$ or $y = 0$. Indeed, if we suppose that $x \neq 0$, then there is $z \in \mathbb{N}$ such that $x = s(z)$. It follows that $0 = x \cdot y = s(z) \cdot y = (z+1) \cdot y = z \cdot y + y$. According to (4) of Proposition 1.5.7, $y = 0$.

(ii) If $x \cdot z = y \cdot z$ and $z \in \mathbb{N}^*$, then $x = y$. Indeed, if we suppose that $x < y$ then there exists $w \in \mathbb{N}^*$ such that $x + w = y$. It follows that $x \cdot z = (x + w) \cdot z$, from where $w \cdot z = 0$, which is absurd because $z, w \in \mathbb{N}^*$. In the case of $y < x$, we proceed in the same way, and then, using the law of trichotomy (see (i) of Remark 1.5.11), it follows that the only remaining possibility is $x = y$.

1.5.2 The Set of Integers

Once we have structured the set of natural numbers as a totally ordered commutative semiring, we can construct the set of integers. We think of an integer as a "difference" between natural numbers. However, since in $\mathbb{N}$, the difference cannot generally be made, we will think of an integer as an ordered pair (a, b) of natural numbers, where a is the minuend and b is the subtrahend, and (a, b) represents the "difference". Since different pairs of natural numbers can have the same "difference", we will introduce an equivalence relation that will identify these pairs.

Definition 1.5.19 Let $(a, b), (c, d) \in \mathbb{N} \times \mathbb{N}$; we say that (a, b) is equivalent with (c, d), and we denote this by $(a, b) \sim (c, d)$, if $a + d = c + b$.

Proposition 1.5.20 $\sim$ *is an equivalence relation on* $\mathbb{N} \times \mathbb{N}$.

Proof Obviously $(a, b) \sim (a, b)$ (reflexivity), and $(a, b) \sim (c, d) \Longrightarrow (c, d) \sim (a, b)$ (symmetry), for every $(a, b), (c, d) \in \mathbb{N} \times \mathbb{N}$. If $(a, b) \sim (c, d)$ and $(c, d) \sim (e, f)$, then $a + d = c + b$ and $c + f = e + d$ from where $a + f + (c + d) = b + e + (c + d)$. According to the cancellation property, $a + f = b + e$ which means that $(a, b) \sim (e, f)$ (transitivity). □

Definition 1.5.21 Let $\mathbb{Z} = \mathbb{N} \times \mathbb{N}/\sim$ be the quotient space of $\mathbb{N} \times \mathbb{N}$ by $\sim$; we recall that $\mathbb{Z} = \{[a, b] : (a, b) \in \mathbb{N} \times \mathbb{N}\}$, where $\alpha = [a, b] = \{(c, d) \in \mathbb{N} \times \mathbb{N} : (c, d) \sim (a, b)\}$ is the equivalence class of the representative (a, b).

$\mathbb{Z}$ is called the set of **integers**.

Remarks 1.5.22

(i) We observe that, for every $a, b, c \in \mathbb{N}$, $[a, b] = [a + c, b + c]$.
(ii) The notation $\mathbb{Z}$ belongs to David Hilbert and comes from the German word "Zahlen" (Numbers). It was first used in 1947 in the book "Algèbre" by Nicolas Bourbaki.

We will organize $\mathbb{Z}$ as a totally ordered commutative ring with unity.

Definition 1.5.23 We will define the addition operation as follows:

$$[a, b] + [c, d] = [a + c, b + d], \text{ for every } [a, b], [c, d] \in \mathbb{Z}.$$

The addition is well-defined, in the sense that it does not depend on representatives; this means that if $(a, b) \sim (a_1, b_1)$ and $(c, d) \sim (c_1, d_1)$, then $(a + c, b + d) \sim (a_1 + c_1, b_1 + d_1)$. Indeed, from the hypotheses, it follows that $a + b_1 = b + a_1$ and $c + d_1 = d + c_1$ and then $a + c + b_1 + d_1 = b + d + a_1 + c_1$.

Theorem 1.5.24 $(\mathbb{Z}, +)$ *is a commutative group.*

Proof Associativity and commutativity of addition are immediate consequences of similar properties of addition on the set of natural numbers.

We also easily observe that $\underline{0} = [0, 0] = \{(a, a) : a \in \mathbb{N}\}$ is the neutral element of addition for integers.

For every $[a, b] \in \mathbb{Z}$, $[b, a] \in \mathbb{Z}$ and $[a, b] + [b, a] = [a + b, a + b] = \underline{0}$. □

Definition 1.5.25 We will say that $-\alpha = [b, a]$ is the **opposite** of the element $\alpha = [a, b]$ and we will denote $[b, a] = -[a, b]$. This definition does not depend on the representatives, in the sense that, if $(a, b) \sim (c, d)$, then $(b, a) \sim (d, a)$. Thus, observe that on the set of integers, we can define the **subtraction** operation as $[a, b] - [c, d] = [a, b] + [d, c] = [a + d, b + c]$.

Definition 1.5.26 We now define the multiplication operation on $\mathbb{Z}$ by:

$$[a, b] \cdot [c, d] = [ac + bd, ad + bc], \text{ for every } [a, b], [c, d] \in \mathbb{Z}.$$

This definition also does not depend on the representatives. Indeed, let $(a, b) \sim (a_1, b_1)$ and $(c, d) \sim (c_1, d_1)$; then

$$a + b_1 = a_1 + b, \qquad (1.30)$$

$$c + d_1 = c_1 + d. \qquad (1.31)$$

By multiplying Eq. (1.30) successively with c and d and adding the results, we obtain:

$$ac + bd + b_1c + a_1d = bc + ad + a_1c + b_1d. \qquad (1.32)$$

1.5 Appendix

Then, multiplying Eq. (1.31) with a_1 and b_1 respectively and adding the results, we obtain:

$$a_1 d_1 + b_1 c_1 + a_1 c + b_1 d = a_1 c_1 + b_1 d_1 + a_1 d + b_1 c. \tag{1.33}$$

Now, by adding Eqs. (1.32) and (1.33) and using the cancellation property (see Theorem 1.5.8), we obtain: $ac + bd + a_1 d_1 + b_1 c_1 = ad + bc + a_1 c_1 + b_1 d_1$, which means that $(ac + bd, ad + bc) \sim (a_1 c_1 + b_1 d_1, a_1 d_1 + b_1 c_1)$.

Remark 1.5.27 $\alpha \cdot \underline{0} = \underline{0}$, for every $\alpha \in \mathbb{Z}$; indeed, if $\alpha = [a, b]$, then $\alpha \cdot \underline{0} = [a, b] \cdot [0, 0] = [0, 0] = \underline{0}$.

Theorem 1.5.28 $(\mathbb{Z}, +, \cdot)$ *is a commutative ring whose only zero divisors is 0* $((\mathbb{Z}, +\cdot)$ *is an integral domain).*

Proof We have shown in the previous theorem that $(\mathbb{Z}, +)$ is a commutative group. Elementary calculations show that the conditions of associativity and commutativity of multiplication, and the distributivity of multiplication over addition are satisfied. If we denote by $\underline{1} = [1, 0] = \{(a + 1, a) : a \in \mathbb{N}\}$, then, for every $\alpha = [a, b] \in \mathbb{Z}$, $\alpha \cdot \underline{1} = [a, b] = \alpha$. Therefore $\underline{1}$ is the neutral element for multiplication.

Let now $\alpha \cdot \beta = \underline{0}$ and we suppose that $\alpha = [a, b] \neq \underline{0}$ and $\beta = [c, d]$. Then $(ac + bd, ad + bc) \sim (0, 0)$ hence $ac + bd = ad + bc$. From $\alpha \neq \underline{0}$ it follows that $a \neq b$. Then $a < b$ or $b < a$. In the case $a < b$ there exists $u \in \mathbb{N}^*$ such that $a + u = b$ and so $ac + ad + ud = ad + ac + uc$ from where $uc = ud$ or $c = d$. Therefore $\beta = \underline{0}$. A similar result in the case $b < a$. □

Proposition 1.5.29 *For every $\alpha, \beta, \gamma \in \mathbb{Z}$, the cancellation laws hold:*

$$\alpha + \beta = \alpha + \gamma \Longrightarrow \beta = \gamma,$$

$$\alpha \cdot \beta = \alpha \cdot \gamma \text{ and } \alpha \neq \underline{0} \Longrightarrow \beta = \gamma.$$

Proof The first implication is obtained by adding $-\alpha$ to both sides of the equation in the hypothesis. For the second implication, let's suppose that $\alpha = [a, b] \neq \underline{0}$; then $a \neq b$ and we suppose that $a < b$; hence there is $u \in \mathbb{N}^*$ such that $a + u = b$. Let now $\beta = [c, d]$ and $\gamma = [e, f]$ two arbitrary numbers in $\mathbb{Z}$. From $\alpha \cdot \beta = \alpha \cdot \gamma$ it follows that $(ac + bd, ad + bc) \sim (ae + bf, af + be)$ or $ac + bd + af + be = ad + bc + ae + bf$. Replacing b in the last equality, we obtain $ac + ad + ud + af + ae + ue = ad + ac + uc + ae + af + uf$, or $u(d + e) = u(c + f)$. According to (ii) of Remark 1.5.18, it follows that $d + e = c + f$, from where $(c, d) \sim (e, f)$, or $\beta = \gamma$.

We will now define an order relation on $\mathbb{Z}$. □

Definition 1.5.30 Let $\alpha = [a, b], \beta = [c, d] \in \mathbb{Z}$; we say that α is **less than or equal** than β, and we denote by $\alpha \leqslant \beta$, if $a + d \leqslant b + c$. α is strictly smaller than β, $\alpha < \beta$, if $\alpha \leqslant \beta$ and $\alpha \neq \beta$; equivalently, $\alpha < \beta$ if and only if $a + d < b + c$.

This definition does not depend on representatives. Indeed, if $\alpha = [a, b] \leqslant [c, d] = \beta$, then $a + d \leqslant b + c$. Let $u \in \mathbb{N}$ such that $a + d + u = b + c$. Now we suppose that $(a, b) \sim (a_1, b_1)$ and $(c, d) \sim (c_1, d_1)$; then $a + b_1 = a_1 + b$ and $c + d_1 = c_1 + d$. If we add the last two equalities, we obtain: $(a_1 + d_1) + (b + c) = (b_1 + c_1) + (a + d)$, and replacing $b + c$, $(a_1 + d_1) + u + a + d = (b_1 + c_1) + a + d$, or $(a_1 + d_1) + u = (b_1 + c_1)$. It follows that $a_1 + d_1 \leqslant b_1 + c_1$ and then $\alpha = [a_1, b_1] \leqslant [c_1, d_1] = \beta$.

Remarks 1.5.31

(i) $\underline{0} = [0, 0] \leqslant [a, b]$ if and only if $b \leqslant a$.
(ii) For every $a \in \mathbb{N}$, $[a, 0] \geqslant \underline{0}$ and $[0, a] \leqslant \underline{0}$.
(iii) For every $[a, b], [c, d] \in \mathbb{Z}$, $[a, b] \leqslant [c, d]$ if and only if $[a, b] - [c, d] = [a, b] + [d, c] = [a + d, b + c] \leqslant \underline{0}$.

Theorem 1.5.32 $(\mathbb{Z}, +, \cdot, \leqslant)$ *is a total ordered commutative ring.*

Proof It is easy to show that $\leqslant$ is an order relation on $\mathbb{Z}$. Let now $\alpha = [a, b]$, $\beta = [c, d] \in \mathbb{Z}$. From statement (i) of Remark 1.5.11, it follows that given the numbers $a + d$ and $b + c \in \mathbb{N}$, there is exactly one of the following situations $a + d < b + c$, $b + c < a + d$ or $a + d = b + c$. These correspond to the following relationships: $\alpha < \beta$, $\beta < \alpha$ or $\alpha = \beta$. Therefore $\leqslant$ is a total order on $\mathbb{Z}$.

Let now $\alpha = [a, b] \leqslant [c, d] = \beta$ and let $\gamma = [e, f] \in \mathbb{Z}$. Then $a + d \leqslant b + c$; we add in both sides of the previous inequality $e + f$ and obtain $a + e + d + f \leqslant c + e + b + f$, which means that $\alpha + \gamma = [a + e, b + f] \leqslant [c + e, d + f] = \beta + \gamma$.

If we assume in addition that $\gamma \geqslant \underline{0}$, then there is $u \in \mathbb{N}$ such that $f + u = e$. Therefore, according to (i) of Remark 1.5.22, $\alpha \cdot \gamma - \beta \cdot \gamma = [ae + bf + cf + de, af + be + ce + df] = [af + au + bf + cf + df + du, af + bf + bu + cf + cu + df] = [(a + d)u, (b + c)u] \leqslant \underline{0}$ and then $\alpha \cdot \gamma \leqslant \beta \cdot \gamma$. □

Remark 1.5.33 A similar calculation to the one in statement Proposition 1.5.29 leads us to the following implications

$$\alpha + \beta \leqslant \alpha + \gamma \Longrightarrow \beta \leqslant \gamma,$$

$$\alpha \cdot \beta \leqslant \alpha \cdot \gamma \text{ and } \alpha > \underline{0} \Longrightarrow \beta \leqslant \gamma.$$

We can now define an embedding of the set of natural numbers into the set of integers and, thus, view $\mathbb{N}$ as a subset of $\mathbb{Z}$. The proof of the following proposition consists of a simple verification of the definitions and, for this reason, we omit it.

Proposition 1.5.34 *The mapping* $i : \mathbb{N} \to \mathbb{Z}$ *defined by* $i(a) = [a, 0]$ *is injective and preserves addition, multiplication, and order relation.*

Remark 1.5.35 Based on the previous proposition, we can identify any natural number a with the equivalence class $[a, 0]$; considering this identification, $\mathbb{N}$ can be viewed as a subset of $\mathbb{Z}$. Moreover, for any $[a, b] \in \mathbb{Z}$, $[a, b] = [a, 0] + [0, b] =$

$[a, 0] - [b, 0] \equiv a - b$. In this way, the intuitive idea of defining integers as differences of natural numbers becomes consistent.

The following theorem shows that $\mathbb{Z}$ is the smallest ring containing natural numbers.

Theorem 1.5.36 $(\mathbb{Z}, +, \cdot)$ *is minimal among commutative rings containing semirings isomorphic to* $\mathbb{N}$.

Proof I have already seen that $(\mathbb{Z}, +, \cdot)$ is a commutative ring and contains a semiring isomorphic to $\mathbb{N}$. Now let $(R, +, \cdot)$ be another commutative ring that contains a copy of $\mathbb{N}$. This means that there exists an injective function $\varphi : \mathbb{N} \to R$ that preserves addition and multiplication. Let $Z = \{\varphi(a) - \varphi(b) : a, b \in \mathbb{N}\} \subseteq R$.

We define the mapping $\Psi : \mathbb{Z} \to Z$ by $\Psi(\alpha) = \varphi(a) - \varphi(b)$, for every $\alpha = [a, b] \in \mathbb{Z}$. The definition does not depend on the representative (a, b) that we choose for the integer number α. Indeed, if (c, d) is another representative for α, then $(a, b) \sim (c, d)$ so that $a + d = b + c$; from this $\varphi(a) + \varphi(d) = \varphi(b) + \varphi(c)$, or equivalently in R, $\varphi(a) - \varphi(b) = \varphi(c) - \varphi(d)$, what means that $\Psi([a, b]) = \Psi([c, d])$.

Now, for every $\alpha = [a, b], \beta = [c, d] \in \mathbb{Z}$, if $\Psi(\alpha) = \Psi(\beta)$, then $\varphi(a) - \varphi(b) = \varphi(c) - \varphi(d)$, or equivalently, $\varphi(a + d) = \varphi(a) + \varphi(d) = \varphi(b) + \varphi(c) = \varphi(b + c)$; it follows that $a + d = b + c$, which means that $(a, b) \sim (c, d)$, and then $\alpha = [a, b] = [c, d] = \beta$. Therefore Ψ is an injection and, since it is obviously surjective, it follows that Ψ is a bijection between $\mathbb{Z}$ and Z.

We will show that Ψ preserves the operations of addition and multiplication. For every $\alpha = [a, b], \beta = [c, d] \in \mathbb{Z}$, $\Psi(\alpha + \beta) = \Psi([a + c, b + d]) = \varphi(a + c) - \varphi(b + d) = (\varphi(a) - \varphi(b)) + (\varphi(c) - \varphi(d)) = \Psi(\alpha) + \Psi(\beta)$.

$\Psi(\alpha \cdot \beta) = \Psi([ac + bd, ad + bc]) = \varphi(ac + bd) - \varphi(ad + bc) = \varphi(a) \cdot \varphi(c) + \varphi(b) \cdot \varphi(d) - \varphi(a) \cdot \varphi(d) - \varphi(b) \cdot \varphi(c) = (\varphi(a) - \varphi(b)) \cdot (\varphi(c) - \varphi(d)) = \Psi(\alpha) \cdot \Psi(\beta)$.

Thus, Ψ is a ring isomorphism, and this shows that any ring containing a copy of $\mathbb{N}$ also contains a copy of $\mathbb{Z}$. □

1.5.3 The Set of Rational Numbers

The integers introduced in the previous subsection allow us to define rational numbers.

We will consider a rational number as a "quotient" of integers. Since in the set of integers, the quotient of two numbers is not always meaningful, we will use the following alternative: we will replace the "quotient" of integers a and b with the ordered pair $(a, b) \in \mathbb{Z} \times \mathbb{Z}^*$ (here will denote $\mathbb{Z}^* = \mathbb{Z} \setminus \{0\}$, a will be the numerator and b the denominator). Since two such pairs can lead to the same rational number, we will identify them through an equivalence relation.

Definition 1.5.37 Let $(a, b), (c, d) \in \mathbb{Z} \times \mathbb{Z}^*$; we say that (a, b) is equivalent with (c, d) if $ad = bc$. We denote this by $(a, b) \sim (c, d)$.

Proposition 1.5.38 $\sim$ *is an equivalence relation on* $\mathbb{Z} \times \mathbb{Z}^*$.

Proof The properties of reflexivity and symmetry are immediate consequences of the definition of the relation $\sim$. Let's prove the transitivity property. If $(a, b) \sim (c, d)$ and $(c, d) \sim (e, f)$, then $ad = bc$ and $cf = de$ and so $adf = bcf = bde$. Because $d \neq 0$ ($d \in \mathbb{Z}^*$) we can apply Proposition 1.5.29 to obtain $af = be$ and then $(a, b) \sim (e, f)$. □

Definition 1.5.39 The quotient space $\mathbb{Q} = \mathbb{Z} \times \mathbb{Z}^*/\sim$ is called the set of **rational numbers**. Therefore $\mathbb{Q} = \{[a, b] : (a, b) \in \mathbb{Z} \times \mathbb{Z}^*\}$, where $[a, b] = \{(c, d) : (a, b) \sim (c, d)\}$.

Remarks 1.5.40

(i) We observe that, for every $a \in \mathbb{Z}$ and $b, c \in \mathbb{Z}^*$, $[a, b] = [ac, bc]$. Particularly, $[a, b] = [-a, -b]$, so that we can always write the denominator of a rational number as a positive number ($b \in \mathbb{N}^*$).
(ii) The notation $\mathbb{Q}$ belongs to Giuseppe Peano and comes from the Italian word "quotziente" (quotient). It was first used also in 1947 in the book "Algèbre" by Nicolas Bourbaki.

We will organize $\mathbb{Q}$ as an ordered field.

Definition 1.5.41 The addition on $\mathbb{Q}$ is defined by:

$$[a, b] + [c, d] = [ad + bc, bd], \text{ for every } [a, b], [c, d] \in \mathbb{Q}.$$

We will make two observations regarding this definition.

Firstly, if $[a, b], [c, d] \in \mathbb{Q}$, then $bd \neq 0$ (from Theorem 1.5.28, $\mathbb{Z}$ is an integral domain), so that $bd \in \mathbb{Z}^*$. Secondly, the above definition does not depend on the choice of representatives. Indeed, if $(a, b) \sim (a_1, b_1)$ and $(c, d) \sim (c_1, d_1)$, then $ab_1 = a_1 b$ and $cd_1 = c_1 d$. By multiplying the first equality by dd_1 and the second one by bb_1, we obtain $adb_1 d_1 = a_1 d_1 bd$ and $bcb_1 d_1 = b_1 c_1 bd$. If we add the last two equalities, we obtain: $(ad + bc)b_1 d_1 = (a_1 d_1 + b_1 c_1)bd$, which means that $(ad + bc, bd) \sim (a_1 d_1 + b_1 c_1, b_1 d_1)$ and then $[a, b] + [c, d] = [a_1, b_1] + [c_1, d_1]$.

Theorem 1.5.42 $(\mathbb{Q}, +)$ *is a commutative group*.

Proof We directly verify, starting from the definition, that addition is commutative and associative. We will define the neutral element for addition as $\underline{0} = [0, 1] = \{(0, a) : a \in \mathbb{Z}^*\}$. Indeed, for every $[a, b] \in \mathbb{Q}$, $[a, b] + [0, 1] = [a, b]$.

The opposite of a rational number $[a, b]$ is $[-a, b]$; indeed, $[a, b] + [-a, b] = [0, b^2] = \underline{0}$. □

Remark 1.5.43 We denote the opposite of $[a, b]$ by $-[a, b] = [-a, b]$ and we define the substraction as $[a, b] - [c, d] = [a, b] + [-c, d] = [ad - bc, bd]$.

Definition 1.5.44 The multiplication is defined by:

$$[a, b] \cdot [c, d] = [ac, bd], \text{ for every } [a, b], [c, d] \in \mathbb{Q}.$$

1.5 Appendix

We remark that $bd \in \mathbb{Z}^*$ and that the above definition does not depend on the choice of representatives of the rational numbers $[a, b]$ and $[c, d]$.

Theorem 1.5.45 $(\mathbb{Q}, +, \cdot)$ *is field.*

Proof It is easy to verify that multiplication is associative, commutative, and distributive over addition. $\underline{1} = [1, 1] = \{(a, a) : a \in \mathbb{Z}^*\}$ is neutral element for multiplication. Indeed, for every $[a, b] \in \mathbb{Q}$, $[a, b] \cdot \underline{1} = [a, b]$. Now, for every $[a, b] \in \mathbb{Q}, [a, b] \neq \underline{0}$, it follows that $a \in \mathbb{Z}^*$. Then $[b, a] \in \mathbb{Q}$ and $[a, b] \cdot [b, a] = [ab, ab] = \underline{1}$. The inverse of the element $[a, b]$ is $[b, a]$, and we denote it by $[a, b]^{-1}$. □

Now let's define an order relation on $\mathbb{Q}$.

Definition 1.5.46 We say that the rational number $[a, b]$ is **positive** if $ab \geq 0$; we denote this by $[a, b] \geq \underline{0}$. The definition does not depend on the choice of representatives of the rational number $[a, b]$; indeed, if $[a, b] \geq \underline{0}$ and if $(a, b) \sim (c, d)$, then $ab \geq 0$ and $ad = bc$. Multiplying the last equality by bd, we obtain: $abd^2 = cdb^2$. Because $d^2, b^2 > 0$ (in any totally ordered commutative ring, the square of any element is positive), and $ab \geq 0$, it follows that $cd \geq 0$. We say that $[a, b] \leq \underline{0}$ if $-[a, b] \geq \underline{0}$, equivalently if $ab \leq 0$.

Let $[a, b], [c, d] \in \mathbb{Q}$; we say that $[a, b]$ is **less than or equal** than $[c, d]$ if $[c, d] - [a, b] = [bc - ad, bd] \geq \underline{0}$, or, equivalently, if $(bc - ad)bd \geq 0$. We denote this by $[a, b] \leq [c, d]$. As we noticed in Remark 1.5.40, the numerator of rational numbers can be chosen to be positive. Therefore, if $b, d \in \mathbb{N}^*$, then $[a, b] \leq [c, d]$ iff $ad \leq bc$.

Theorem 1.5.47 *The relation $\leq$ is a total order on $\mathbb{Q}$ and $(\mathbb{Q}, +, \cdot, \leq)$ is an ordered field.*

Proof Reflexivity and antisymmetry of the relation $\leq$ are easily verified; let us demonstrate that $\leq$ is transitive. Let us suppose that $[a, b] \leq [c, d]$ and $[c, d] \leq [e, f]$ where $b, d, f \in \mathbb{N}^*$; then $ad \leq bc$ and $cf \leq ed$. We multiply the first inequality by f, the second one by b, and we obtain: $afd \leq bcf$ and $bcf \leq bed$, from where $afd \leq bed$; using Remark 1.5.33 we deduce that $af \leq be$ which means that $[a, b] \leq [e, f]$. Let now $[a, b], [c, d] \in \mathbb{Q}$ with $b, d \in \mathbb{N}^*$. Since the order relation on $\mathbb{Z}$ is total, $ad \leq bc$ or $bc \leq ad$, which implies that $[a, b] \leq [c, d]$ or $[c, d] \leq [a, b]$. Thus, $\leq$ is a total order relation on $\mathbb{Q}$.

Let us show that it is compatible with addition and multiplication. Let $[a, b], [c, d], [e, f] \in \mathbb{Q}$ with $b, d, f \in \mathbb{N}^*$, and $[a, b] \leq [c, d]$; then $ad \leq bc$. Since $[a, b] + [e, f] = [af + be, bf]$, it follows that $(af + be)df = adf^2 + bdef \leq bcf^2 + bdef = bf(cf + de)$, which means that $[a, b] + [e, f] \leq [c, d] + [e, f] = [cf + de, df]$.

If, in addition, we assume that $[e, f] \geq \underline{0}$, then $ef \geq 0$ and so $aedf \leq bcef$ which means that $[a, b] \cdot [e, f] = [ae, bf] \leq [ce, df] = [c, d] \cdot [e, f]$.

Therefore $(\mathbb{Q}, +, \cdot, \leq)$ is an ordered field. □

Remarks 1.5.48

(i) The mapping $j : \mathbb{Z} \to \mathbb{Q}$ defined by $j(a) = [a, 1]$ is injective and preserves addition, multiplication, and order relation; therefore it is an embedding of $\mathbb{Z}$ in $\mathbb{Q}$. Therefore we can identify any integer a with the equivalence class $[a, 1]$; considering this identification, $\mathbb{Z}$ can be viewed as a subset of $\mathbb{Q}$. Moreover, for any $[a, b] \in \mathbb{Q}$, $[a, b] = [a, 1] \cdot [1, b] = [a, 1] \cdot [b, 1]^{-1} \equiv \frac{a}{b}$. In this way, the intuitive idea of defining integers as a quotient of integers becomes consistent.

(ii) If we compose the function i defined in Proposition 1.5.34 with the function j mentioned above, we obtain an embedding of the set of natural numbers into $\mathbb{Q}$. Thus, $\mathbb{N}$ can be thought of as a subset of $\mathbb{Q}$ and therefore $\mathbb{N} \subseteq \mathbb{Z} \subseteq \mathbb{Q}$.

(iii) A similar proof to that of Theorem 1.5.36 leads us to: $\mathbb{Q}$ is minimal among the fields containing rings isomorphic with $\mathbb{Z}$.

Proposition 1.5.49 $(\mathbb{Q}, +, \cdot, \leqslant)$ *is an Archimedean ordered field.*

Proof Theorem 1.5.47 guarantees that $(\mathbb{Q}, +, \cdot, \leqslant)$ is an ordered field. Let now $\alpha, \beta \in \mathbb{Q}$ with $\alpha > \underline{0}$; there exist $a, b \in \mathbb{Z}$ such that $\beta \cdot \alpha^{-1} = [a, b] \in \mathbb{Q}$. According to (i) of Remark 1.5.40, we can suppose that $b \in \mathbb{N}^*$. If $a \in \mathbb{Z} \setminus \mathbb{N}^*$, then there is $n = 1 \in \mathbb{N}$ such that $nb > a$. If $a \in \mathbb{N}^*$, then there is $n = a + 1 \in \mathbb{N}$ such that $nb = ab + b > ab \geqslant a$. In both cases $[n, 1] - [a, b] = [nb - a, b] > \underline{0}$, what it trains $n \cdot \alpha \equiv [n, 1] \cdot \alpha > \beta$. □

Proposition 1.5.50 *For every* $\alpha = [a, b] \in \mathbb{Q}$ *there exists* $k \in \mathbb{Z}$ *such that* $k \equiv [k, 1] \leqslant \alpha = [a, b] < [k + 1, 1] \equiv k + 1$. *This number k is called the* **integer part** *of the rational number α and is denoted by* $k = \lfloor \alpha \rfloor$.

Proof We can suppose that $b \in \mathbb{N}^*$.

Let $A = \{a - kb : k \in \mathbb{Z}, a \geqslant kb\} \subseteq \mathbb{N}$.

If $a \in \mathbb{N}$, then there is $k = 0 \in \mathbb{Z}$ such that $a \geqslant kb$. If $a \in \mathbb{Z} \setminus \mathbb{N}$, then, according to Proposition 1.5.49, there exists $n \in \mathbb{N}$ such that $nb > -a$. We denote $k = -n \in \mathbb{Z}$ and then $kb = -nb < a$.

It is observed that, in all cases $A \subseteq \mathbb{N}$, $A \neq \emptyset$. According to Theorem 1.5.12, there exists a smallest element in the set A, denoted as x_0. Let $k_0 \in \mathbb{Z}$ such that $x_0 = a - k_0 b$.

Since $x_0 \in A$, $k_0 b \leqslant a$ and then $k_0 \equiv [k_0, 1] \leqslant [a, b] = \alpha$. Since x_0 is the smallest element in A, and since $x = a - (k_0 + 1)b < x_0$, it follows that $x \notin A$. Therefore $a < (k_0 + 1)b$, from where $\alpha = [a, b] < [k_0 + 1, 1] \equiv k_0 + 1$. □

References

1. Bachman, G. (1964). *Introduction to p-adic numbers and valuation theory*. New York: Academic Press.
2. Baker, A. (2011). *An introduction to p-adic numbers and p-adic analysis*. http://www.maths.gla.ac.uk/ajb

References

3. Bercovich, V. (2009). *Non-archimedean analytic spaces, Lecture at the Advanced School on p-adic Analysis and Applications*, ICTP, Trieste, 31 August–11 September, 2009.
4. Bosch, S., Güntzer, U., & Remmert, R. (1984). *Non-Archimedean analysis*. Berlin: Springer-Verlag.
5. Gallistel, C. R., Gelman, R., & Cordes, S. (2006). The cultural and evolutionary history of the real numbers. In S. C. Levinson & J. Pierre (Eds.), *Evolution and culture: A Fyssen Foundation symposium* (pp. 247–274). Cambridge, MA: MIT Press.
6. Gelbaum, B. R., & Olmsted, J. M. H. (1992). *Counterexamples in analysis*. Mineola, NY: Dover Publications, Inc.
7. Gowers, T. (2017). *What is so wrong with thinking of real numbers as infinite decimals?* Cambridge University, Department of Pure Mathematics and Mathematical Statistics. https://www.dpmms.cam.ac.uk/~wtg10/decimals.html
8. Halbeisen, L., & Krapf, R. (2020). *Gödel's theorems and Zermelo's axioms. A firm foundation of mathematics*. Birkhäuser, Springer Nature Switzerland AG.
9. Jacobson, N. (2009). *Basic algebra II* (2nd ed.). Mineola, NY: Dover Publications Inc.
10. Katz, K. U., & Katz, M. G. (2012). Stevin numbers and reality. *Foundations of Science, 17*, 109–123.
11. Peano, G. (1889). *Arithmetices principia: nova methodo exposita*. Torino: Fratres Bocca.
12. Robert, A. M. (2000). *A course in p-adic analysis, Graduate texts in mathematics* (Vol. 198). New York: Springer-Verlag.
13. Stevin, S. (1608). *Disme: The art of tents or decimal aritmetike*. London.
14. Weis, I. (2015). The real numbers. A survey of constructions. *Rocky Mountain Journal of Mathematics, 45*, 737–762.

Chapter 2
Recurrences

A **recurrence** is a relationship between the terms of a sequence that allows us, recursively, to determine all the terms of the sequence, i.e. an equation of the form:

$$\begin{cases} F(n, x_n, x_{n+1}, \cdots, x_{n+p}) = 0, \text{ for every } n \in \mathbb{N} \\ x_0 = a_0, x_1 = a_1, \cdots, x_{p-1} = a_{p-1} \end{cases}, \qquad (*)$$

where $p \in \mathbb{N}^*$, $F : \mathbb{N} \times \mathbb{R}^{p+1} \to \mathbb{R}$ is a known function, and $a_0, a_1, \cdots, a_{p-1} \in \mathbb{R}$.

Such recurrences frequently arise in problems involving the approximation of solutions to equations, the calculation of values of special functions, discrete models in economics, biology, etc.

If in $(*)$ we substitute $n = 0$, then we obtain $F(0, x_0, x_1, \cdots, x_p) = 0$ from which, in theory, we can express x_p in terms of $a_0, a_1, \cdots, a_{p-1}$. Then, if we let n take on values $1, 2, \cdots$ in turn, we can obtain all the terms of the sequence. Of course, this depends on the form of the function F. This is, in fact, one of the goals of the recurrence theory: to find the general form of the sequence (x_n), when this is possible. In most cases, this is not possible, so we settle for studying the behavior of the sequence defined by the recurrence $(*)$ (boundedness, monotonicity, limit).

In the case where F is a linear function of the variables $x_n, \cdots, x_{n+p}$, equation $(*)$ is called a linear recurrence. In this case, there are methods to obtain an explicit formula for the sequence (x_n); these methods will be presented in the first paragraph of this section. Here we only focus on solving linear recurrences; a thorough study of stability and asymptotic behavior, as well as applications in the theory of differential equations, can be found in [6] and [9].

Nonlinear recurrences form an extremely broad class for which there are no general solving techniques. For this reason, we will limit ourselves in the second paragraph to only a few special cases. After briefly reviewing some cases where nonlinear recurrences can be reduced to linear ones, we will focus on two notable nonlinear recurrences, both of which have a quadratic convergence rate.

The first is Newton's method for finding the roots of functions. A remarkable particular case of this method is represented by the ancient method of Heron for extracting square roots, known as the Babylonian algorithm.

The second is the well-known algorithm of the arithmetic-geometric mean (the (A, G) algorithm). The last three subsections of the paragraph are dedicated to double nonlinear algorithms, of which the (A, G) is a part. Gauss's and Legendre's proofs for the (A, G) algorithm's limit are given. The elliptic integrals of the first and second kind are presented, as well as Legendre's formula and its applications to the approximation of the length of the lemniscate and the ellipse.

2.1 Linear Recurrences

The general form of a linear recurrence is:

$$f(n+p) + a_1(n) \cdot f(n+p-1) + \cdots + a_p(n) \cdot f(n) = a(n), \text{ for every } n \in \mathbb{N}, \quad (R)$$

where $p \in \mathbb{N}^*, a_0, a_1, \cdots, a_p, a$ are known sequences of real numbers, and $f = (x_n)_{n \in \mathbb{N}}$ is an unknown sequence. We will show that there are general techniques for solving these types of equations.

First-order linear recurrences can be solved using elementary methods. Indeed, the general form of such a recurrence is:

$$x_{n+1} = a_n \cdot x_n + b_n, \text{ for every } n \in \mathbb{N}.$$

For any $n \in \mathbb{N}$, we will denote the product $a_0 a_1 \cdots a_n = A_n$. Then

$$x_1 = a_0 \cdot x_0 + b_0 = a_0 \left(x_0 + \frac{b_0}{a_0} \right) = A_0 \left(x_0 + \frac{b_0}{A_0} \right),$$
$$x_2 = a_1 (a_0 x_0 + b_0) + b_1 = a_0 a_1 \left(x_0 + \frac{b_0}{a_0} + \frac{b_1}{a_0 a_1} \right) = A_1 \left(x_0 + \frac{b_0}{A_0} + \frac{b_1}{A_1} \right).$$

If we suppose that

$$x_n = A_{n-1} \left(x_0 + \sum_{k=0}^{n-1} \frac{b_k}{A_k} \right), \text{ then}$$

$$x_{n+1} = a_n \cdot x_n + b_n = a_n \cdot A_{n-1} \left(x_0 + \sum_{k=0}^{n-1} \frac{b_k}{A_k} \right) + b_n =$$

$$= A_n \left(x_0 + \sum_{k=0}^{n-1} \frac{b_k}{A_k} \right) + A_n \cdot \frac{b_n}{A_n} = A_n \left(x_0 + \sum_{k=0}^{n} \frac{b_k}{A_k} \right).$$

2.1 Linear Recurrences

Then the general solution of the first-order linear recurrence is

$$x_n = A_{n-1}\left(x_0 + \sum_{k=0}^{n-1} \frac{b_k}{A_k}\right), \text{ for every } n \in \mathbb{N}^*.$$

However, to address general recurrences of type (R), a more complicated mathematical apparatus is required.

2.1.1 Difference Calculus

The difference operator is the substitute of differential operator in the differential calculus.

Definition 2.1.1 Let S be the set of all sequences of real numbers $f : \mathbb{N} \to \mathbb{R}$, $f(n) = x_n$, for every $n \in \mathbb{N}$. We define the **difference operator** $\Delta : S \to S$ by:

$$\Delta f(n) = f(n+1) - f(n) = x_{n+1} - x_n, \text{ for all } n \in \mathbb{N}.$$

It is immediately noticeable that the difference operator is linear:

$$\Delta(f+g) = \Delta f + \Delta g, \Delta(\alpha f) = \alpha \Delta f, \text{ for every } f, g \in S, \text{ and every } \alpha \in \mathbb{R}.$$

Now we define iterated finite differences of higher order of f by:

$$\Delta^0 f(n) = f(n) = x_n,$$
$$\Delta^1 f(n) = \Delta f(n) = f(n+1) - f(n) = x_{n+1} - x_n,$$
$$\Delta^2 f(n) = \Delta(\Delta^1 f(n)) = f(n+2) - 2f(n+1) + f(n) =$$
$$= x_{n+2} - 2x_{n+1} + x_n.$$
$$\Delta^p f(n) = \Delta(\Delta^{p-1} f(n)), \text{ for every } p, n \in \mathbb{N}$$

Lemma 2.1.2 Let $f : \mathbb{N} \to \mathbb{R}$ be a sequence of real numbers. Then, for every $p, n \in \mathbb{N}$, we have

$$\Delta^p f(n) = \sum_{k=0}^{p}(-1)^k \binom{p}{k} \cdot f(n+p-k), \text{ for every } p, n \in \mathbb{N}. \quad (*)$$

(Here, for every $0 \leqslant k \leqslant p$, $\binom{p}{k} = \dfrac{p!}{k!(p-k)!}$ are the binomial coefficients.)

Proof $(*)$ is obviously satisfied for $p = 0$.

If we suppose that $\Delta^{p-1} f(n) = \sum_{k=0}^{p-1} (-1)^k \binom{p-1}{k} \cdot f(n+p-1-k)$, then

$$\Delta^p f(n) = \Delta(\Delta^{p-1} f(n)) = \sum_{k=0}^{p-1} (-1)^k \binom{p-1}{k} f(n+p-k) -$$

$$- \sum_{k=0}^{p-1} (-1)^k \binom{p-1}{k} f(n+p-k-1) = \binom{p-1}{0} f(n+p) +$$

$$+ \sum_{k=1}^{p-1} (-1)^k \binom{p-1}{k} f(n+p-k) - \underbrace{\sum_{k=0}^{p-2} (-1)^k \binom{p-1}{k} f(n+p-k-1)}_{k \to k-1} -$$

$$(-1)^{p-1} \binom{p-1}{p-1} f(n) =$$

$$= \binom{p}{0} f(n+p) + \sum_{k=1}^{p-1} (-1)^k \left(\binom{p-1}{k} + \binom{p-1}{k-1} \right) f(n+p-k) +$$

$$+ (-1)^p \binom{p}{p} f(n) = \sum_{k=0}^{p} (-1)^k \binom{p}{k} \cdot f(n+p-k).$$

□

Remarks 2.1.3

(i) A simple calculation leads us to:

$$\Delta^p (\Delta f) = \Delta^{p+1} f, \text{ and } \Delta^p (\Delta^q f) = \Delta^{p+q} f.$$

(ii) Let f_p be a polynomial of degree p in the variable $n \in \mathbb{N}$ (that is, for every $n \in \mathbb{N}$, $f_p(n) = a_0 n^p + a_1 n^{p-1} + \cdots + a_{p-1} n + a_p$, where $a_0, \cdots a_p \in \mathbb{R}$); then $\Delta^{p+1} f_p(n) = 0$, for every $n \in \mathbb{N}$.

We will prove this fact by induction on p. For $p = 0$, $\Delta^0(a_0) = a_0$, and then $\Delta^1(a_0) = 0$. Now we suppose that $\Delta^p g_{p-1}(n) = 0$, for every polynomial g_{p-1} of degree $p-1$; then $\Delta^{p+1} f_p(n) = \Delta^p(\Delta f_p k(n)) = \Delta^p(f_p(n+1) - f_p(n)) = 0$ ($f_p(n+1) - f_p(n)$ is a polynomial of degree $p-1$ in the variable n).

2.1 Linear Recurrences

(iii) We can obtain a dual formula to formula (∗), that is, we can express the translates of a sequence f using the difference operator:

$$f(n+1) = x_{n+1} = \Delta^1 f(n) + \Delta^0 f(n),$$
$$f(n+2) = x_{n+2} = \Delta^2 f(n) + 2\Delta^1 f(n) + \Delta^0 f(n),$$

and inductively we obtain

$$f(n+p) = x_{n+p} = \sum_{k=0}^{p} \binom{p}{k} \cdot \Delta^k f(n), \text{ for every } p, n \in \mathbb{N}. \quad (**)$$

We can now use the formula (∗∗) to substitute the translates of the sequence f into the linear recurrence (R); we then transform the recurrence (R) into linear difference equation:

$$\Delta^p f(n) + b_1(n) \cdot \Delta^{p-1} f(n) + \cdots + b_p(n) \cdot \Delta^0 f(n) = a(n), \text{ for every } n \in \mathbb{N}. \quad (D)$$

2.1.2 Homogeneous Linear Recurrences

In this paragraph, we aim to present methods for obtaining solutions to the equation (R) or, equivalently, the difference equation (D). We will begin by studying the homogeneous linear equations. These are obtained by replacing the sequence $(a_n)_n$ in (R) or (D) with the constant zero sequence. The homogeneous linear recurrence is therefore:

$$f(n+p) + a_1(n) \cdot f(n+p-1) + \cdots + a_p(n) \cdot f(n) = 0, \text{ for every } n \in \mathbb{N}, \quad (RO)$$

and the associated homogeneous difference equation is

$$\Delta^p f(n) + b_1(n) \cdot \Delta^{p-1} f(n) + \cdots + b_p(n) \cdot \Delta^0 f(n) = 0, \text{ for every } n \in \mathbb{N}. \quad (DO)$$

Remark 2.1.4 The order of the homogeneous linear recurrence is determined by p. If we assume that there exists $n_0 \in \mathbb{N}$ such that $a_p(n_0) = 0$, then the order p of the homogeneous recurrence can be reduced. Indeed, in this case, if we assume that $a_{p-1}(n) \neq 0$ for all $n \in \mathbb{N}$, then by setting $n = n_0$ in (RO), we obtain:

$$f(n_0 + 1) = F_1(f(n_0 + 2), \cdots, f(n_0 + p)),$$

where F_1 is a linear map of its $p - 1$ variables.

In (RO), we now set $n = n_0 + 1$, and replace $f(n_0 + 1)$ with the value obtained above, giving:

$$f(n_0 + 2) = F_2(f(n_0 + 3), \cdots, f(n_0 + p + 1)),$$

where F_2 is linear in all its $p - 1$ variables, and so on.

After n steps, we obtain

$$f(n_0 + n) = F_n(f(n_0 + 1 + n), \cdots, f(n_0 + p - 1 + n)).$$

If we perform the translation $f(n_0 + n) = g(n)$ in the last equation, we obtain

$$g(n) = F_n(g(n + 1), \cdots, g(n + p - 1))$$

which is a homogeneous linear recurrence of order $p - 1$.

To avoid such a situation, in what follows we will only consider homogeneous linear recurrences of order p with $a_p(n) \neq 0$ for all $n \in \mathbb{N}$.

Theorem 2.1.5 *Let $p \in \mathbb{N}^*$, let $a_1, a_2, \cdots, a_p$ be p sequences of real numbers such that $a_p(n) \neq 0$, for every $n \in \mathbb{N}$, and let the homogeneous linear recurrence:*

$$f(n+p) + a_1(n) \cdot f(n+p-1) + \cdots + a_p(n) \cdot f(n) = 0, \text{ for every } n \in \mathbb{N}. \quad (RO)$$

The set of solutions of the equation (RO),

$$V = \{f : \mathbb{N} \to \mathbb{R} | f \text{ satisfies } (RO)\}$$

is a real vector space of dimension p.

Proof It is evident that V is a real linear space with respect to the usual operations of addition and scalar multiplication of sequences; let $\dim V$ be its dimension.

We observe that a sequence $f \in V$ is perfectly determined if we know its first p values. Indeed, by knowing $f(0), f(1), \cdots, f(p-1)$, we determine $f(p)$ by setting n to 0 in (RO); then, by setting $n = 1$, we determine $f(p+1)$, and so on.

Thus, we can build p solutions to the equation (RO), $f^0, f^1, \cdots, f^{p-1} \in V$, by choosing the first p values of each of them in the following manner: for every $k \in \{0, \cdots, p-1\}$,

$$f^k(k) = 1 \text{ and } f^k(i) = 0, \text{ for every } i \in \{0, \cdots, p-1\} \setminus \{k\}.$$

Let $\lambda_0 \cdot f^0 + \cdots + \lambda_{p-1} \cdot f^{p-1} = \underline{0}$ be a null linear combination of these sequences; it follows in particular that

$$\lambda_0 \cdot f^0(n) + \cdots + \lambda_{p-1} \cdot f^{p-1}(n) = 0, \text{ for every } n = 0, \cdots, p-1. \quad (2.1)$$

2.1 Linear Recurrences

But, for any $n \in \{0, \cdots, p-1\}$, $f^k(n) = 0$, for every $k \neq n$ and $f^n(n) = 1$, from where $\lambda_n = 0$, for any $n = 0, \cdots, p-1$ and then $f^0, \cdots, f^{p-1}$ are linearly independent.

It follows from the above that:

$$\dim V \geqslant p. \tag{2.2}$$

Let now $g^0, \cdots, g^{p-1}, g^p \in V$, $(p+1)$ solutions of the equation (RO) and let $\lambda_0, \cdots, \lambda_p$ such that

$$\sum_{i=0}^{p} \lambda_i \cdot g^i = \underline{0}. \tag{2.3}$$

Then

$$\sum_{i=0}^{p} \lambda_i \cdot g^i(n) = 0, \text{ for every } n = 0, \cdots, p-1. \tag{2.4}$$

Equation (2.4) is a linear and homogeneous algebraic system that consists of p equations with $p+1$ unknowns: $\lambda_0, \cdots, \lambda_p$. Therefore, this system also admits non-trivial solutions; therefore there exist $\bar{\lambda}_0, \cdots, \bar{\lambda}_p$, not all zero, such that:

$$\sum_{i=0}^{p} \bar{\lambda}_i \cdot g^i(n) = 0, \text{ for every } n = 0, \cdots, p-1. \tag{2.5}$$

We will prove by induction that (2.5) is true for any $n \in \mathbb{N}$. Equation (2.5) is verified for $n = 0, \cdots, p-1$.

We suppose that $m \geqslant p-1$ and that (2.5) holds for $n = 0, 1, \cdots, m$; using the fact that g^i, $i = 0, \cdots, p$, satisfy (RO), so that, for every $i = 0, \cdots, p$, and every $n \in \mathbb{N}$,

$$g^i(n+p) = -\sum_{j=1}^{p} a_j(n) \cdot g^i(n+p-j),$$

we obtain:

$$\sum_{i=0}^{p} \bar{\lambda}_i \cdot g^i(m+1) = \sum_{i=0}^{p} \bar{\lambda}_i \cdot g^i((m+1-p)+p) =$$

$$= -\sum_{i=0}^{p} \sum_{j=1}^{p} \bar{\lambda}_i \cdot a_j(m+1-p) \cdot g^i(m+1-j) =$$

$$= -\sum_{j=1}^{p} \left[a_j(m+1-p) \cdot \sum_{i=0}^{p} \bar{\lambda}_i \cdot g^i(m+1-j) \right].$$

But, for every $j = 1, \cdots, p$, $m + 1 - j \leqslant m$ and so, from the inductive hypothesis, $\sum_{i=0}^{p} \bar{\lambda}_i \cdot g^i(m + 1 - j) = \underline{0}$. It follows that (2.5) it is verified for $m + 1$.

Since (2.5) is true for any $n \in \mathbb{N}$, it follows that

$$\sum_{i=0}^{p} \bar{\lambda}_i \cdot g^i = \underline{0}.$$

Therefore any $p + 1$ vectors, $g^0, \cdots, g^p$, are linearly dependent.

It follows that

$$\dim V < p + 1. \qquad (2.6)$$

From (2.2) and (2.6) $\dim V = p$. □

Definition 2.1.6 Let $f^0, f^1, \cdots, f^{p-1} \in V$; the determinant

$$D[f^0, \cdots, f^{p-1}](n) = \begin{vmatrix} f^0(n) & f^1(n) & \cdots & f^{p-1}(n) \\ f^0(n+1) & f^1(n+1) & \cdots & f^{p-1}(n+1) \\ \cdots & \cdots & \cdots & \cdots \\ f^0(n+p-1) & f^1(n+p-1) & \cdots & f^{p-1}(n+p-1) \end{vmatrix}$$

$n \in \mathbb{N}$, is called the **Casorati determinant** associated to the functions $f^0, \cdots, f^{p-1}$. This determinant plays an important role in establishing the linear dependence of vectors $f^0, f^1, \cdots, f^{p-1}$. First, we will establish a calculation formula for the Casorati determinant.

Lemma 2.1.7 For every $f^0, \cdots, f^{p-1} \in V$ and every $n \in \mathbb{N}^*$,

$$D[f^0, \cdots, f^{p-1}](n) = (-1)^{np} a_p(0) a_p(1) \cdots a_p(n-1) \cdot D[f^0, \cdots, f^{p-1}](0).$$

Proof Since $f^0, \cdots, f^{p-1}$ satisfy the relationship (RO), for every $n \geqslant 1$, and for any $k = 0, \cdots, p - 1$,

$$f^k(n + p - 1) = -a_1(n-1) \cdot f^k(n + p - 2) - \cdots - a_p(n-1) \cdot f^k(n-1).$$

2.1 Linear Recurrences

By substituting the previous relationships into the last row of the Casorati determinant and taking into account the properties of determinants, we obtain, for every $n \geq 1$:

$$D[f^0, \cdots, f^{p-1}](n) = \begin{vmatrix} f^0(n) & \vdots & f^{p-1}(n) \\ f^0(n+1) & \vdots & f^{p-1}(n+1) \\ \cdots & \vdots & \cdots \\ f^0(n+p-2) & \vdots & f^{p-1}(n+p-2) \\ -a_p(n-1) \cdot f^0(n-1) & \vdots & -a_p(n-1) \cdot f^{p-1}(n-1) \end{vmatrix}$$

$$= -a_p(n-1) \cdot (-1)^{p-1} \cdot D[f^0, \cdots, f^{p-1}](n-1).$$

From here:

$$D[f^0, \cdots, f^{p-1}](n) = (-1)^{pk} a_p(n-1) \cdots a_p(n-k) \cdot D[f^0, \cdots, f^{p-1}](n-k) =$$

$$= (-1)^{pn} a_p(n-1) \cdots a_p(0) \cdot D[f^0, \cdots, f^{p-1}](0).$$

□

Corollary 2.1.8 $D[f^0, \cdots, f^{p-1}](0) = 0$ *if and only if, for every* $n \in \mathbb{N}$, $D[f^0, \cdots, f^{p-1}](n) = 0$.

Proposition 2.1.9 *The vectors* $f^0, \cdots, f^{p-1} \in V$ *are linearly independent if and only if* $D[f^0, \cdots, f^{p-1}](0) \neq 0$ *(equivalently, iff* $D[f^0, \cdots, f^{p-1}](n) \neq 0$, *for all* $n \in \mathbb{N}$).

Proof First, we assume that $f^0, \cdots, f^{p-1}$ are not linearly independent, so there exist $\lambda_0, \cdots, \lambda_{p-1} \in \mathbb{R}$, not all zero, such that $\sum_{k=0}^{p-1} \lambda_k \cdot f^k = \underline{0}$. Without loss of generality, we can assume that $\lambda_0 \neq 0$; then:

$$\lambda_0 \cdot D[f^0, \cdots, f^{p-1}](0) = \begin{vmatrix} \lambda_0 \cdot f^0(0) & f^1(0) & \cdots & f^{p-1}(0) \\ \lambda_0 \cdot f^0(1) & f^1(1) & \cdots & f^{p-1}(1) \\ \cdots & \cdots & & \cdots \\ \lambda_0 \cdot f^0(p-1) & f^1(p-1) & \cdots & f^{p-1}(p-1) \end{vmatrix}.$$

If we multiply the second column of the determinant above by $\lambda_1, \cdots$, the p-th column by λ_{p-1}, and add everything to the first column, we obtain only zeros in the first column. Therefore: $\lambda_0 \cdot D[f^0, \cdots, f^{p-1}](0) = 0$ and so $D[f^0, \cdots, f^{p-1}](0) = 0$.

Conversely, if $f^0, \cdots, f^{p-1}$ are linearly independent, the system:

$$\lambda_0 \cdot f^0(k) + \cdots + \lambda_{p-1} \cdot f^{p-1}(k) = 0, k = 0, \cdots, p-1$$

has only the trivial solution $\lambda_0 = \lambda_1 = \cdots = \lambda_{p-1} = 0$, which implies that its determinant is non-zero. But the determinant of the system above is: $D[f^0, \cdots, f^{p-1}](0)$. □

Remark 2.1.10 If we can determine p linearly independent vectors $f^1, \cdots, f^p \in V$ then the general solution of the homogeneous recurrence (RO) is:

$$f = c_1 \cdot f^1 + c_2 \cdot f^2 + \cdots + c_p \cdot f^p, \quad c_1, c_2, \cdots, c_p \in \mathbb{R}.$$

In a particular case, that of homogeneous equations with constant coefficients, we can determine p linearly independent vectors in V.

2.1.3 Equations with Constant Coefficients

Let $p \in \mathbb{N}^*, a_1, \cdots, a_p \in \mathbb{R}$ with $a_p \neq 0$; the equation

$$f(n+p) + a_1 \cdot f(n+p-1) + \cdots + a_p \cdot f(n) = 0, \text{ for every } n \in \mathbb{N}. \quad (RO')$$

is called a homogeneous linear recurrence with constant coefficients; it is associated with a homogeneous difference equation with constant coefficients of the form:

$$\Delta^p f(n) + b_1 \cdot \Delta^{p-1} f(n) + \cdots + b_p \cdot \Delta^0 f(n) = 0, \text{ for every } n \in \mathbb{N}, \quad (DO')$$

where $b_1, \cdots, b_p \in \mathbb{R}$.

We seek solutions of the form $f(n) = \lambda^n, n \in \mathbb{N}$ for the equation (RO'); imposing the condition that f satisfies (RO'), we obtain:

$$\boxed{\lambda^p + a_1 \cdot \lambda^{p-1} + \cdots + a_p = 0.} \quad (EC)$$

The equation (EC) is called the **characteristic equation** associated with the homogeneous linear equation with constant coefficients (RO'). This equation admits p real or complex solutions.

We will study each case that may occur separately.

I. We assume that the equation (EC) has distinct real roots $\lambda_1, \cdots, \lambda_p$. In this case, for every $k = 1, \cdots, p$, $f^k(n) = \lambda_k^n, n \in \mathbb{N}$, define p linearly independent vectors in V. Indeed, the Casorati determinant for this set of vectors is

$$D[f^1, \cdots, f^p](0) = \begin{vmatrix} 1 & 1 & \cdots & 1 \\ \lambda_1 & \lambda_2 & \cdots & \lambda_p \\ \cdots & \cdots & \cdots & \cdots \\ \lambda_1^{p-1} & \lambda_2^{p-1} & \cdots & \lambda_p^{p-1} \end{vmatrix} = \prod_{1 \leq k < l \leq p} (\lambda_l - \lambda_k) \neq 0.$$

2.1 Linear Recurrences

According to Proposition 2.1.9, $f^1, \cdots, f^p$ form a set of linearly independent vectors, and thus the general solution of the homogeneous recurrence is:

$$f(n) = \sum_{k=1}^{p} c_k \cdot \lambda_k^n, n \in \mathbb{N}, c_1, \cdots, c_p \in \mathbb{R}.$$

II. Let us assume that the roots $\lambda_1, \cdots, \lambda_p$ are distinct, but among them there is also a complex conjugate pair: $\lambda_{1,2} = R \cdot e^{\pm i\alpha} = R(\cos\alpha \pm i\sin\alpha)$, where $i = \sqrt{-1}$. In this case the set $\{f^1, f^2, \cdots, f^p\} \subseteq V$, where $f^1(n) = R^n \cos n\alpha$, $f^2(n) = R^n \sin n\alpha$, $f^3(n) = \lambda_3^n, \cdots, f^p(n) = \lambda_p^n$, is linearly independent. Indeed, the Casorati determinant for this set is:

$$D[f^1, \cdots, f^p](0) = \begin{vmatrix} 1 & 0 & 1 & \vdots & 1 \\ \mathbb{R}\cos\alpha & R\sin\alpha & \lambda_3 & \vdots & \lambda_p \\ \cdots & \cdots & \cdots & \vdots & \cdots \\ \mathbb{R}^{p-1}\cos(p-1)\alpha & R^{p-1}\sin(p-1)a & \lambda_3^{p-1} & \vdots & \lambda_p^{p-1} \end{vmatrix}$$

In the determinant above, we multiply the second column by $i = \sqrt{-1}$ and add it to the first column:

$$D[f^1, \cdots, f^p](0) = \begin{vmatrix} 1 & 0 & 1 & \cdots & 1 \\ \lambda_1 & R\sin\alpha & \lambda_3 & \cdots & \lambda_p \\ \cdots & \cdots & \cdots & \cdots \\ \lambda_1^{p-1} & R^{p-1}\sin(p-1)\alpha & \lambda_3^{p-1} & \cdots & \lambda_p^{p-1} \end{vmatrix}.$$

We now multiply the second column by $-2i$ and add to it the first column:

$$(-2i)D[f^1, \cdots, f^p](0) = \begin{vmatrix} 1 & 1 & \cdots & 1 \\ \lambda_1 & \lambda_2 & \cdots & \lambda_p \\ \cdots & \cdots & \cdots \\ \lambda_1^{p-1} & \lambda_2^{p-1} & \cdots & \lambda_p^{p-1} \end{vmatrix} =$$

$$= \prod_{1 \leqslant k < l \leqslant p} (\lambda_l - \lambda_k) \neq 0.$$

Proposition 2.1.9 thus ensures the independence of the vectors; the general solution will be in this case:

$$f(n) = c_1 \cdot R^n \cos n\alpha + c_2 \cdot R^n \sin n\alpha + \sum_{k=3}^{p} c_k \cdot \lambda_k^n, n \in \mathbb{N}, c_1, \cdots, c_p \in \mathbb{R}.$$

If the equation (EC) has multiple pairs of complex conjugate roots, the same procedure is followed for each pair.

III. Let us now consider the case of multiple roots.

If λ_1 is a multiple root of order s for the characteristic equation, we need to replace the first s vectors (which are equal to each other) with new ones in the system $\lambda_1^n = \lambda_2^n = \cdots = \lambda_s^n, \lambda_{s+1}^n, \cdots, \lambda_p^n$; we search for new solutions of the form $f(n) = g(n) \cdot \lambda_1^n, n \in \mathbb{N}$. By replacing in the equation (RO') and simplifying by λ_1^n, we obtain:

$$g(n+p) \cdot \lambda_1^p + a_1 \cdot g(n+p-1) \cdot \lambda_1^{p-1} + \cdots + a_p \cdot g(n) = 0, \text{ for all } n \in \mathbb{N}. \quad (2.7)$$

We now apply the relationships $(**)$ given in (iii) of Remark 2.1.3 to the sequence g and obtain, for every $n \in \mathbb{N}$, and every $k = 0, \cdots, p$,

$$g(n+k) = \sum_{j=0}^{k} \binom{k}{j} \Delta^j g(n).$$

We now substitute these values into the Eqs. (2.7)

$$\sum_{j=0}^{p} \binom{p}{j} \Delta^j g(n) \cdot \lambda_1^p + a_1 \sum_{j=0}^{p-1} \binom{p-1}{j} \Delta^j g(n) \cdot \lambda_1^{p-1} + \cdots +$$

$$+ a_p \sum_{j=0}^{0} \binom{0}{j} \Delta^j g(n) = 0,$$

or, after grouping the terms:

$$\left(\binom{p}{0} \lambda_1^p + a_1 \binom{p-1}{0} \lambda_1^{p-1} + \cdots + a_p \right) \cdot \Delta^0 g(n) + \quad (2.8)$$

$$+ \left(\binom{p}{1} \lambda_1^p + a_1 \binom{p-1}{1} \lambda_1^{p-1} + \cdots + a_{p-1} \lambda_1 \right) \cdot \Delta^1 g(n) + \cdots +$$

$$+ \left(\binom{p}{p-1} \lambda_1^p + a_1 \binom{p-1}{p-1} \lambda_1^{p-1} \right) \cdot \Delta^{p-1} g(n) + \binom{p}{p} \lambda_1^p \cdot \Delta^p g(n) = 0.$$

Let now $P(\lambda) = \lambda^p + a_1 \lambda^{p-1} + \cdots + a_p$ be the characteristic polynomial; then the relationship (2.8) can be rewritten as:

$$P(\lambda_1) \cdot \Delta^0 g(n) + \lambda_1 P'(\lambda_1) \cdot \Delta^1 g(n) + \cdots + \quad (2.9)$$

$$+ \frac{\lambda_1^{p-1}}{(p-1)!} P^{(p-1)}(\lambda_1) \cdot \Delta^{p-1} g(n) + \frac{\lambda_1^p}{p!} P^{(p)}(\lambda_1) \cdot \Delta^p g(n) = 0.$$

2.1 Linear Recurrences

Since λ_1 is a multiple root of order s, $P(\lambda_1) = P'(\lambda_1) = \cdots = P^{(s-1)}(\lambda_1) = 0$, and then the relationship (2.9) becomes:

$$\frac{\lambda_1^s}{s!} P^{(s)}(\lambda_1) \cdot \Delta^s g(n) + \frac{\lambda_1^{s+1}}{(s+1)!} P^{(s+1)}(\lambda_1) \cdot \Delta^{s+1} g(n) + \cdots + \quad (2.10)$$

$$+ \frac{\lambda_1^p}{p!} P^{(p)}(\lambda_1) \cdot \Delta^p g(n) = 0.$$

Relationship (2.10) is clearly verified if g is a polynomial of degree less than s because, in this case, the finite differences of g of order greater than or equal to s are identically zero (see (ii) of Remark 2.1.3). In particular, we can choose:

$$g_1(n) = 1, g_2(n) = n, \cdots, g_s(n) = n^{s-1}.$$

In this way, we replace the equal terms $\lambda_1^n, \lambda_2^n, \cdots, \lambda_s^n$ with the sequences: $\lambda_1^n, n\lambda_1^n, \cdots, n^{s-1}\lambda_1^n$.

We still need to prove linear independence. We will do this in the general case of many multiple roots for the characteristic equation. Let us assume that the characteristic equation (EC) has multiple roots:

λ_1 with multiplicity order s_1,
λ_2 with multiplicity order s_2,
$\cdots$
λ_q with multiplicity order s_q.

Obviously, $s_1 + \cdots s_q = p$.

We replace each multiple root with a sequence of solutions as shown above, and thus obtain the system of solutions of the characteristic equation:

$$\begin{cases} \lambda_1^n, n\lambda_1^n, \cdots, n^{s_1-1}\lambda_1^n, \\ \lambda_2^n, n\lambda_2^n, \cdots, n^{s_2-1}\lambda_2^n, \\ \cdots \\ \lambda_q^n, n\lambda_q^n, \cdots, n^{s_q-1}\lambda_q^n. \end{cases} \quad (S)$$

Theorem 2.1.11 *The p solutions of the system (S) above form a system of linearly independent vectors in V.*

Proof Let us consider a null linear combination of these p vectors:

$$(c_{11}\lambda_1^n + c_{21}n\lambda_1^n + \cdots + c_{s_11}n^{s_1-1}\lambda_1^n) + \cdots + \quad (2.11)$$

$$+ (c_{1q}\lambda_q^n + c_{2q}n\lambda_q^n + \cdots + c_{s_qq}n^{s_q-1}\lambda_q^n) = 0.$$

We will first order the terms appearing in Eq. (2.11) according to the following criteria:

We assume first that the roots of the characteristic equation are ordered by the magnitude, such that:

$$|\lambda_1| \geq |\lambda_2| \geq \cdots \geq |\lambda_q|.$$

If we have roots with equal maximum modulus, let's say

$$|\lambda_1| = \cdots = |\lambda_r| = R > |\lambda_{r+1}| \geq \cdots \geq |\lambda_q|,$$

then we order the first ones by their multiplicity, meaning we assume that:

$$s_1 \geq s_2 \geq \cdots \geq s_r.$$

Let us assume, even in this last case, that we could have equal maximal multiplicities, and let us denote:

$$s_1 = s_2 = \cdots = s_t = s > s_{t+1} \geq \cdots \geq s_r.$$

We will say that the terms $n^{s-1}\lambda_1^n, n^{s-1}\lambda_2^n, \cdots, n^{s-1}\lambda_t^n$ are **dominant** in relationship (2.11).

The idea of the proof is as follows: we will first show that all dominant terms in relationship (2.11) have null coefficients; then we will reorder relationship (2.11), highlighting the next dominant terms, and repeat the reasoning until we obtain that all coefficients in relationship (2.11) are null. From here, it will follow that the system of vectors (S) is linearly independent.

Let us observe that a term $n^k \lambda_j^n$ is not dominant if: $j > r$ or if $j \leq r$ and $k < s - 1$. Let us also observe that there exists an $n_0 \in \mathbb{N}$ such that for any term $n^k \lambda_j^n$ that is not dominant:

$$\frac{|n^k \lambda_j^n|}{n^{s-1} R^n} \leq \frac{1}{n}, \text{ for every } n \geq n_0. \quad (*)$$

Indeed, if $j > r$ $|\lambda_j| < R$ and then

$$n \cdot \frac{|n^k \lambda_j^n|}{n^{s-1} R^n} = n^{k-s+2} \cdot \left(\frac{|\lambda_j|}{R}\right)^n \to 0$$

and thus we can find an $n_0 \in \mathbb{N}$ such that, for every $n \geq n_0$,

$$n \cdot \frac{|n^k \lambda_j^n|}{n^{s-1} R^n} = n^{k-s+2} \cdot \left(\frac{|\lambda_j|}{R}\right)^n < 1,$$

which immediately leads to the relationship $(*)$; since k and j take on finite sets of values, n_0 can be chosen independently.

2.1 Linear Recurrences   77

If $j \leqslant r$ and $k < s - 1$, then $|\lambda_j| = R$ and so

$$\frac{|n^k \lambda_j^n|}{n^{s-1} R^n} = \frac{n^k}{n^{s-1}} \leqslant \frac{n^{s-2}}{n^{s-1}} = \frac{1}{n}.$$

Thus, $(*)$ is demonstrated in all possible cases.

Let us also note that in the relationship (2.11) we have a finite number of terms, and therefore there exists a number $M > 0$ such that:

$$|c_{kj}| \leqslant M, \text{ for every } j = 1, \cdots, q, \text{ and every } k = 1, \cdots, s_j. \qquad (**)$$

Let us return to relationship (2.11) where we keep in the left-hand side only the dominant terms:

$$c_{s1} n^{s-1} \lambda_1^n + \cdots + c_{st} n^{s-1} \lambda_t^n = -\left(\sum{'} c_{kj} n^{k-1} \lambda_j^n\right), \qquad (2.12)$$

where $\sum{'}$ indicates the sum of all terms in the left-hand side of relationship (2.11) except for the dominant ones.

Since $|\lambda_1| = |\lambda_2| = \cdots = |\lambda_t| = R$, it follows that

$$\lambda_1 = R e^{i\alpha_1}, \lambda_2 = R e^{i\alpha_2}, \cdots, \lambda_t = R e^{i\alpha_t},$$

where $i = \sqrt{-1}$ and $\alpha_1, \alpha_2, \cdots, \alpha_t \in [0, 2\pi[$ are distinct pairwise because $\lambda_1, \lambda_2, \cdots, \lambda_t$ are distinct pairwise.

We divide relationship (2.12) by $n^{s-1} R^n$:

$$c_{s1} e^{i\alpha_1 n} + \cdots + c_{st} e^{i\alpha_t n} = -\left(\sum{'} c_{kj} \frac{n^k \lambda_j^n}{n^{s-1} R^n}\right). \qquad (2.13)$$

In relationship (2.13) we take the absolute value of both terms, bound the absolute value of the second term using the sum of absolute values, and using conditions $(*)$, $(**)$; then there exists $n_0 \in \mathbb{N}$ such that:

$$|c_{s1} e^{i\alpha_1 n} + \cdots + c_{st} e^{i\alpha_t n}| \leqslant p \cdot \frac{M}{n}, \text{ for every } n \geqslant n_0. \qquad (2.14)$$

Let

$$D(n) = \begin{vmatrix} e^{i\alpha_1 n} & e^{i\alpha_2 n} & \cdots & e^{i\alpha_t n} \\ e^{i\alpha_1 (n+1)} & e^{i\alpha_2 (n+1)} & \cdots & e^{i\alpha_t (n+1)} \\ \cdots & \cdots & \cdots \\ e^{i\alpha_1 (n+t-1)} & e^{i\alpha_2 (n+t-1)} & \cdots & e^{i\alpha_t (n+t-1)} \end{vmatrix} =$$

$$= e^{i\alpha_1 n} e^{i\alpha_2 n} \cdots e^{i\alpha_t n} \begin{vmatrix} 1 & 1 & \cdots & 1 \\ e^{i\alpha_1} & e^{i\alpha_2} & \cdots & e^{i\alpha_t} \\ \cdots & \cdots & \cdots & \cdots \\ e^{i\alpha_1(t-1)} & e^{i\alpha_2(t-1)} & \cdots & e^{i\alpha_t(t-1)} \end{vmatrix}.$$

Then

$$|D(n)| = \prod_{1 \leq j < k \leq t} \left| e^{i\alpha_k} - e^{i\alpha_j} \right|. \tag{2.15}$$

Since $\alpha_1, \alpha_2, \cdots, \alpha_t$ are pairwise distinct, it follows from (2.15) that $|D(n)|$ is a strictly positive number independent of n.

On the other hand, if we multiply the first column of the determinant $D(n)$ by c_{s1} and then multiply the second column of the determinant by c_{s2} and add it to the first, $\cdots$, and multiply the t-th column by c_{st} and add it to the first column, we obtain:

$$c_{s1} \cdot D(n) = \begin{vmatrix} A_1 & e^{i\alpha_2 n} & \cdots & e^{i\alpha_t n} \\ A_2 & e^{i\alpha_2(n+1)} & \cdots & e^{i\alpha_t(n+1)} \\ \cdots & \cdots & & \cdots \\ A_t & e^{i\alpha_2(n+t-1)} & \cdots & e^{i\alpha_t(n+t-1)} \end{vmatrix},$$

where, for every $k = 1, \cdots, t$, $A_k = c_{s1} \cdot e^{i\alpha_1(n+k-1)} + \cdots + c_{st} \cdot e^{i\alpha_t(n+k-1)}$. From relationship (2.14), $|A_k| \leq \frac{pM}{n}$, for all $n \geq n_0$, and since all the other elements of the above determinant have absolute value equal to 1, we obtain by applying the definition of determinants:

$$|c_{s1}| \cdot |D(n)| \leq t! \cdot \frac{pM}{n}, \text{ for every } n \geq n_0.$$

Taking into account (2.15), $|D(n)|$ is constant and strictly positive, and therefore the above relationship can only occur if $c_{s1} = 0$.

By a similar reasoning, it can be shown that the other coefficients of the dominant terms in (2.11) are also zero.

As we have already announced, the dominant terms in (2.11) disappear and, by a reordering process, we consider the dominant terms among the remaining ones. We continue the reasoning until we obtain all the coefficients in relationship (2.11) equal to zero. □

Remark 2.1.12 If one of the multiple roots is complex, then we will proceed as in case **II**.

Let's assume, for example, that $\lambda_1 = R \cdot e^{i\alpha} = R(\cos\alpha + i\sin\alpha)$ (where $i = \sqrt{-1}$) is a multiple root of order s_1; then it is clear that $\lambda_2 = R \cdot e^{-i\alpha} = R(\cos\alpha -$

2.1 Linear Recurrences

$i \sin \alpha$) will have the same multiplicity order $s_2 = s_1$. In this case, the first two rows of the system of solutions (S) will be replaced with:

$$\begin{cases} R^n \cos n\alpha, nR^n \cos n\alpha, \cdots, n^{s_1-1} R^n \cos n\alpha, \\ \mathbb{R}^n \sin n\alpha, nR^n \sin n\alpha, \cdots, n^{s_2-1} R^n \sin n\alpha. \end{cases}$$

2.1.4 Fibonacci Sequence

An immediate application of the theory of linear homogeneous recurrences is finding the general form of the Fibonacci sequence. In 1202, the Italian mathematician Leonardo di Pisa, better known as Fibonacci, published a book called "Liber Abaci", which presented sequences that satisfy the following second-order linear homogeneous recurrence:

$$F(n+2) = F(n+1) + F(n), \text{ for all } n \in \mathbb{N}; F(0) = 0, F(1) = 1.$$

This equation is a linear recurrence with constant coefficients. Its characteristic equation is:

$$\lambda^2 - \lambda - 1 = 0.$$

Its roots are:

$$\lambda_1 = \frac{1+\sqrt{5}}{2}, \lambda_2 = \frac{1-\sqrt{5}}{2} = -\frac{1}{\lambda_1}.$$

We observe that $\lambda_1 = \phi$ is the golden ratio. The golden ratio, or as Euclid called it, "the extreme and mean ratio", is the ratio in which a point C divides a segment AB such that $\dfrac{AC}{CB} = \dfrac{AB}{AC}$.

$$\underset{\underset{a}{}}{A \bullet} \quad\quad\quad \underset{\underset{b}{}}{C \bullet} \quad\quad\quad B \bullet$$

If we denote, as in the figure above, $AC = a$ and $CB = b$, then

$$\frac{a}{b} = \frac{a+b}{a},$$

or equivalently,
$$\left(\frac{a}{b}\right)^2 - \frac{a}{b} - 1 = 0,$$

from where $\phi = \frac{a}{b} = \frac{1+\sqrt{5}}{2}$ ($\phi > 0$).

According to **I** of previous subsection, the general solution of the Fibonacci equation is:

$$F(n) = c_1 \left(\frac{1+\sqrt{5}}{2}\right)^n + c_2 \left(\frac{1-\sqrt{5}}{2}\right)^n, \text{ for every } n \in \mathbb{N}, c_1, c_2 \in \mathbb{R}.$$

If we use the initial conditions $F(0) = 0$, $F(1) = 1$, we obtain $c_1 = \frac{1}{\sqrt{5}}$ and $c_2 = -\frac{1}{\sqrt{5}}$, and thus we obtain Binet's formula:

$$F(n) = \frac{1}{\sqrt{5}} \left(\frac{1+\sqrt{5}}{2}\right)^n - \frac{1}{\sqrt{5}} \left(\frac{1-\sqrt{5}}{2}\right)^n =$$

$$= \frac{1}{\sqrt{5}} \left(\phi^n - (-\phi)^{-n}\right), \text{ for every } n \in \mathbb{N}.$$

Although the numbers $F(n)$ are called **Fibonacci numbers**, it appears that they were known long before by Indian mathematicians (see [13]).

We conclude this subsection with a result first observed by Kepler but proved only a century later.

Proposition 2.1.13 $\frac{F(n+1)}{F(n)} \xrightarrow[n \to +\infty]{} \phi.$

Proof For every $n \in \mathbb{N}$,

$$\frac{F(n+1)}{F(n)} = \frac{\left(\frac{1-\sqrt{5}}{2}\right)^{n+1} - \left(\frac{1+\sqrt{5}}{2}\right)^{n+1}}{\left(\frac{1-\sqrt{5}}{2}\right)^n - \left(\frac{1+\sqrt{5}}{2}\right)^n} = \frac{\left(\frac{1-\sqrt{5}}{1+\sqrt{5}}\right)^{n+1} - 1}{\left(\frac{1-\sqrt{5}}{1+\sqrt{5}}\right)^n - 1} \cdot \frac{1+\sqrt{5}}{2} \to \frac{1+\sqrt{5}}{2},$$

because $\left(\frac{1-\sqrt{5}}{1+\sqrt{5}}\right)^n = (-1)^n \cdot \left(\frac{3-\sqrt{5}}{2}\right)^n \to 0$. Therefore $\frac{F(n+1)}{F(n)} \to \frac{1+\sqrt{5}}{2} = \phi$. □

The Fibonacci numbers and the golden ratio have many other interesting properties for mathematicians (such as their connection to fractal geometry) as well as for other fields such as crystallography, biology, and the visual arts. Entire books are dedicated to exploring these properties. For interested readers, we will mention just two references: [5, 8].

2.1.5 Non-homogeneous Linear Equations

Proposition 2.1.14 *Let be a non-homogeneous linear recurrence:*

$$f(n+p)+a_1(n)\cdot f(n+p-1)+\ldots+a_p(n)\cdot f(n) = a(n), \text{ for all } n \in \mathbb{N}, \quad (R)$$

where $a_p(n) \neq 0$, for every $n \in \mathbb{N}$, and let

$$f(n+p)+a_1(n)\cdot f(n+p-1)+\ldots+a_p(n)\cdot f(n) = 0, \text{ for all } n \in \mathbb{N} \quad (RO)$$

be the associated homogeneous linear recurrence.

If $f^1, \cdots f^p$ are p linearly independent solutions of the homogeneous equation (RO) and if f^* is a particular solution of the non-homogeneous equation (R), then the general solution of the equation (R) is given by

$$f = c_1 \cdot f^1 + \cdots + c_p \cdot f^p + f^*, \; c_1, \cdots, c_p \in \mathbb{R}.$$

Proof It is evident that, for any $c_1, \cdots, c_p \in \mathbb{R}$, $\sum_{k=1}^{p} c_k \cdot f^k + f^*$ satisfies the equation (R).

Let f be an arbitrary solution of the equation (R):

$$f(n+p)+a_1(n)\cdot f(n+p-1)+\ldots+a_p(n)\cdot f(n) = a(n), \text{ for all } n \in \mathbb{N}. \quad (1)$$

Since f^* is also a solution of (R),

$$f^*(n+p)+a_1(n)\cdot f^*(n+p-1)+\ldots+a_p(n)\cdot f^*(n) = a(n), \text{ for all } n \in \mathbb{N}. \quad (2)$$

We subtract (2) from (1) and obtain, for every $n \in \mathbb{N}$:

$$(f - f^*)(n+p) + a_1(n) \cdot (f - f^*)(n+p-1) + \ldots + a_p(n) \cdot (f - f^*)(n) = 0.$$

So $f - f^*$ is a solution of (RO), and therefore there exist $c_1, \cdots, c_p \in \mathbb{R}$ such that $(f - f^*) = \sum_{k=1}^{p} c_k f^k$, and so $f = \sum_{k=1}^{p} c_k f^k + f^*$. □

From the previous proposition it follows that, to solve a non-homogeneous linear recurrence, it is sufficient to find a particular solution of it (assuming, of course, that we can solve the associated homogeneous equation).

The method of finding such a particular solution is called the **method of variation of parameters** or Lagrange's method.

Assume that $f^1, \cdots, f^p$ are p linearly independent solutions of the homogeneous equation (RO); we know that the general solution of the homogeneous equation is $\sum_{k=1}^{p} c_k f^k$.

We will look for a particular solution of the non-homogeneous equation (R) by considering that $c_1, \cdots, c_p$ are not constants but sequences that we will determine.

Therefore, we try to determine a particular solution of the non-homogeneous equation of the form

$$f^*(n) = \sum_{k=1}^{p} c_k(n) f^k(n), \; n \in \mathbb{N}.$$

To determine the p unknown sequences $c_1(n), \cdots, c_p(n)$, we will impose on f^* the condition to satisfy the equation (R) and we will add $p - 1$ additional conditions that will lead to the following compatible system:

$$\begin{cases} (1) \sum_{k=1}^{p} c_k(n+p) f^k(n+p) + a_1(n) \sum_{k=1}^{p} c_k(n+p-1) f^k(n+p-1) + \\ \quad + \cdots + a_p(n) \sum_{k=1}^{p} c_k(n) f^k(n) = a(n) \\ (2) \sum_{k=1}^{p} c_k(n+1) f^k(n+1) = \sum_{k=1}^{p} c_k(n) f^k(n+1) \\ (3) \sum_{k=1}^{p} c_k(n+2) f^k(n+2) = \sum_{k=1}^{p} c_k(n) f^k(n+2) \\ \cdots \\ (p) \sum_{k=1}^{p} c_k(n+p-1) f^k(n+p-1) = \sum_{k=1}^{p} c_k(n) f^k(n+p-1). \end{cases}$$

We will solve this system by transforming it into a system of difference equations.

First, in equation (p), we substitute $n \mapsto n+1$, and we obtain the first term from Eq. (1):

$$\sum_{k=1}^{p} c_k(n+p) f^k(n+p) = \sum_{k=1}^{p} c_k(n+1) f^k(n+p) =$$

$$= \sum_{k=1}^{p} \Delta c_k(n) f^k(n+p) + \sum_{k=1}^{p} c_k(n) f^k(n+p).$$

The next terms of Eq. (1) are modified accordingly using equations (p), $(p-1), \cdots, (2)$ in this order, and we obtain:

$$\sum_{k=1}^{p} \Delta c_k(n) f^k(n+p) + \sum_{k=1}^{p} c_k(n) f^k(n+p) + a_1(n) \sum_{k=1}^{p} c_k(n) f^k(n+p-1) +$$

$$+ \cdots + a_p(n) \sum_{k=1}^{p} c_k(n) f^k(n) = a(n),$$

2.1 Linear Recurrences

or

$$\sum_{k=1}^{p} \Delta c_k(n) f^k(n+p) +$$

$$+ \sum_{k=1}^{p} c_k(n) \left[f^k(n+p) + a_1(n) f^k(n+p-1) + \cdots + a_p(n) f^k(n) \right] = a(n).$$

Since $f^1, \cdots, f^p$ are solutions of the homogeneous equation (RO), the square brackets in the above relationship vanish and thus Eq. (1) becomes:

$$\sum_{k=1}^{p} \Delta c_k(n) f^k(n+p) = a(n). \tag{1'}$$

Equation (2) can be rewritten as:

$$\sum_{k=1}^{p} \Delta c_k(n) f^k(n+1) = 0. \tag{2'}$$

If we replace n with $n+1$ in Eq. (2) and substitute it into Eq. (3), we get:

$$\sum_{k=1}^{p} \Delta c_k(n) f^k(n+2) = 0. \tag{3'}$$

We will replace n with $n+1$ in Eq. (3) and substitute in Eq. (4) to obtain:

$$\sum_{k=1}^{p} \Delta c_k(n) f^k(n+3) = 0, \tag{4'}$$

$$\cdots$$

$$\sum_{k=1}^{p} \Delta c_k(n) f^k(n+p-1) = 0. \tag{p'}$$

Finally, we rewrite the above system placing Eq. (1') last:

$$\begin{cases} \sum_{k=1}^{p} \Delta c_k(n) f^k(n+1) = 0 \\ \sum_{k=1}^{p} \Delta c_k(n) f^k(n+2) = 0 \\ \sum_{k=1}^{p} \Delta c_k(n) f^k(n+3) = 0 \\ \cdots \\ \sum_{k=1}^{p} \Delta c_k(n) f^k(n+p-1) = 0 \\ \sum_{k=1}^{p} \Delta c_k(n) f^k(n+p) = a(n). \end{cases} \tag{S}$$

The system (S) has p equations with p unknowns: $\Delta c_1(n), \cdots, \Delta c_p(n)$. The determinant of the system is the Casorati determinant $D[f^1, \cdots, f^p](n+1)$ associated to the system of p linearly independent vectors in the vector space of solutions of the homogeneous equation (RO); by Proposition 2.1.9, this determinant is nonzero, and thus the system (S) admits a unique solution: $\Delta c_1(n), \cdots, \Delta c_p(n)$, where, for every $k = 1, \cdots, p$,

$$\Delta c_k(n) = \frac{\begin{vmatrix} f^1(n+1) & \vdots & f^{k-1}(n+1) & 0 & f^{k+1}(n+1) & \vdots & f^p(n+1) \\ \vdots & \vdots & \vdots & \vdots & \vdots & \vdots & \vdots \\ f^1(n+p) & \vdots & f^{k-1}(n+p) & a(n) & f^{k+1}(n+p) & \vdots & f^p(n+p) \end{vmatrix}}{D[f^1, \ldots, f^p](n+1)}$$

or

$$\Delta c_k(n) = (-1)^{p+k} a(n) \frac{D_k(n)}{D(n)},$$

where we have denoted $D(n) = D[f^1, \cdots, f^p](n+1)$ and $D_k(n) =$

$$= \begin{vmatrix} f^1(n+1) & \vdots & f^{k-1}(n+1) & f^{k+1}(n+1) & \vdots & f^p(n+1) \\ \vdots & \vdots & \vdots & \vdots & \vdots & \vdots \\ f^1(n+p-1) & \vdots & f^{k-1}(n+p-1) & f^{k+1}(n+p-1) & \vdots & f^p(n+p-1) \end{vmatrix}$$

We can then determine, up to an additive constant, the unknown sequences $c_1(n), \cdots, c_p(n)$; therefore, for every $k = 1, \cdots, p$, and every $n \in \mathbb{N}^*$,

$$c_k(n) = c_k(0) + \sum_{l=0}^{n-1} \Delta c_k(l) = c_k(0) + (-1)^{p+k} \sum_{l=0}^{n-1} a(l) \frac{D_k(l)}{D(l)}.$$

A particular solution for the non-homogeneous equation (R) will be:

$$\sum_{k=1}^{p} c_k(n) \cdot f^k(n) = \sum_{k=1}^{p} c_k(0) f^k(n) + \sum_{k=1}^{p} (-1)^{p+k} \left(\sum_{l=0}^{n-1} a(l) \frac{D_k(l)}{D(l)} \right) f^k(n).$$

But since $\sum_{k=1}^{p} c_k(0) f^k(n)$ is a solution of the homogeneous equation, we can take as a particular solution:

$$f^*(n) = \sum_{k=1}^{p} (-1)^{p+k} \left(\sum_{l=0}^{n-1} a(l) \frac{D_k(l)}{D(l)} \right) f^k(n).$$

2.1 Linear Recurrences

Then, according to Proposition 2.1.14, the general solution of the non-homogeneous equation is:

$$f(n) = \sum_{k=1}^{p} c_k f^k(n) + \sum_{k=1}^{p} (-1)^{p+k} \left(\sum_{l=0}^{n-1} a(l) \frac{D_k(l)}{D(l)} \right) f^k(n).$$

Remarks 2.1.15

(i) The method of variation of parameters can be successfully applied to non-homogeneous linear equations whose associated homogeneous equations have constant coefficients. Indeed, in this case, the characteristic equation associated with the homogeneous equation allows finding a linearly independent system of solutions for the homogeneous equation and thus finding a particular solution to the non-homogeneous equation.

(ii) From a practical point of view, the method of variation of parameters is quite difficult to apply. In certain particular cases, we may find particular solutions more easily.

Theorem 2.1.16 *Consider the non-homogeneous linear recurrence with constant coefficients:*

$$f(n+p) + a_1 \cdot f(n+p-1) + \ldots + a_p \cdot f(n) = Q(n)e^{\alpha n}, \text{ for every } n \in \mathbb{N},$$

where Q is a polynomial of degree q, and $\alpha \in \mathbb{R}$; let

$$C(\lambda) = \lambda^p + a_1 \lambda^{p-1} + \cdots + a_p$$

be the characteristic polynomial associated with the homogeneous equation

$$f(n+p) + a_1 \cdot f(n+p-1) + \ldots + a_p \cdot f(n) = 0.$$

Then, a particular solution of the non-homogeneous equation is of the form:

$$f^*(n) = R(n)e^{\alpha n},$$

where R is a polynomial of degree $r = q + s$, and s is the order of multiplicity of the root e^α for the polynomial C (if e^α is not a root of C, then $s = 0$).

Proof We try, for the non-homogeneous equation, a particular solution of the form $f^*(n) = R(n)e^{\alpha n}$, where R is a polynomial of degree r to be determined. Imposing the condition that the non-homogeneous equation is satisfied by f^*, we obtain:

$$R(n+p)e^{\alpha(n+p)} + a_1 R(n+p-1)e^{\alpha(n+p-1)} + \cdots + a_p R(n)e^{\alpha n} = Q(n)e^{\alpha n}. \tag{2.18}$$

We simplify Eq. (2.18) with $e^{\alpha n}$ and obtain:

$$R(n+p)e^{\alpha p} + a_1 R(n+p-1)e^{\alpha(p-1)} + \cdots + a_p R(n) = Q(n). \quad (2.19)$$

For every $k = 0, \cdots, p$, we replace in (2.19), $R(n+k) = \sum_{j=0}^{k} \binom{k}{j} \Delta^j R(n)$ (see the relationships (**) in (iii) of Remark 2.1.3) and we obtain

$$\sum_{j=0}^{p} \binom{p}{j} \Delta^j R(n) e^{\alpha p} + a_1 \sum_{j=0}^{p-1} \binom{p-1}{j} C_{p-1}^{j} \Delta^j R(n) e^{\alpha(p-1)} + \quad (2.20)$$

$$+ \cdots + a_p \sum_{j=0}^{0} \binom{j}{j} \Delta^j R(n) = Q(n).$$

If in (2.20) we rearrange the terms we obtain:

$$\left[\binom{p}{0} e^{\alpha p} + a_1 \binom{p-1}{0} e^{\alpha(p-1)} + \cdots + a_p \binom{0}{0} \right] \Delta^0 R(n) + \quad (2.21)$$

$$+ \left[\binom{p}{1} e^{\alpha p} + a_1 \binom{p-1}{1} e^{\alpha(p-1)} + \cdots + a_{p-1} \binom{1}{1} e^{\alpha} \right] \Delta^1 R(n) +$$

$$+ \cdots + \left[\binom{p}{k} e^{\alpha p} + a_1 \binom{p-1}{k} e^{\alpha(p-1)} + \cdots + a_{p-k} \binom{k}{k} e^{\alpha k} \right] \Delta^k R(n) + \cdots$$

$$+ \left[\binom{p}{p-1} e^{\alpha p} + a_1 \binom{p-1}{p-1} e^{\alpha(p-1)} \right] \Delta^{p-1} R(n) + \binom{p}{p} e^{\alpha p} \Delta^p R(n) = Q(n).$$

Taking into account the form of the characteristic polynomial $C(\lambda)$ and its derivatives, we can rewrite Eq. (2.21) as follows:

$$C(e^{\alpha}) \Delta^0 R(n) + C'(e^{\alpha}) e^{\alpha} \Delta^1 R(n) + \cdots + \frac{1}{k!} C^{(k)}(e^{\alpha}) e^{\alpha k} \Delta^k R(n) + \quad (2.22)$$

$$+ \cdots + \frac{1}{p!} C^{(p)}(e^{\alpha}) e^{\alpha p} \Delta^p R(n) = Q(n).$$

Since e^{α} is a multiple root of order s for the polynomial C, we have $C(e^{\alpha}) = C'(e^{\alpha}) = \cdots = C^{(s-1)}(e^{\alpha}) = 0$. Therefore, from Eq. (2.22), we are left with

$$\frac{1}{s!} C^{(s)}(e^{\alpha}) e^{\alpha s} \Delta^s R(n) + \cdots + \frac{1}{p!} C^{(p)}(e^{\alpha}) e^{\alpha p} \Delta^p R(n) = Q(n). \quad (2.23)$$

2.1 Linear Recurrences

In Eq. (2.23), $\Delta^s R(n), \cdots, \Delta^p R(n)$ are polynomials of degree $r-s, \cdots, r-p$ respectively. In order to identify the coefficients of R, we need $r - s = q$ or equivalently $r = q+s$. It is worth noting that if we substitute $R(n) = \sum_{j=0}^{r} A_j n^{r-j}$ into (2.23), then $\Delta^s R(n) = \Delta^s \left(\sum_{j=0}^{r-s} A_j n^{r-j} \right)$. Thus, in the left-hand side of Eq. (2.23), there are $q + 1$ unknowns that need to be determined by matching the coefficients with those of the polynomial Q. $\square$

Remarks 2.1.17

(i) The above demonstration implies that the unknown polynomial R should be sought in the form

$$R(n) = A_0 n^{q+s} + A_1 n^{q+s-1} + \cdots + A_q n^s.$$

In the case where e^α is not a root of the polynomial C, we have $s = 0$, and therefore R will be a polynomial of the same degree as Q.

(ii) If we are looking for a particular solution to the non-homogeneous equation

$$f(n+p) + a_1 \cdot f(n+p-1) + \ldots + a_p \cdot f(n) = Q(n), \forall n \in \mathbb{N},$$

where Q is a polynomial of degree q, we can apply the result from the previous theorem with $\alpha = 0$. The conclusion is that the particular solution should be sought in the form of a polynomial of degree q if 1 is not a solution of the characteristic equation.

If 1 is a multiple root of order s of the characteristic equation, then the particular solution Q should be sought in the form:

$$R(n) = A_0 n^{q+s} + A_1 n^{q+s-1} + \cdots + A_q n^s.$$

Example 2.1.18 Consider the non-homogeneous recurrence

$$f(n+2) - 2f(n+1) + f(n) = 12n^2, \text{ for every } n \in \mathbb{N}.$$

The associated homogeneous equation is

$$f(n+2) - 2f(n+1) + f(n) = 0, \text{ for every } n \in \mathbb{N},$$

whose characteristic equation is $\lambda^2 - 2\lambda + 1 = 0$. The characteristic equation has a double root $\lambda_1 = \lambda_2 = 1$. The general solution of the homogeneous equation is therefore of the form $c_1 \lambda_1^n + c_2 n \lambda_1^n = c_1 + c_2 n$. In our case $\alpha = 0$, and thus $e^\alpha = 1$ is a double root of the characteristic equation, so $q = s = 2$. According to point (ii) of Remark 2.1.17, we will look for a particular solution of the nonhomogeneous equation in the form $f^*(n) = A_0 n^4 + A_1 n^3 + A_2 n^2$. If we impose the condition that f^* satisfies the nonhomogeneous equation, after identifying the coefficients, we obtain $A_0 = 1$, $A_1 = -4$, $A_2 = 5$. In the end, the particular solution is $f^*(n) =$

$n^4 - 4n^3 + 5n^2$, and the general solution of the nonhomogeneous equation is $f(n) = n^4 - 4n^3 + 5n^2 + c_2 n + c_1$, for every $n \in \mathbb{N}$.

2.2 Nonlinear Recurrences

As noted in the introduction to this chapter, nonlinear recurrences (that is, those recurrences of the form (∗), where F is nonlinear) form an extremely vast class for which we do not have methods to find the general form of the sequence $(x_n)_n$. Actually, in the special cases we will analyze, we will not present the general form of sequences defined by such recurrences, but rather study only their convergence.

In a first subsection, we will provide two examples of nonlinear recurrences that can be reduced to linear recurrences, thus allowing the application of the techniques discussed in the previous section.

The second subsection deals with nonlinear recurrences defined by contractions.

In the third subsection, we will present Newton's method, which allows us to iteratively approximate the zeros of a function. An immediate application of this method is represented by the Heron's algorithm for extracting square roots.

The last three subsections will be dedicated to double algorithms generated by means and, in particular, the well-known algorithm of Gauss (the arithmetic-geometric mean algorithm).

2.2.1 Linearizable Nonlinear Recurrences

A **homogeneous recurrence** is a recurrence of the form:

$$f\left(\frac{x_{n+1}}{x_n}, n\right) = 0, \text{ for every } n \in \mathbb{N}, \qquad (2.24)$$

where f is a polynomial in its first variable:

$$f(y, n) = a_0(n) y^p + a_1(n) y^{p-1} + \cdots + a_{p-1}(n) y + a_p(n), \text{ for every } n \in \mathbb{N},$$

and $a_0, \cdots a_p$ are sequences of real numbers.

We note that we are only considering sequences of real numbers that satisfy the above recurrence. Thus, let $y_1(n), \cdots, y_k(n)$ be the real roots of the equation $a_0(n) y^p + a_1(n) y^{p-1} + \cdots + a_{p-1}(n) y + a_p(n) = 0$. Therefore $f(y, n) = \prod_{i=1}^{k}(y - y_i(n)) \cdot g(y, n)$, where $g(\cdot, n)$ is a polynomial that has only complex roots.

A sequence $(x_n)_n \subset \mathbb{R}$ satisfies Eq. (2.24) if and only if it is a solution of one of the first-order linear homogeneous equations of the form

$$x_{n+1} = y_i(n) x_n, i = 1, \cdots, k.$$

2.2 Nonlinear Recurrences

The solutions to these equations have been presented in the introduction of Sect. 2.1.

Thus, the solutions of recurrence (2.24) are the sequences:

$$(x_n^i)_n, \text{ where, for every } i = 1, \cdots, k, \ x_n^i = \prod_{j=0}^{n-1} y_i(j) \cdot x_0^i, n \in \mathbb{N}.$$

A **Riccati recurrence** is an equation of the form:

$$x_{n+1} \cdot x_n + a_n x_{n+1} + b_n x_n + c_n = 0, n \in \mathbb{N}, \tag{2.25}$$

where $(a_n)_n$, $(b_n)_n$, and $(c_n)_n$ are known sequences of real numbers.

If we substitute $x_n = \dfrac{y_{n+1}}{y_n} - a_n$ in Eq. (2.25) and simplify through y_{n+1}, we obtain the second-order homogeneous linear equation

$$y_{n+2} + (b_n - a_{n+1})y_{n+1} + (c_n - a_n b_n)y_n = 0. \tag{2.26}$$

If $c_n \neq a_n b_n$ for any $n \in \mathbb{N}$, and if we can determine two linearly independent solutions, $(y_n^1)_n$ and $(y_n^2)_n$, of Eq. (2.26), then the general solution of Eq. (2.26) is given by: $y_n = c^1 y_n^1 + c^2 y_n^2$, for every $n \in \mathbb{N}$ (see Theorem 2.1.5 and Remark 2.1.10).

We will now present an example of a Riccati recurrence.

Example 2.2.1 Consider the Riccati recurrence

$$x_{n+1} \cdot x_n + x_{n+1} + x_n - 1 = 0, x_1 = 0. \tag{2.27}$$

Let's find the solution $(x_n)_n$ of this recurrence and then determine $\lim_n x_n$.

By making the substitution $x_n = \dfrac{y_{n+1}}{y_n} - 1$, we obtain the second-order homogeneous linear recurrence $y_{n+2} - y_{n+1} - y_n = 0$.

In Sect. 2.1.4, we found the general solution of this recurrence to be:

$$y_n = c_1 \left(\frac{1+\sqrt{5}}{2}\right)^n + c_2 \left(\frac{1-\sqrt{5}}{2}\right)^n, \text{ for all } n \in \mathbb{N}.$$

Since $x_1 = 0$, it follows that $y_2 = y_1$, which leads to $c_1 + c_2 = 0$. Therefore, the solution of recurrence (2.27) is

$$x_n = \frac{\left(\frac{1+\sqrt{5}}{2}\right)^{n+1} - \left(\frac{1-\sqrt{5}}{2}\right)^{n+1}}{\left(\frac{1+\sqrt{5}}{2}\right)^n - \left(\frac{1-\sqrt{5}}{2}\right)^n} - 1, \text{ for every } n \in \mathbb{N}.$$

If we observe that $a = \frac{1+\sqrt{5}}{2} > 1$ and that $\frac{1-\sqrt{5}}{2} = -\frac{1}{a}$, then we obtain

$$x_n = \frac{1}{a} \cdot \frac{a^{2n+2} - (-1)^{n+1}}{a^{2n} - (-1)^n} - 1,$$

and then $\lim_n x_n = a - 1 = \frac{\sqrt{5}-1}{2}$.

2.2.2 Contractions

In this subsection, we will address the recurrences generated by contractions.

Definition 2.2.2 Let $I \subset \mathbb{R}$ be an interval and $f : I \to \mathbb{R}$ be a real-valued function. A point $c \in I$ is called a **fixed point** for f if $f(c) = c$.

Let $0 < k < 1$; f is a k-**contraction** on I if:

$$|f(x) - f(y)| \leq k \cdot |x - y|, \text{ for every } x, y \in I.$$

f is a **contraction** on I if there is $0 < k < 1$ such the f is a k-contraction.

Remark 2.2.3 A contraction on I has at most one fixed point on I. Indeed if f is a k-contraction ($0 < k < 1$) and if $c_1, c_2 \in I$ are two fixed points for f, then $|c_1 - c_2| = |f(c_1) - f(c_2)| \leq k \cdot |c_1 - c_2|$, and this inequality can occur if and only if $c_1 = c_2$.

Theorem 2.2.4 (The Contraction Mapping Principle) *Let $J \subseteq \mathbb{R}$ be a closed interval and let $f : J \to J$ be a contraction on J. Then f has a unique fixed point in the interval J.*

Proof Since f is a contraction on J, there exists a constant $k \in]0, 1[$ such that $|f(x) - f(y)| \leq k|x - y|$, for all $x, y \in J$.

Let $x_0 \in J$ be an arbitrary point. We will define $x_1 = f(x_0) \in J$, $x_2 = f(x_1) \in J$, and in general, $x_{n+1} = f(x_n) \in J$, for every $n \in \mathbb{N}$. Now we will show that the sequence $(x_n)_n$ thus constructed is a Cauchy sequence.

Let $n \in \mathbb{N}$ arbitrary; then

$$|x_{n+1} - x_n| = |f(x_n) - f(x_{n-1})| \leq k \cdot |x_n - x_{n-1}| = k \cdot |f(x_{n-1}) - f(x_{n-2})| \leq$$

$$\leq k^2 \cdot |x_{n-1} - x_{n-2}| \leq \cdots \leq k^n \cdot |x_1 - x_0|.$$

Using the previous inequalities, for any $n, p \in \mathbb{N}$, we obtain:

$$|x_{n+p} - x_n| \leq |x_{n+p} - x_{n+p-1}| + \cdots + |x_{n+1} - x_n| \leq \qquad (*)$$

$$\leq \left(k^{n+p-1} + \cdots + k^n\right) \cdot |x_1 - x_0| = \frac{k^n - k^{n+p}}{1 - k} \cdot |x_1 - x_0| < k^n \cdot \frac{1}{1-k} \cdot |x_1 - x_0|.$$

2.2 Nonlinear Recurrences

Since $0 < k < 1$, for every $\varepsilon > 0$, there exists $n_\varepsilon \in \mathbb{N}$ such that

$$|x_{n+p} - x_n| < k^n \cdot \frac{1}{1-k} \cdot |x_1 - x_0| < \varepsilon, \text{ for every } n \geq n_\varepsilon \text{ and every } p \in \mathbb{N}.$$

Since $(x_n)_n$ is a Cauchy sequence, there exists $c \in \mathbb{R}$ such that $x_n \to c$. Since J is a closed interval, we have $c \in J$. We observe that f, being a contraction on J, is a continuous function on J, and therefore $f(x_n) \to f(c)$. Finally, taking the limit in the recurrence relationship $x_{n+1} = f(x_n)$, we obtain $c = f(c)$.

In the previous remark, we showed that a contraction can have at most one fixed point. Therefore, it follows that c is the unique fixed point of the function f. □

Remarks 2.2.5

(i) The sequence $(x_n)_n$ constructed in the previous theorem's proof is called the sequence of **successive approximations** starting from the point x_0.

We note that this sequence is a solution to the nonlinear recurrence equation

$$x_{n+1} = f(x_n), n \in \mathbb{N}.$$

Thus, the previous theorem states that if f is a contraction, then this recurrence defines a convergent sequence towards the unique solution of the equation $f(x) = x$.

(ii) If we take the limit as $p \to \infty$ in equation $(*)$ from the previous theorem's proof, we obtain the following assessment of the convergence rate of the sequence $(x_n)_n$:

$$|c - x_n| \leq \frac{k^n}{1-k} \cdot |x_1 - x_0|, \text{ for every } n \in \mathbb{N}.$$

The following proposition is a sort of reciprocal of the contraction mapping principle.

Proposition 2.2.6 *Let $c \in \mathbb{R}$, $\delta > 0$ and let $f : [c-\delta, c+\delta] \to \mathbb{R}$ be a contraction with the fixed point c ($f(c) = c$). Then $f([c-\delta, c+\delta]) \subseteq [c-\delta, c+\delta]$.*

Proof Let us suppose that $0 < k < 1$ and that f is a k-contraction on $[c-\delta, c+\delta]$ that has a fixed point c. For every $x \in [c-\delta, c+\delta]$, $|f(x) - c| = |f(x) - f(c)| \leq k \cdot |x - c| \leq k\delta < \delta$. It follows that $f(x) \in [c-\delta, c+\delta]$ and so $f([c-\delta, c+\delta]) \subseteq [c-\delta, c+\delta]$. □

2.2.3 Newton's Method

Newton's method, also known as the Newton-Raphson method, is used for the algorithmic determination of the zeros of a function. It can be described as a special class of nonlinear recurrence.

Let's assume that $I \subset \mathbb{R}$ is an interval and that $f : I \to \mathbb{R}$ is a differentiable function on I with a zero at $c \in I$. Under the assumption that the derivative of the function f does not vanish in a neighborhood of c, we aim to present an algorithm that converges to c. The idea is as follows: we start with an initial point $x_0 \in I$; at $(x_0, f(x_0))$, we construct the tangent to the graph of f and denote by x_1 the point where it intersects the x-axis. Then we denote by x_2 the point where the tangent to the graph of f at $(x_1, f(x_1))$ intersects the x-axis, and so on. In the figure below, we have outlined these initial steps of the algorithm.

The equation of the tangent to the graph of function f at point $(x_0, f(x_0))$ is given by $y - f(x_0) = f'(x_0)(x - x_0)$. From this equation, we can find the intersection point with the x-axis, which gives us the value of x_1. So, we have $x_1 = x_0 - \dfrac{f(x_0)}{f'(x_0)}$. Similarly, we obtain $x_2 = x_1 - \dfrac{f(x_1)}{f'(x_1)}$ and, in the general case,

$$x_{n+1} = x_n - \frac{f(x_n)}{f'(x_n)}, \text{ for all } n \in \mathbb{N}. \qquad (*)$$

We will say that the algorithm defined in relation $(*)$ is **Newton's algorithm**.

If J is a neighborhood of c where the derivative of f does not vanish, and if we define $g : J \to \mathbb{R}$ by $g(x) = x - \dfrac{f(x)}{f'(x)}$, for $x \in J$, then for any $n \in \mathbb{N}$, $x_{n+1} = g(x_n)$.

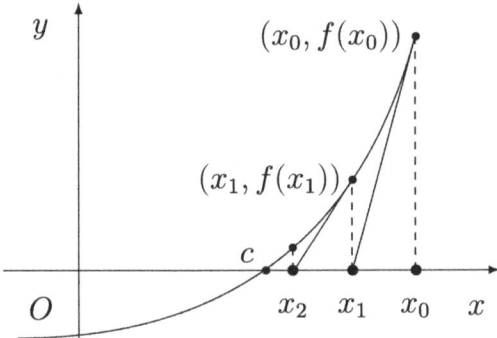

Here are some possible questions regarding the sequence $(x_n)_n$ defined as mentioned: (1) Are there any restrictions on the choice of the neighborhood J for the sequence $(x_n)_n$ to be well-defined? We observe that in order to define x_{n+1}, it is necessary for $x_n \in J$. Therefore, it should be that $g(J) \subseteq J$. (2) Are there any conditions on the function f and on the start point that guarantee the convergence of $(x_n)_n$? (3) In the case where the sequence $(x_n)_n$ is convergent, is its limit equal to c? (4) If the answer to the second question is affirmative, what is the convergence rate?

As we will see in the following the convergence of the sequence $(x_n)_n$ depends on various factors, such as the behavior of the function f and the initial point x_0. In

2.2 Nonlinear Recurrences

general, there is no guarantee that the sequence will always converge for any choice of x_0. Convergence may depend on the properties of the function and the specific starting point. The following theorem provides some answers to the aforementioned questions.

Theorem 2.2.7 *Let $I \subseteq \mathbb{R}$ be an interval, c be an interior point of I and let $f : I \to \mathbb{R}$ such that:*

(a) $f(c) = 0$.
(b) f is differentiable and f' is Lipschitz continuous on I.
(c) $f'(c) \neq 0$.

Then, for every $k \in]0, 1[$ there exists $\delta > 0$ such that:

(1) $[c - \delta, c + \delta] \subseteq I$.
(2) $f'(x) \neq 0$, for every $x \in [c - \delta, c + \delta]$.
(3) $g([c - \delta, c + \delta]) \subseteq [c - \delta, c + \delta]$, where $g(x) = x - \dfrac{f(x)}{f'(x)}$.

For every $x_0 \in [c - \delta, c + \delta]$, the recurrence starting by x_0: $x_{n+1} = g(x_n) = x_n - \dfrac{f(x_n)}{f'(x_n)}$, for all $n \in \mathbb{N}$, defines a sequence $(x_n)_n \subseteq [c - \delta, c + \delta]$ convergent to c with the following properties:

(4) $|x_n - c| \leq k^n \cdot |x_0 - c|$, for every $n \in \mathbb{N}$.
(5) *There exists $K > 0$ such that $|x_{n+1} - c| \leq K|x_n - c|^2$, for every $n \in \mathbb{N}$.*

Proof Since f' is Lipschitz continuous, there exists $L > 0$ such that

$$|f'(x) - f'(y)| \leq L \cdot |x - y|, \text{ for every } x, y \in I. \tag{2.28}$$

Let's prove that

$$|f(x) - f(y) - f'(y)(x - y)| \leq \frac{L}{2} \cdot |x - y|^2, \text{ for every } x, y \in I. \tag{2.29}$$

Indeed, for every $x, y \in I$,

$$f(x) - f(y) = \int_0^1 \frac{d}{dt}[f(y + t(x - y))] dt = (x - y) \cdot \int_0^1 f'(y + t(x - y)) dt,$$

from where, using (2.28),

$$|f(x) - f(y) - f'(y)(x - y)| \leq |x - y| \cdot \int_0^1 |f'(y + t(x - y)) - f'(y)| dt \leq$$

$$\leq |x - y| \cdot \int_0^1 Lt|x - y| dt = \frac{L}{2} \cdot |x - y|^2.$$

Since f' is Lipschitz continuous, it is continuous on I, and since $f'(c) \neq 0$, there exists $\delta_1 > 0$ such that

$$f'(x) \neq 0, \text{ for every } x \in [c - \delta_1, c + \delta_1] \subseteq I. \quad (2.30)$$

Since $\dfrac{1}{f'}$ is continue on the closed interval $[c - \delta_1, c + \delta_1]$, there exists $M > 0$ such that

$$\frac{1}{|f'(x)|} \leqslant M, \text{ for every } x \in [c - \delta_1, c + \delta_1]. \quad (2.31)$$

As f is continue in c and $f(c) = 0$, there exists $\delta < \min\{\delta_1, \frac{k}{2LM}\}$ such that

$$|f(x)| < \frac{k}{2LM^2}, \text{ for every } x \in [c - \delta, c + \delta] \subseteq [c - \delta_1, c + \delta_1] \subseteq I. \quad (2.32)$$

Let's consider the function $g : [c - \delta, c + \delta] \to \mathbb{R}$ defined by

$$g(x) = x - \frac{f(x)}{f'(x)}, x \in [c - \delta, c + \delta].$$

For every $x, y \in [c - \delta, c + \delta]$,

$$|g(x) - g(y)| = \left| x - y - \frac{f(x)}{f'(x)} + \frac{f(y)}{f'(y)} \right| = \left| \frac{f'(y)(x-y) + f(y)}{f'(y)} - \frac{f(x)}{f'(x)} \right| =$$

$$= \left| \frac{f'(y)(x-y) + f(y) - f(x)}{f'(y)} + \frac{f(x)[f'(x) - f'(y)]}{f'(x)f'(y)} \right| \leqslant$$

$$\leqslant \frac{|f(x) - f(y) - f'(y)(x-y)|}{|f'(y)|} + \frac{|f(x)| \cdot |f'(x) - f'(y)|}{|f'(x)| \cdot |f'(y)|} \overset{(2.28),(2.29),(2.31)}{\leqslant}$$

$$\overset{(2.28),(2.29),(2.31)}{\leqslant} \frac{LM}{2} \cdot (x-y)^2 + |f(x)| \cdot LM^2 \cdot |x - y| \overset{(2.32)}{<} \frac{LM}{2} \cdot 2\delta \cdot |x - y| +$$

$$+ \frac{k}{2LM^2} \cdot LM^2 \cdot |x - y| < LM \cdot \frac{k}{2LM} \cdot |x - y| + \frac{k}{2} \cdot |x - y| = k \cdot |x - y|.$$

It follows that g is a k-contraction on $[c - \delta, c + \delta]$ and c is a fixed point of g ($g(c) = c - \dfrac{f(c)}{f'(c)} = c$). According to Proposition 2.2.6, $g([c - \delta, c + \delta]) \subseteq [c - \delta, c + \delta]$. It follows that, for any $x_0 \in [c - \delta, c + \delta]$, $x_1 = x_0 + \dfrac{f(x_0)}{f'(x_0)} = g(x_0) \in [c - \delta, c + \delta]$.

2.2 Nonlinear Recurrences

Inductively, assuming that $x_n \in [c-\delta, c+\delta]$, then $x_{n+1} = x_n + \dfrac{f(x_n)}{f'(x_n)} = g(x_n) \in [c-\delta, c+\delta]$. Therefore $(x_n)_n \subseteq [c-\delta, c+\delta]$.

The sequence $(x_n)_n$ is the sequence of successive approximations starting from x_0 constructed with the contraction g. Taking into account Theorem 2.2.4 and Remarks 2.2.5, $x_n \to c$. Moreover, for every $n \in \mathbb{N}$,

$$|x_n - c| = |g(x_{n-1}) - g(c)| \leqslant k \cdot |x_{n-1} - c| = k \cdot |g(x_{n-2}) - g(c)| \leqslant$$

$$\leqslant k^2 \cdot |x_{n-2} - c| \leqslant \cdots \leqslant k^n \cdot |x_0 - c|.$$

We still need to prove conclusion (5) of the theorem.

Using the mean value theorem, for every $n \in \mathbb{N}$, there exists c_n between c and x_n such that $f(x_n) - f(c) = f'(c_n)(x_n - c)$ or

$$f(x_n) = f'(c_n) \cdot (x_n - c). \tag{2.33}$$

From (2.33) we obtain, for every $n \in \mathbb{N}$,

$$|x_{n+1} - c| = |(x_{n+1} - x_n) + (x_n - c)| = \left| -\frac{f(x_n)}{f'(x_n)} + \frac{f(x_n)}{f'(c_n)} \right| =$$

$$= \frac{|f(x_n)| \cdot |f'(x_n) - f'(c_n)|}{|f'(x_n)| \cdot |f'(c_n)|} = \frac{|f'(c_n)| \cdot |x_n - c| \cdot |f'(x_n) - f'(c_n)|}{|f'(x_n)| \cdot |f'(c_n)|} \overset{(2.28),(2.32)}{\leqslant}$$

$$\overset{(2.28),(2.32)}{\leqslant} ML \cdot |x_n - c|^2.$$

Therefore, there exists $K = ML > 0$ such that

$$|x_{n+1} - c| \leqslant K \cdot |x_n - c|^2, \text{ for every } n \in \mathbb{N}.$$

$\square$

Remarks 2.2.8

(i) The inequality (5) obtained in the conclusion of the previous theorem states that the convergence rate to c of the sequence (x_n) is quadratic. This means that at each new step, the number of exact decimal places found for c in the previous step is doubled. In fact, based on this inequality, we can obtain, for every $n \in \mathbb{N}$:

$$|x_n - c| \leqslant K \cdot |x_{n-1} - c|^2 \leqslant K \cdot K^2 \cdot |x_{n-2} - c|^{2^2} \leqslant \cdots \leqslant K \cdot K^2 \cdots K^{2^{n-1}} \cdot |x_0 - c|^{2^n}$$

$$= K^{2^n - 1} \cdot |x_0 - c|^{2^n} \leqslant (LM)^{2^n - 1} \cdot \delta^{2^n} < (LM)^{2^n - 1} \cdot \left(\frac{k}{2LM} \right)^{2^n} = \frac{1}{K} \cdot \left(\frac{k}{2} \right)^{2^n}.$$

We observe that quadratic convergence implies exponential convergence.

(ii) Obviously, the smaller k is, the smaller the neighborhood $[c - \delta, c + \delta]$ in which the iterations of the recurrence are located. However, as we will see from the following example, in practice, it is possible to conveniently increase this interval.

Example 2.2.9 Let's consider the function $f : \mathbb{R} \to \mathbb{R}$ defined by $f(x) = x^3 - 2x + 2$. This function has a single real zero $c \in]-2, -1[$. If we attempt to start Newton's algorithm with $x_0 = 0$, we obtain an oscillating sequence $(x_n)_n$, where $x_{2n} = 0$ and $x_{2n+1} = 1$, for every $n \in \mathbb{N}$, which is clearly divergent. In this case
$$g(x) = x - \frac{f(x)}{f'(x)} = 2 \cdot \frac{x^3 - 1}{3x^2 - 2}, \text{ and } g'(x) = \frac{f(x) \cdot f''(x)}{[f'(x)]^2} = \frac{6x(x^3 - 2x + 2)}{(3x^2 - 2)^2}.$$
We remark that, for every $x \in]-\infty, c[$, $g'(x) > 0$ and so g is an increasing function on the interval $]-\infty, c[$. Therefore, for every $x \in]-\infty, c[$, $x < g(x) < g(c) = c$; hence $g(]-\infty, c[) \subseteq]-\infty, c[$.

It follows that the sequence starting from any point $x_0 \in]-\infty, c[$, $x_{n+1} = g(x_n)$, is well constructed and $(x_n)_n \subseteq]-\infty, c[$. This sequence converges monotonically to c, and through a similar method to that used to establish conclusion (5) in the previous theorem, it can be shown that its convergence is quadratic.

As seen from this example, the choice of the starting point in Newton's algorithm is crucial. In general, this point should be chosen as close as possible to the root c that we want to approximate.

Example 2.2.10 Even in simple cases, if the conditions of the previous theorem are not satisfied, the recurrence defined by Newton's algorithm may not converge. Let's consider, for example, the function $f : \mathbb{R} \to \mathbb{R}$, $f(x) = |x|^\alpha$. The function f is differentiable on $\mathbb{R} \setminus \{0\}$ if $0 < \alpha < 1$; if $\alpha \geq 1$, then f is differentiable on $\mathbb{R}$. If we choose the initial point $x_0 > 0$, then $x_n = (-1)^n \cdot \left(\frac{1-\alpha}{\alpha}\right)^n \cdot x_0$, for every $n \in \mathbb{N}$. In the case where $0 < \alpha < \frac{1}{2}$, $x_{2n} \to +\infty$, and $x_{2n+1} \to -\infty$. If $\alpha = \frac{1}{2}$, then $x_n = (-1)^n \cdot x_0$, for every $n \in \mathbb{N}$. Finally, if $\alpha \geq \frac{1}{2}$, then $x_n \to 0$.

2.2.4 Heron's Method

An immediate application of Newton's method allows obtaining the square root of a number with high accuracy. Let $a > 0$; the square root of a is the positive zero of the function $f : \mathbb{R} \to \mathbb{R}$, defined by $f(x) = x^2 - a$. Let the function $g :]0, +\infty[\to \mathbb{R}$, defined by $g(x) = x - \frac{f(x)}{f'(x)} = \frac{x^2 + a}{2x}$. We remark that $g(x) \geq \sqrt{a}$, for every $x > 0$. The Newton's recurrence is $x_{n+1} = g(x_n) = \frac{x_n^2 + a}{2x_n}$, for every $n \in \mathbb{N}$. Then $x_{n+1} \geq \sqrt{a}$, for every $n \in \mathbb{N}$, and so $x_{n+1} - x_n = \frac{a - x_n^2}{2x_n} \leq 0$, for every

2.2 Nonlinear Recurrences

$n \in \mathbb{N}^*$, and then $\sqrt{a} \leqslant x_n \leqslant x_1$, for every $n \in \mathbb{N}^*$. It follows that $(x_n)_{n \geqslant 1}$ is a monotonically decreasing sequence that is bounded. This means that the Newton recurrence converges regardless of the choice of the starting point $x_0 > 0$. If we denote the limit of the sequence as $l > 0$, then follows $l = \dfrac{l^2 + a}{2l}$, which implies that $l = \sqrt{a}$.

A simple calculation shows that, for every $n \in \mathbb{N}$,

$$|x_{n+1} - \sqrt{a}| = x_{n+1} - \sqrt{a} = \frac{(x_n - \sqrt{a})^2}{2x_n} \leqslant \frac{1}{2\sqrt{a}} \cdot (x_n - \sqrt{a})^2.$$

Therefore, the convergence rate of the sequence $(x_n)_n$ to $\sqrt{a}$ is quadratic.

Surprisingly, this algorithm was known long before Newton, by Heron of Alexandria in the first century AD. Heron provides the first explicit description of this algorithm: Suppose we want to calculate the square root of a number a. We start with an initial approximation $x_0 > \sqrt{a}$. Then, $\dfrac{a}{x_0} < \sqrt{a}$. The arithmetic mean of x_0 and $\dfrac{a}{x_0}$, $x_1 = \dfrac{1}{2} \cdot \left(x_0 + \dfrac{a}{x_0} \right) > \sqrt{a}$ is an approximation of $\sqrt{a}$ that is better than $\dfrac{a}{x_0}$ and x_0. We continue in the same manner and construct an even better approximation $x_2 = \dfrac{1}{2} \cdot \left(x_1 + \dfrac{a}{x_1} \right) > \sqrt{a}$.

At step $n + 1$, we obtain the approximation

$$x_{n+1} = \frac{1}{2} \cdot \left(x_n + \frac{a}{x_n} \right) = \frac{x_n^2 + a}{2x_n},$$

which is exactly the Newton's recurrence. We say that this is the Heron's algorithm.

It is worth mentioning that very accurate approximations of square roots were found long before. For example, on the YBC 7289 tablet (approximately between 1800 and 1600 BCE) from the Yale Babylonian Collection, a five-decimal approximation of the square root of 2 appears. This value was used for calculating the diagonal of a square. Analyzing this tablet, as well as a series of coefficient lists found on other Babylonian tablets, D. Fowler and E. Robson attempt to convince us that these results are based on a similar algorithm to Heron's algorithm (see [7]). This is one of the reasons why Heron's algorithm is also called the Babylonian algorithm.

We can use Newton's method to approximate higher-order roots as well. Let $a > 0$ and let $p \in \mathbb{N}^*$. The p-th root of a is the positive zero of the function $f : \mathbb{R} \to \mathbb{R}$, defined by $f(x) = x^p - a$. Let the function $g :]0, +\infty[\to \mathbb{R}$ defined by $g(x) = x - \dfrac{f(x)}{f'(x)} = x - \dfrac{x^p - a}{px^{p-1}} = \dfrac{p-1}{p} x + \dfrac{a}{px^{p-1}}$. The Newton's recurrence in this case is: $x_{n+1} = g(x_n) = \dfrac{p-1}{p} x_n + \dfrac{a}{px_n^{p-1}}$.

Let's show that $x_n \geqslant \sqrt[p]{a}$, for every $n \in \mathbb{N}^*$. We write x_{n+1} in the following convenient form $x_{n+1} = x_n + \dfrac{a - x_n^p}{p x_n^{p-1}} = x_n \left(1 + \dfrac{a - x_n^p}{p x_n^p}\right)$ and by raising both sides to the power of p, we obtain $x_{n+1}^p = x_n^p \cdot \left(1 + \dfrac{a - x_n^p}{p x_n^p}\right)^p$. Now, we can use Bernoulli's inequality $\left(\dfrac{a - x_n^p}{p x_n^p} > -1\right)$:

$$x_{n+1}^p \geqslant x_n^p \left(1 + p \cdot \dfrac{a - x_n^p}{p x_n^p}\right) = a, \text{ from where } x_{n+1} \geqslant \sqrt[p]{a}, \text{ for all } n \in \mathbb{N}^*.$$

It follows that $x_{n+1} - x_n = \dfrac{a - x_n^p}{p x_n^{p-1}} \leqslant 0$, for every $n \in \mathbb{N}^*$, from where $\sqrt[p]{a} \leqslant x_n \leqslant x_1$, for every $n \in \mathbb{N}^*$. It follows that $(x_n)_{n \geqslant 1}$ is a monotonically decreasing sequence that is bounded. This means that the Newton recurrence converges regardless of the choice of the starting point $x_0 > 0$.

Obviously $\lim_n x_n = \sqrt[p]{a}$.

2.2.5 Double Algorithms Generated by Means

In this subsection, we will analyze the behavior of double recurrences of the form

$$\begin{cases} x_{n+1} = f(x_n, y_n), & x_0 = x \\ y_{n+1} = g(x_n, y_n), & y_0 = y \end{cases}, n \in \mathbb{N},$$

where f and g are means. We are searching for conditions in which the sequences $(x_n)_n$ and $(y_n)_n$, as defined, have the same limit $\varphi(x, y)$, and this limit can be determined. The main source of inspiration for writing this subsection was the book [12]. Information provided by [3, 4, 10, 11] was also utilized.

Let's first define the concept of mean.

Definition 2.2.11 A **mean** is a function $f :]0, +\infty[\times]0, +\infty[\to]0, +\infty[$, that possesses the following properties:

(1) $x < f(x, y) < y$, for every $x, y > 0$ with $x < y$.
(2) $f(\lambda x, \lambda y) = \lambda \cdot f(x, y)$, for every $x, y, \lambda > 0$.
(3) f is a continuous function on $]0, +\infty[\times]0, +\infty[$.

We will denote by $\mathcal{M}$ the set of all means.
We say that $f \in \mathcal{M}$ is **symmetric** if

(4) $f(x, y) = f(y, x)$, for every $x, y > 0$.

Remark 2.2.12 We observe that, for every $f \in \mathcal{M}$ and every $x > 0$, $f(x, x) = x$. Indeed, if $f \in \mathcal{M}$ and $x > 0$, then we choose arbitrarily two sequences

$(x_n)_n, (y_n)_n \subseteq]0, +\infty[$ such that $x_n < y_n$, for every $n \in \mathbb{N}$ and $x_n \to x$, $y_n \to x$. From 1), $x_n < f(x_n, y_n) < y_n$, for every $n \in \mathbb{N}$ and, from 3), $x \leqslant f(x, x) \leqslant x$.

Definition 2.2.13 On the set $\mathcal{M}$, we will define the following transitive relationship: let $f, g \in \mathcal{M}$; we say that $f \prec g$ if $f(x, y) < g(x, y)$, for every $x, y > 0$ with $x \neq y$. From the previous remark, $f(x, x) = x = g(x, x)$, for every $x > 0$.

In the following, we will provide several examples of symmetric and non-symmetric means.

Examples 2.2.14

(1) The **power means**. For every $r \in \mathbb{R} \setminus \{0\}$ we define $M_r :]0, +\infty[\times]0, +\infty[\to]0, +\infty[$ by

$$M_r(x, y) = \left(\frac{x^r + y^r}{2} \right)^{\frac{1}{r}}, \text{ for every } x, y > 0.$$

It is easy to verify that, for any $r \in \mathbb{R}^*$, M_r is a symmetric mean. Moreover, the power means have the following properties:

(i) $\lim_{r \to 0} M_r(x, y) = \sqrt{xy}$, for every $x, y > 0$. Then let $M_0(x, y) = \sqrt{xy}$.
(ii) $\lim_{r \to -\infty} M_r(x, y) = \min\{x, y\}$, $\lim_{r \to +\infty} M_r(x, y) = \max\{x, y\}$, for every $x, y > 0$.
(iii) $M_r \prec M_s$, for every $r < s$.

Indeed, for every $x, y > 0$,

(i) $\lim_{r \to 0} M_r(x, y) = \lim_{r \to 0} \left(\frac{x^r + y^r}{2} \right)^{\frac{1}{r}} = \exp \left[\lim_{r \to 0} \frac{\log(x^r + y^r) - \log 2}{r} \right] = $

$= \exp \left(\lim_{r \to 0} \frac{x^r \log x + y^r \log y}{x^r + y^r} \right) = \exp \left(\frac{\log(xy)}{2} \right) = \sqrt{xy}.$

(ii) Since M_r is symmetric, it is sufficient to consider the case $x < y$.

$$\lim_{r \to -\infty} M_r(x, y) = x \cdot \lim_{r \to -\infty} \left(\frac{1 + \left(\frac{y}{x}\right)^r}{2} \right)^{\frac{1}{r}} = x = \min\{x, y\}.$$

The limit at $+\infty$ can be computed in a similar manner.
(iii) Using Bernoulli's inequality, it can be easily shown that:

$$a^b > ab + 1 - b, \text{ for every } a > 0, a \neq 1, b > 1. \quad (*)$$

Let $x, y > 0$ be arbitrarily chosen such that $x \neq y$.

First let $0 < r < s$ and $M_r = M_r(x, y)$, $M_s = M_s(x, y)$. Then

$$\left(\frac{M_s}{M_r}\right)^s = \frac{1}{2}\left[\left(\frac{x}{M_r}\right)^s + \left(\frac{y}{M_r}\right)^s\right] = \frac{1}{2}\left[\left(\frac{x^r}{M_r^r}\right)^{\frac{s}{r}} + \left(\frac{y^r}{M_r^r}\right)^{\frac{s}{r}}\right] \overset{(*)}{\geq}$$

$$\overset{(*)}{\geq} \frac{1}{2}\left(\frac{s}{r} \cdot \frac{x^r}{M_r^r} + 1 - \frac{s}{r} + \frac{s}{r} \cdot \frac{y^r}{M_r^r} + 1 - \frac{s}{r}\right) = 1$$

(we use the inequality $(*)$ with $b = \frac{s}{r} > 1$, $a = \frac{x^r}{M_r^r}, \frac{y^r}{M_r^r}, a > 0, a \neq 1$.)
It follows that $M_r < M_s$.
If $0 = r < s$, then

$$\left(\frac{M_s}{M_0}\right)^s = \frac{1}{2}\left[\left(\frac{x}{\sqrt{xy}}\right)^s + \left(\frac{y}{\sqrt{xy}}\right)^s\right] = \frac{1}{2}\left[\left(\frac{x}{y}\right)^{\frac{s}{2}} + \left(\frac{y}{x}\right)^{\frac{s}{2}}\right] > 1$$

$\left(\left(\frac{x}{y}\right)^{\frac{s}{2}} \neq 1\right)$. Therefore $M_0 < M_s$.

The remaining possible cases can be reduced to the previous ones by taking into account the fact that, for every $r < 0$, $M_r(x, y) = \frac{1}{M_{-r}(x^{-1}, y^{-1})}$.

Thus, we have demonstrated that $M_r \prec M_s$, for every $r, s \in \mathbb{R}, r < s$.
Some particular cases of power means are as follows:

(a) **Arithmetic mean**: $A(x, y) = M_1(x, y) = \frac{x+y}{2}$.
(b) **Geometric mean**: $G(x, y) = M_0(x, y) = \sqrt{xy}$.
(c) **Harmonic mean**: $H(x, y) = M_{-1}(x, y) = \frac{2xy}{x+y}$.

Actually, the arithmetic, geometric, and harmonic means are commonly referred to as Pythagorean means.

(d) **Quadratic mean**: $Q(x, y) = M_2(x, y) = \sqrt{\frac{x^2+y^2}{2}}$.

Property (iii) mentioned above leads us to the following relationships: $H \prec G \prec A \prec Q$.

(2) Another example of a symmetric mean is the **contraharmonic mean** discovered by Eudoxus of Cnidos in the fourth century BCE, defined by

$$C(x, y) = \frac{x^2 + y^2}{x + y}, \text{ for every } x, y > 0.$$

It is easily shown that C is a symmetric mean and that $Q \prec C$.

(3) **The logarithmic mean**: $L :]0, +\infty[\times]0, +\infty[\to]0, +\infty[$,

$$L(x, y) = \begin{cases} \frac{x-y}{\log x - \log y}, & x \neq y, \\ x, & x = y. \end{cases} = \begin{cases} \frac{x-y}{2\,\text{artanh}\,\frac{x-y}{x+y}}, & x \neq y, \\ x, & x = y. \end{cases}$$

2.2 Nonlinear Recurrences

(4) **The Neuman–Sándor mean**: $NS :]0, +\infty[\times]0, +\infty[\to]0, +\infty[$,

$$NS(x, y) = \begin{cases} \dfrac{x-y}{2 \operatorname{arsinh} \frac{x-y}{x+y}}, & x \neq y, \\ x, & x = y. \end{cases}$$

It can easily be shown that $G \prec L \prec A \prec NS \prec Q$.

We will now present some examples of non-symmetric means.

(5) $B :]0, +\infty[\times]0, +\infty[\to]0, +\infty[$, $B(x, y) = \sqrt{\dfrac{x+y}{2} \cdot y}$.

(6) **The Schwab-Borchardt mean**: $SB :]0, +\infty[\times]0, +\infty[\to]0, +\infty[$,

$$SB(x, y) = \begin{cases} \dfrac{\sqrt{y^2 - x^2}}{\arccos \frac{x}{y}}, & 0 < x < y, \\ x, & x = y \\ \dfrac{\sqrt{x^2 - y^2}}{\operatorname{arcosh} \frac{x}{y}}, & 0 < y < x. \end{cases}$$

Let us consider two means, $f, g \in \mathcal{M}$, and the double recurrence constructed using them:

$$\begin{cases} x_{n+1} = f(x_n, y_n), \ x_0 = x \\ y_{n+1} = g(x_n, y_n), \ y_0 = y \end{cases}, n \in \mathbb{N},$$

In the following theorem, we present conditions under which the sequences $(x_n)_n$ and $(y_n)_n$ defined in this manner have the same limit $\varphi(x, y)$.

Theorem 2.2.15 *If $f, g \in \mathcal{M}$, $f \prec g$, then, for every $x, y > 0$, there exist $\lim_n x_n = \varphi(x, y) = \lim_n y_n$. Moreover, $\varphi \in \mathcal{M}$, $f \prec \varphi \prec g$ and φ satisfies the property of invariance with respect to f and g:*

$$\varphi(x, y) = \varphi(f(x, y), g(x, y)), \text{ for every } x, y > 0. \tag{I}$$

Proof Let $x, y > 0$. If $x = y$, then $x_1 = f(x, x) = x = g(x, x) = y_1$. By induction, we find that, for every $n \in \mathbb{N}$, $x_n = x = y_n$. Therefore $\lim_n x_n = \lim_n y_n = x \equiv \varphi(x, x)$.

If $x \neq y$, then $x_1 = f(x, y) < g(x, y) = y_1, x_2 = f(x_1, y_1) < g(x_1, y_1) = y_2$. If we suppose that $x_n < y_n$, then $x_{n+1} = f(x_n, y_n) < g(x_n, y_n) = y_{n+1}$. Then the sequence $(x_n)_{n \geqslant 1}$ is increasing and bounded and $(y_n)_{n \geqslant 1}$ is decreasing and bounded. Therefore, there exist $l, L > 0$ such that $\lim_n x_n = l \leqslant L = \lim_n y_n$. If we suppose that $l < L$, then $l < f(l, L) = \lim_n f(x_n, y_n) = \lim_n x_{n+1} = l$, which leads to a contradiction. Therefore $l = L$ and so $\lim_n x_n \equiv \varphi(x, y) \equiv \lim_n y_n$.

Let's show that φ is a mean. For every $x, y > 0$ with $x < y$,

$$x < x_1 < \cdots < x_n < \varphi(x, y) < y_n < \cdots < y_1 < y.$$

For every $x, y > 0$, $\begin{cases} x_1 = f(x, y) \equiv f_1(x, y), \\ y_1 = g(x, y) \equiv g_1(x, y). \end{cases}$ It follows that

$$\begin{cases} x_2 = f(f(x, y), g(x, y)) \equiv f_2(x, y), \\ y_2 = g(f(x, y), g(x, y)) \equiv g_2(x, y). \end{cases}$$

If we suppose that $\begin{cases} x_n = f_n(x, y), \\ y_n = g_n(x, y), \end{cases}$ it follows that

$$\begin{cases} x_{n+1} = f(x_n, y_n) = f(f_n(x, y), g_n(x, y)) \equiv f_{n+1}(x, y), \\ y_{n+1} = g(x_n, y_n) = g(f_n(x, y), g_n(x, y)) \equiv g_{n+1}(x, y). \end{cases}$$

It can be easily proven by induction that, for every $\lambda > 0$, and for every $n \in \mathbb{N}^*$, $f_n(\lambda x, \lambda y) = \lambda \cdot f_n(x, y)$. It follows that

$$\varphi(\lambda x, \lambda y) = \lim_n f_n(\lambda x, \lambda y) = \lambda \cdot \lim_n f_n(x, y) = \lambda \cdot \varphi(x, y).$$

Let's show that φ is continuous on $]0, +\infty[\times]0, +\infty[$. Let $x, y > 0$ be two arbitrary points. We have observed that $x_n \uparrow \varphi(x, y)$ and $y_n \downarrow \varphi(x, y)$. Then, for every $\varepsilon > 0$ there exists $n_0 \in \mathbb{N}^*$ such that $y_{n_0} - x_{n_0} < \frac{\varepsilon}{2}$.

As we see above, for every $n \in \mathbb{N}$, $\begin{cases} x_n = f_n(x, y), \\ y_n = g_n(x, y), \end{cases}$ where f_n and g_n are obtained through repeated compositions of functions f and g. It follows that, for every $n \in \mathbb{N}$, f_n and g_n are continuous at (x, y). Then let $\delta > 0$ be such that, for every $x', y' > 0$ with $|x - x'| < \delta, |y - y'| < \delta$,

$$|x_{n_0} - x'_{n_0}| = |f_{n_0-1}(x, y) - f_{n_0-1}(x', y')| < \frac{\varepsilon}{2}, \text{ and}$$
$$|y_{n_0} - y'_{n_0}| = |g_{n_0-1}(x, y) - g_{n_0-1}(x', y')| < \frac{\varepsilon}{2}.$$

(x', y') is the starting point for the double recurrence $\begin{cases} x'_{n+1} = f(x'_n, y'_n), x'_0 = x', \\ y'_{n+1} = g(x'_n, y'_n), y'_0 = y', \end{cases}$
and therefore $x'_1 < x'_2 < \cdots < x'_{n_0} < \varphi(x', y') < y'_{n_0} < \cdots < y'_2 < y'_1$.
Therefore $\varphi(x, y) - \varphi(x', y') < y_{n_0} - x'_{n_0} = (y_{n_0} - x_{n_0}) + (x_{n_0} - x'_{n_0}) < \frac{\varepsilon}{2} + \frac{\varepsilon}{2}$,
$\varphi(x', y') - \varphi(x, y) < y'_{n_0} - x_{n_0} = (y_{n_0} - x_{n_0}) + (y'_{n_0} - y_{n_0}) < \frac{\varepsilon}{2} + \frac{\varepsilon}{2}$.

It follows that φ is continuous in (x, y) and, since $x, y > 0$ are arbitrary, φ is continuous on $]0, +\infty[\times]0, +\infty[$; therefore $\varphi \in \mathcal{M}$.

Let now $x, y > 0, x \neq y$. Then $f(x, y) = x_1 < \varphi(x, y) < g(x, y)$, so that $f \prec \varphi \prec g$.

Finally, let's show that φ satisfies the property of invariance (I) with respect to f and g. Let $x, y > 0$ and let $(x_n)_n, (y_n)_n$ be the double recurrence starting from (x, y), and $(x'_n)_n, (y'_n)_n$ be the double recurrence starting from (x', y'), where $x' = f(x, y) = x_1, y' = g(x, y) = y_1$. We remark that, for every $n \in \mathbb{N}$, $x'_n =$

2.2 Nonlinear Recurrences

$x_{n+1}, y'_n = y_{n+1}$. Taking into account that $x'_n \uparrow \varphi(x', y') \downarrow y'_n$, it follows that $\varphi(x', y') = \varphi(x, y)$, from where $\varphi(x, y) = \varphi(f(x, y), g(x, y))$. □

Definition 2.2.16 Let $f, g \in \mathcal{M}$ with $f \prec g$ and let

$$\begin{cases} x_{n+1} = f(x_n, y_n), & x_0 = x \\ y_{n+1} = g(x_n, y_n), & y_0 = y \end{cases}, n \in \mathbb{N},$$

be the double recurrence constructed using f, g. We say that the limit of this recurrence, φ, is **generated by the means** f and g and we denote this by $\varphi = (f, g)$.

Remark 2.2.17 The limit of the double algorithm generated by f and g, φ, is the unique mean that satisfies the property of invariance relative to f and g. Indeed, let $\psi \in \mathcal{M}$ be a mean that satisfies the property of invariance relative to f and g: $\psi(x, y) = \psi(f(x, y), g(x, y))$, for every $x, y > 0$. Then $\psi(x, y) = \psi(x_1, y_1) = \psi(f(x_1, y_1), g(x_1, y_1)) = \psi(x_2, y_2) = \cdots = \psi(x_n, y_n)$, for every $n \in \mathbb{N}$. Since ψ is continuous, $\psi(x_n, y_n) \to \psi(\varphi(x, y), \varphi(x, y)) = \varphi(x, y)$ and so $\psi(x, y) = \varphi(x, y)$, for every $x, y > 0$.

This observation allows us, in some cases, to determine the limit of a double algorithm by highlighting a mean that satisfies the property of invariance.

Examples 2.2.18

(1) Let's determine the limit of the double recurrence constructed using the harmonic mean H and the arithmetic mean A.
From Examples 2.2.14 we know that $H \prec A$. Let $x, y > 0$; $x_1 = \dfrac{2xy}{x+y}$, $y_1 = \dfrac{x+y}{2}$. Therefore $x_1 \cdot y_1 = x \cdot y$ and so $\sqrt{x_1 y_1} = \sqrt{xy}$ or $G(H(x, y), A(x, y)) = G(x, y)$. It follows that the geometric mean satisfies the property of invariance relative to H and A. Taking into account the previous remark, $G = (H, A)$.

Let's further observe here that Heron's method for square root extraction (see Sect. 2.2.4) can be obtained as a particular case of the algorithm generated by H and A. Indeed, let's initiate the double recurrence with the starting data $x = 1, y = a > 0$. We have already observed that $x_1 \cdot y_1 = x \cdot y = a$ and similarly, for every $n \in \mathbb{N}$, $x_n \cdot y_n = x \cdot y = a$, which implies $x_n = \dfrac{a}{y_n}$. Then, for every $n \in \mathbb{N}$,

$$y_{n+1} = \frac{x_n + y_n}{2} = \frac{1}{2}\left(y_n + \frac{a}{y_n}\right),$$

which is actually the Heron's algorithm.

(2) In the following example, we will study a case where Theorem 2.2.15 cannot be applied. However, we will still be able to find the general form of the sequences that form the double recurrence, as well as their common limit.

Let $x, y > 0$ and let the double recurrence

$$\begin{cases} x_{n+1} = A(x_n, y_n) = \dfrac{x_n + y_n}{2} & , x_0 = x, \\ y_{n+1} = B(x_n, y_n) = \sqrt{\dfrac{x_n + y_n}{2} \cdot y_n} & , y_0 = y. \end{cases}$$

(see (5) of Example 2.2.14).

A and B are not comparable (if $x < y$, then $A(x, y) < B(x, y)$ and if $y < x$, then $B(x, y) < A(x, y)$). We will show that $\lim_n x_n = \lim_n y_n = SB(x, y)$ where SB is the Schwab-Borchardt mean (see (6) in Example 2.2.14).

(a) If $x < y$, then there exists an unique $\alpha \in]0, \frac{\pi}{2}[$ such that $\cos\alpha = \frac{x}{y}$. For every $n \in \mathbb{N}$, we denote by $r_n = \frac{x_n}{y_n} < 1$, then r_n satisfies the recurrence $r_{n+1} = \sqrt{\frac{1+r_n}{2}}$. For every $n \in \mathbb{N}$, let $\alpha_n \in]0, \frac{\pi}{2}[$ such that $r_n = \cos\alpha_n$. Then $\cos\alpha_{n+1} = \cos\frac{\alpha_n}{2}$, from where $\alpha_{n+1} = \frac{\alpha_n}{2}$, for every $n \in \mathbb{N}$ and then $\alpha_n = \frac{\alpha}{2^n}$. It follows that $x_n = y_n \cdot \cos\frac{\alpha}{2^n}$ and then $y_{n+1} = \sqrt{\frac{x_n+y_n}{1} \cdot y_n} = y_n \cdot \cos\frac{\alpha}{2^{n+1}}$. The last relationship leads us to $y_n = y \cos\frac{\alpha}{2} \cos\frac{\alpha}{2^2} \cdots \cos\frac{\alpha}{2^n} = y \cdot \frac{\sin\alpha}{2^n \sin\frac{\alpha}{2^n}}$, from where $\lim_n y_n = y \cdot \frac{\sin\alpha}{\alpha} = \frac{\sqrt{y^2-x^2}}{\arccos\frac{x}{y}} = \lim_n x_n$.

(b) If $y < x$, then there exits an unique $\alpha > 0$ such that $\cosh\alpha = \frac{x}{y} > 1$. Similar calculations to those in point (a) lead us to

$$y_n = y \cdot \frac{\sinh\alpha}{2^n \sinh\frac{\alpha}{2^n}} \to \frac{\sqrt{x^2 - y^2}}{\operatorname{arcosh}\frac{x}{y}} \text{ and } x_n = y_n \cdot \cosh\frac{\alpha}{2^n} \to \frac{\sqrt{x^2 - y^2}}{\operatorname{arcosh}\frac{x}{y}}.$$

It follows that $\lim_n x_n = \lim_n y_n = SB(x, y)$, for every $x, y > 0$.

2.2.6 The Arithmetic-Geometric Mean

The double recurrence consisting of the arithmetic mean and the geometric mean (the (A, G) algorithm) is one of the most important nonlinear algorithms. Its connections with one of the deepest theories, that of elliptic functions and integrals, have led us to dedicate this subsection to it.

The arithmetic-geometric mean appeared in a memoir by Lagrange published in 1784–1785. As a teenager, Carl Friedrich Gauss rediscovered the algorithm. In 1816, Gauss stated in a letter to H.C. Schumacher that he independently discovered the arithmetic-geometric mean in 1791 at the age of 14. However, his major contribution, which included an elegant integral representation of the limit, came 7–9 years later. At the age of 22–23, Gauss wrote a paper describing his discoveries regarding the arithmetic-geometric mean. Like many of his research works, this

2.2 Nonlinear Recurrences

paper was only published after his death. Gauss clearly attached great importance to the (A, G) algorithm, as evidenced by numerous references made in his journal between 1799 and 1800. Some of these references are quite vague, and it is possible that we have not yet fully uncovered Gauss's discoveries in this regard.

We present Gauss's theorem with two proofs: one by Gauss himself and the other by Legendre, which utilizes the Landen transformation.

In the preparation of this subsection, valuable sources of information were the works [1, 2].

Let's assume that $a, b > 0$; the algorithm of arithmetic-geometric means is:

$$\begin{cases} a_{n+1} = \dfrac{a_n + b_n}{2} = A(a_n, b_n), \ a_0 = a, \\ b_{n+1} = \sqrt{a_n b_n} = G(a_n, b_n), \quad b_0 = b, \end{cases} \quad n \in \mathbb{N}. \qquad (A-G)$$

From (1) of Example 2.2.14 $G \prec A$, and then, Theorem 2.2.15 leads us to the conclusion that there exists a symmetric mean $M \in \mathcal{M}$ such that $\lim_n a_n = M(a, b) = \lim_n b_n$, $G \prec M \prec A$ and M satisfies the property of invariance with respect to A and G:

$$M(a, b) = M(A(a, b), G(a, b)) = M\left(\frac{a+b}{2}, \sqrt{ab}\right), \text{ for every } a, b > 0.$$

We will denote $M(a, b)$ as the **arithmetic-geometric mean** of the numbers a and b.

Gauss provides four numerical examples to demonstrate the fast convergence speed of the arithmetic-geometric mean algorithm. We will present just one of them:

Let $a = 1$ and $b = 0, 8$; then:

$a_1 = 0, 9$ $\qquad\qquad\qquad\ b_1 = 0, 894427190999915\ldots$
$a_2 = 0, \underline{897213595499957}\ldots\ b_2 = 0, \underline{897209268732734}\ldots$
$a_3 = 0, \underline{897211432116346}\ldots\ b_3 = 0, \underline{897211432113738}\ldots$
$a_4 = 0, \underline{897211432115042}\ldots\ b_4 = 0, \underline{897211432115042}\ldots$

The proposition below illustrates the exponential convergence speed of this algorithm.

Proposition 2.2.19 *Let $a, b > 0$; the sequences (a_n) and (b_n) generated by the arithmetic-geometric mean algorithm $(A - G)$ satisfy the inequalities:*

$$\begin{matrix} a_n - M(a, b) \\ M(a, b) - b_n \end{matrix} \leqslant a_n - b_n \leqslant 8b \cdot \left(\frac{a-b}{8b}\right)^{2^n}, \text{ for every } n \in \mathbb{N}. \qquad (*)$$

Proof The first inequality is a consequence of the fact that, for every $n \in \mathbb{N}$, $b_n \leqslant M(a, b) \leqslant a_n$.

Let now $n \in \mathbb{N}^*$; then

$$a_n - b_n = \frac{a_{n-1} + b_{n-1}}{2} - \sqrt{a_{n-1} \cdot b_{n-1}} = \frac{(\sqrt{a_{n-1}} - \sqrt{b_{n-1}})^2}{2} =$$

$$= \frac{(a_{n-1} - b_{n-1})^2}{2(\sqrt{a_{n-1}} + \sqrt{b_{n-1}})^2} \leqslant \frac{(a_{n-1} - b_{n-1})^2}{2 \cdot 2^2 \cdot b},$$

from where

$$a_n - b_n \leqslant \frac{(a_{n-1} - b_{n-1})^2}{8b}. \tag{2.34}$$

In inequality (2.34), we replace n with $n - 1$ and obtain:

$$a_{n-1} - b_{n-1} \leqslant \frac{(a_{n-2} - b_{n-2})^2}{8b}, n \geqslant 2. \tag{2.35}$$

From (2.34) and (2.35), we obtain through repeated iterations:

$$a_n - b_n \leqslant \frac{(a_{n-2} - b_{n-2})^{2^2}}{(8b) \cdot (8b)^2} \leqslant \frac{(a_{n-3} - b_{n-3})^{2^3}}{(8b) \cdot (8b)^2 \cdot (8b)^{2^2}} \leqslant \cdots \leqslant \frac{(a-b)^{2^n}}{(8b)^{1+2+2^2+\ldots+2^{n-1}}} =$$

$$= 8b \cdot \left(\frac{a-b}{8b}\right)^{2^n}.$$

$\square$

From (2.34) of the previous proof, we deduce that $(a_n - b_n)_n$ converges quadratically to zero.

We will show that the arithmetic-geometric mean can be represented using a complete elliptic integral of the first kind.

Definition 2.2.20 The integral

$$K(x) = \int_0^{\frac{\pi}{2}} \frac{dt}{\sqrt{1 - x^2 \sin^2 t}}, |x| < 1$$

is called the **complete elliptic integral of the first kind**.

Theorem 2.2.21 (Gauss 1799; published in 1818) *For every* $x \in]-1, 1[$,

$$M(1+x, 1-x) = \frac{\pi}{2 \cdot K(x)} = \frac{\pi}{2 \cdot \int_0^{\frac{\pi}{2}} \frac{dt}{\sqrt{1 - x^2 \sin^2 t}}}.$$

2.2 Nonlinear Recurrences

Proof

Gauss Demonstration Firstly, we will show how Gauss arrived at the integral representation of the arithmetic-geometric mean. Because M is symmetric, we observe that, for every $x \in]-1, 1[$,

$$M(1+(-x), 1-(-x)) = M(1-x, 1+x) = M(1+x, 1-x).$$

Therefore, M is an even function. If we assume that $x \mapsto \dfrac{1}{M(1+x, 1-x)}$ is an analytic function, then we can write

$$\frac{1}{M(1+x, 1-x)} = \sum_{k=0}^{\infty} A_k x^{2k}. \tag{2.36}$$

We make the substitution $x = \dfrac{2t}{1+t^2}$; since M is invariant with respect to A and G, we obtain:

$$M(1+x, 1-x) = M\left(1 + \frac{2t}{1+t^2}, 1 - \frac{2t}{1+t^2}\right) = M\left(\frac{(1+t)^2}{1+t^2}, \frac{(1-t)^2}{1+t^2}\right)$$

$$= \frac{1}{1+t^2} M\left((1+t)^2, (1-t)^2\right) =$$

$$= \frac{1}{1+t^2} \cdot M\left(\frac{(1+t)^2 + (1-t)^2}{2}, \sqrt{(1+t)^2(1-t)^2}\right) =$$

$$= \frac{1}{1+t^2} \cdot M\left(1+t^2, 1-t^2\right).$$

By substituting into (2.36), we obtain $\dfrac{1+t^2}{M(1+t^2, 1-t^2)} = \sum_{k=0}^{\infty} A_k \left(\dfrac{2t}{1+t^2}\right)^{2k}$, from which, using again (2.36), we obtain the relationship

$$\left(1+t^2\right) \sum_{k=0}^{\infty} A_k t^{4k} = \sum_{k=0}^{\infty} A_k \left(\frac{2t}{1+t^2}\right)^{2k}, \text{ or}$$

$$\sum_{k=0}^{\infty} A_k t^{4k} = \sum_{k=0}^{\infty} A_k 2^{2k} t^{2k} \left(1+t^2\right)^{-2k-1}. \tag{2.37}$$

We recall the formula for the binomial series: for every $\alpha \in \mathbb{R}$,

$$(1+x)^\alpha = 1 + \sum_{n=1}^\infty \frac{\alpha(\alpha-1)\cdots(\alpha-n+1)}{n!} x^n, \text{ for all } x \in]-1, 1[.$$

Therefore, for every $t \in]-1, 1[: (1+t^2)^{-2k-1} =$

$$= 1 + \sum_{n=1}^\infty \frac{(2k+1)(2k+2)\cdots(2k+n)}{n!} (-1)^n t^{2n} =$$

$$= 1 + \sum_{n=1}^\infty \binom{2k+n}{n} (-1)^n t^{2n}.$$

Returning to Eq. (2.37), we obtain:

$$\sum_{k=0}^\infty A_k t^{4k} = \sum_{k=0}^\infty \left[A_k 2^{2k} t^{2k} \left(1 + \sum_{n=1}^\infty (-1)^n \binom{2k+n}{n} t^{2n} \right) \right] =$$

$$= \sum_{k=0}^\infty A_k 2^{2k} t^{2k} \left[1 - \binom{2k+1}{1} t^2 + \cdots + (-1)^n \binom{2k+n}{n} t^{2n} + \cdots \right] =$$

$$= \sum_{k=0}^\infty \left[A_k 2^{2k} t^{2k} - \binom{2k+1}{1} A_k 2^{2k} t^{2k+2} + \cdots + \right.$$

$$\left. + (-1)^n \binom{2k+n}{n} A_k 2^{2k} t^{2k+2n} + \cdots + \right] =$$

$$= (A_0 - A_0 t^2 + A_0 t^4 - A_0 t^6 + \cdots + (-1)^n A_0 t^{2n} + \cdots) +$$

$$+ \left[A_1 2^2 t^2 - \binom{3}{1} A_1 2^2 t^4 + \binom{4}{2} A_1 2^2 t^6 + \cdots \right.$$

$$\left. + (-1)^n A_1 \binom{n+2}{n} 2^2 t^{2n+2} + \cdots \right] +$$

$$+ \left[A_2 2^4 t^4 - \binom{5}{1} A_2 2^4 t^6 + \binom{6}{2} A_2 2^4 t^8 + \cdots \right.$$

$$\left. + (-1)^n \binom{n+4}{n} A_2 2^4 t^{2n+4} + \cdots \right] + \cdots$$

2.2 Nonlinear Recurrences

Therefore

$$\sum_{k=0}^{\infty} A_k t^{4k} = A_0 + \left(-A_0 + 2^2 A_1\right) t^2 + \left[A_0 - \binom{3}{1} 2^2 A_1 + 2^4 A_2\right] \cdot t^4 + \quad (2.38)$$

$$+ \left[-A_0 + \binom{4}{2} A_1 2^2 - \binom{5}{1} A_2 2^4 + A_3 2^6\right] \cdot t^6 + \cdots$$

In (2.36), we substitute $x = 0$,

$$A_0 = \frac{1}{M(1,1)} = 1.$$

Then, we identify the coefficients in (2.38) and obtain:

$$0 = -A_0 + 2^2 A_1 \Rightarrow A_1 = \frac{1}{2^2} = \left(\frac{1}{2}\right)^2,$$

$$A_1 = A_0 - \binom{3}{1} 2^2 A_1 + 2^4 A_2 \Rightarrow A_2 = \frac{9}{2^6} = \left(\frac{1 \cdot 3}{2 \cdot 4}\right)^2$$

$$0 = -A_0 + \binom{4}{2} A_1 2^2 - \binom{5}{1} C_5^1 A_2 2^4 + A_3 2^6 \Rightarrow A_3 = \frac{25}{2^8} = \left(\frac{1 \cdot 3 \cdot 5}{2 \cdot 4 \cdot 6}\right)^2.$$

Analogously, we obtain

$$A_k = \left[\frac{1 \cdot 3 \cdots (2k-1)}{2 \cdot 4 \cdots (2k)}\right]^2 = \left[\frac{(2k-1)!!}{(2k)!!}\right]^2, \text{ for every } k \in \mathbb{N}^*.$$

Returning to (2.36):

$$\frac{1}{M(1+x, 1-x)} = 1 + \left(\frac{1}{2}\right)^2 x^2 + \left(\frac{1 \cdot 3}{2 \cdot 4}\right)^2 x^4 + \left(\frac{1 \cdot 3 \cdot 5}{2 \cdot 4 \cdot 6}\right)^2 x^6 + \cdots$$

or

$$\frac{1}{M(1+x, 1-x)} = 1 + \sum_{k=1}^{\infty} \left[\frac{(2k-1)!!}{(2k)!!}\right]^2 \cdot x^{2k}. \quad (2.39)$$

Now we will expand the integral $K(x)$ in a series. For this, we will expand the integrand of $K(x)$ using the binomial series and integrate the resulting series term by term.

$$\left(1 - x^2 \sin^2 t\right)^{-\frac{1}{2}} = 1 + \left(-\frac{1}{2}\right)\left(-x^2 \sin^2 t\right) + \frac{\left(-\frac{1}{2}\right)\left(-\frac{1}{2}-1\right)}{2!} x^4 \sin^4 t + \cdots +$$

$$+ \frac{\left(-\frac{1}{2}\right)\left(-\frac{1}{2}-1\right)\cdots\left(-\frac{1}{2}-1-k\right)}{k!} (-1)^k x^{2k} \sin^{2k} t + \cdots =$$

$$= 1 + \sum_{k=1}^{\infty} (-1)^k \left(\frac{1}{2}\right)^k \frac{1 \cdot 3 \cdots (2k-1)}{k!} (-1)^k x^{2k} \sin^{2k} t =$$

$$= 1 + \sum_{k=1}^{\infty} \frac{1 \cdot 3 \cdots (2k-1)}{2^k k!} x^{2k} \sin^{2k} t.$$

So, by integrating the series above:

$$K(x) = \int_0^{\pi/2} \left[1 + \sum_{k=1}^{\infty} \frac{(2k-1)!!}{2^k k!} x^{2k} \sin^{2k} t\right] dt =$$

$$= \frac{\pi}{2} + \int_0^{\pi/2} \sum_{k=1}^{\infty} \frac{(2k-1)!!}{2^k k!} x^{2k} \sin^{2k} t\, dt$$

or

$$K(x) = \frac{\pi}{2} + \sum_{k=1}^{\infty} \frac{(2k-1)!!}{2^k k!} x^{2k} \int_0^{\pi/2} \sin^{2k} t\, dt. \qquad (2.40)$$

Let's calculate now

$$I_k = \int_0^{\pi/2} \sin^{2k} t\, dt = \int_0^{\pi/2} (-\cos t)' \sin^{2k-1} t\, dt =$$

$$= -\cos t \sin^{2k-1} t \Big|_0^{\pi/2} + (2k-1) \int_0^{\pi/2} \cos^2 t \sin^{2k-2} t\, dt =$$

$$= (2k-1)(I_{k-1} - I_k).$$

2.2 Nonlinear Recurrences

Hence: $I_k = \frac{2k-1}{2k} I_{k-1}$ and $I_0 = \frac{\pi}{2}$. It follows that

$$I_k = \frac{(2k-1)}{(2k)} \frac{\pi}{2}.$$

Therefore, returning to (2.40):

$$K(x) = \frac{\pi}{2} + \sum_{k=1}^{\infty} \frac{(2k-1)!!}{2^k k!} x^{2k} \frac{(2k-1)!! \pi}{(2k)!!} \frac{\pi}{2} = \frac{\pi}{2} + \sum_{k=1}^{\infty} \frac{\pi}{2} \frac{[(2k-1)!!]^2}{[2k!!]^2} x^{2k}$$

Hence, from (2.39)

$$K(x) = \frac{\pi}{2} \cdot \frac{1}{M(1+x, 1-x)}$$

or

$$M(1+x, 1-x) = \frac{\pi}{2K(x)}.$$

Legendre Demonstration Once we know the integral representation of the arithmetic-geometric mean, we have several options available to verify that the double algorithm converges to this limit.

The following proof was obtained by Adrien-Marie Legendre in 1825, using an idea from the English mathematician John Landen (1719–1790).

Let $(a_n)_n$ and $(b_n)_n$ be the sequences of arithmetic means and geometric means, respectively, that converge to the common limit $M(a, b)$. Let's denote, for every $n \in \mathbb{N}$, $x_n = \frac{\sqrt{a_n^2 - b_n^2}}{a_n}$; then $x_n \to 0$. In addition, let $x_0 = \frac{\sqrt{a^2 - b^2}}{a} = x$ and, for $n \in \mathbb{N}$, let:

$$x_{n+1} = \frac{\sqrt{a_{n+1}^2 - b_{n+1}^2}}{a_{n+1}} = \frac{\sqrt{\left(\frac{a_n+b_n}{2}\right)^2 - a_n b_n}}{\frac{a_n+b_n}{2}} = \frac{a_n - b_n}{a_n + b_n} \quad (2.41)$$

In the complete elliptic integral of the first kind, $K(x) = \int_0^{\frac{\pi}{2}} \frac{dt}{\sqrt{1 - x^2 \sin^2 t}}$, we make the substitution:

$$\tan s = \frac{\sin 2t}{x_1 + \cos 2t} \Leftrightarrow x_1 \sin s = \sin(2t - s), \quad (2.42)$$

substitution known as the **Landen transformation** (here $x_1 = \frac{\sqrt{a_1^2 - b_1^2}}{a_1} = \frac{a-b}{a+b}$).

By differentiating Eq. (2.42), we obtain:

$$\frac{ds}{\cos^2 s} = \frac{2(1 + x_1 \cos 2t)}{(x_1 + \cos 2t)^2} dt. \qquad (2.43)$$

Let's calculate it now $\dfrac{ds}{\sqrt{1 - x_1^2 \sin^2 s}}$:

$$\frac{ds}{\sqrt{1 - x_1^2 \sin^2 s}} = \frac{\cos^2 s}{\sqrt{1 - x_1^2 \sin^2 s}} \cdot \frac{ds}{\cos^2 s} =$$

$$= \frac{1}{\sqrt{1 + \tan^2 s} \cdot \sqrt{1 + (1 - x_1^2) \tan^2 s}} \cdot \frac{2(1 + x_1 \cos 2t)}{(x_1 + \cos 2t)^2} dt =$$

$$= \frac{1}{\sqrt{1 + \frac{\sin^2 2t}{(x_1 + \cos 2t)^2}} \cdot \sqrt{1 + (1 - x_1^2)\frac{\sin^2 2t}{(x_1 + \cos 2t)^2}}} \cdot \frac{2(1 + x_1 \cos 2t)}{(x_1 + \cos 2t)^2} dt =$$

$$= \frac{2dt}{\sqrt{(x_1 + 1)^2 - 2x_1 \cdot 2\sin^2 t}} = \frac{2dt}{(x_1 + 1) \cdot \sqrt{1 - \frac{4x_1}{(x_1+1)^2} \sin^2 t}}.$$

From (2.41) we obtain

$$\frac{4x_1}{(x_1 + 1)^2} = \frac{4 \cdot \frac{a-b}{a+b}}{\left(\frac{a-b}{a+b} + 1\right)^2} = \frac{a^2 - b^2}{a^2} = x^2.$$

By substituting in the above equality, we obtain

$$\frac{ds}{\sqrt{1 - x_1^2 \sin^2 s}} = \frac{2dt}{(x_1 + 1) \cdot \sqrt{1 - x^2 \sin^2 t}}$$

or

$$\frac{dt}{\sqrt{1 - x^2 \sin^2 t}} = \frac{x_1 + 1}{2} \cdot \frac{ds}{\sqrt{1 - x_1^2 \sin^2 s}}. \qquad (2.44)$$

2.2 Nonlinear Recurrences

Since $t \in \left[0, \frac{\pi}{2}\right]$, $2t \in [0, \pi]$, and therefore there exists a unique $t_0 \in \left[0, \frac{\pi}{2}\right]$ such that $\cos(2t_0) = -x_1$. Then:

$$K(x) = \int_0^{\frac{\pi}{2}} \frac{dt}{\sqrt{1-x^2\sin^2 t}} = \int_0^{t_0} \frac{dt}{\sqrt{1-x^2\sin^2 t}} + \int_{t_0}^{\frac{\pi}{2}} \frac{dt}{\sqrt{1-x^2\sin^2 t}}$$

We observe that as t approaches t_0 from below ($t \uparrow t_0$), $\cos(2t) \downarrow \cos(2t_0) = -x_1$ and then $\tan s = \frac{\sin(2t)}{x_1+\cos(2t)} \to +\infty$ so that $s \uparrow \frac{\pi}{2}$, and when t approaches t_0 from above ($t \downarrow t_0$), $\cos(2t) \uparrow \cos(2t_0) = -x_1$ so that $\tan s = \frac{\sin(2t)}{x_1+\cos(2t)} \to -\infty$ and then $s \downarrow -\frac{\pi}{2}$.

Therefore

$$K(x) = \int_0^{\frac{\pi}{2}} \frac{1+x_1}{2} \frac{ds}{\sqrt{1-x_1^2\sin^2 s}} + \int_{-\frac{\pi}{2}}^0 \frac{1+x_1}{2} \frac{ds}{\sqrt{1-x_1^2\sin^2 s}} =$$

$$= \frac{1+x_1}{2} \cdot \int_{-\frac{\pi}{2}}^{\frac{\pi}{2}} \frac{ds}{\sqrt{1-x_1^2\sin^2 s}} = (x_1+1) \cdot K(x_1).$$

After n iterations, we deduce that:

$$K(x) = (1+x_1)(1+x_2) \cdots (1+x_n) K(x_n). \qquad (2.45)$$

Using the relationship (2.44) again, we have $1 + x_k = \frac{2a_{k-1}}{a_{k-1}+b_{k-1}} = \frac{a_{k-1}}{a_k}$, and then relationship (2.45) can be written as

$$K(x) = \frac{a}{a_n} \cdot K(x_n), \text{ for every } n \in \mathbb{N}. \qquad (2.46)$$

Taking the limit in relationship (2.46), we obtain,

$$K(x) = \frac{a}{M(a,b)} \cdot \frac{\pi}{2}. \qquad (2.47)$$

On the other hand, since M is an invariant mean with respect to A and G,

$$M(1+x, 1-x) = M\left(\frac{1+x+1-x}{2}, \sqrt{(1+x)(1-x)}\right) =$$

$$= M\left(1, \sqrt{1-x^2}\right) = M\left(1, \frac{b}{a}\right) = M\left(\frac{a}{a}, \frac{b}{a}\right) = \frac{1}{a}M(a,b),$$

which, based on the equality (2.47), represents the conclusion of theorem. $\square$

Remark 2.2.22 The Landen transformation was introduced in 1771. There are multiple versions of the Landen transformation. Most often, the Landen transformation is expressed through an equality between two differentials in the form of relationship (2.44) from the previous proof.

The result from Theorem 2.2.21 allows obtaining $M(a, b)$ for any positive numbers a and b.

Corollary 2.2.23

$$M(a,b) = \frac{\pi}{2 \cdot \int_0^{\frac{\pi}{2}} \frac{dt}{\sqrt{a^2 \sin^2 t + b^2 \cos^2 t}}}, \text{ for every } a, b > 0.$$

Proof Let

$$I(a,b) = \int_0^{\frac{\pi}{2}} \left(a^2 \sin^2 t + b^2 \cos^2 t\right)^{-\frac{1}{2}} dt = \int_0^{\frac{\pi}{2}} \left(a^2 \cos^2 t + b^2 \sin^2 t\right)^{-\frac{1}{2}} dt =$$

$$= \int_0^{\frac{\pi}{2}} \left[a^2 - (a^2 - b^2) \sin^2 t\right]^{-\frac{1}{2}} dt = \frac{1}{a} \int_0^{\frac{\pi}{2}} \left[1 - \frac{a^2 - b^2}{a^2} \sin^2 t\right]^{-\frac{1}{2}} dt =$$

$$= \frac{1}{a} \int_0^{\frac{\pi}{2}} \left[1 - x^2 \sin^2 t\right]^{-\frac{1}{2}} dt = \frac{1}{a} K(x), \text{ where } x = \frac{\sqrt{a^2 - b^2}}{a}.$$

The equality in the statement is a consequence of relationship (2.47) in the proof of the preceding theorem. □

2.2.7 The Perimeter of the Lemniscate and the Ellipse

In this subsection, we will introduce elliptic integral of the second kind and we prove Legendre's formula that connects the two types of integrals. The results we establish will be used to derive formulas for calculating the perimeter of Bernoulli's lemniscate and the ellipse.

Definition 2.2.24 The integral

$$E(x) = \int_0^{\frac{\pi}{2}} \sqrt{1 - x^2 \sin^2 t} \, dt, \, |x| < 1$$

is called the **complete elliptic integral of the second kind**.

2.2 Nonlinear Recurrences

We would like to remind you that the complete elliptic integral of the first kind is given by:

$$K(x) = \int_0^{\frac{\pi}{2}} \frac{dt}{\sqrt{1 - x^2 \sin^2 t}}, \ |x| < 1.$$

In the following, we present a formula that connects the two types of elliptic integrals, known as **Legendre's formula**.

Theorem 2.2.25 *Let $x \in]0, 1[$ and let $x' = \sqrt{1 - x^2}$; then:*

$$K(x) E(x') + K(x') E(x) = K(x) K(x') + \frac{\pi}{2}. \tag{L}$$

Proof Let $c = x^2$ and $c' = 1 - c$; then the formula (L) is rewritten as follows:

$$k(c) \cdot e(1 - c) + k(1 - c) \cdot e(c) = k(c) \cdot k(1 - c) + \frac{\pi}{2}, 0 < c < 1, \tag{l}$$

where $k(c) = \int_0^{\frac{\pi}{2}} \frac{dt}{\sqrt{1 - c \sin^2 t}}$, and $e(c) = \int_0^{\frac{\pi}{2}} \sqrt{1 - c \sin^2 t} \, dt$.

Let the function $l :]0, 1[\to \mathbb{R}$ be defined by:

$$l(c) = e(c) \cdot k(1 - c) + e(1 - c) \cdot k(c) - k(c) \cdot k(1 - c).$$

l is a differentiable function on $]0, 1[$; let's calculate its derivative.

$$e'(c) = \int_0^{\frac{\pi}{2}} \frac{-\sin^2 t}{2\sqrt{1 - c \sin^2 t}} dt = \frac{1}{2c} \int_0^{\frac{\pi}{2}} \frac{1 - c \sin^2 t - 1}{\sqrt{1 - c \sin^2 t}} dt,$$

from where

$$e'(c) = \frac{1}{2c} [e(c) - k(c)]. \tag{2.48}$$

$$k'(c) = -\frac{1}{2c} k(c) + \frac{1}{2c} \int_0^{\frac{\pi}{2}} \frac{dt}{(1 - c \sin^2 t)^{\frac{3}{2}}}. \tag{2.49}$$

To evaluate the second term in Eq. (2.49), we observe that:

$$\left(\frac{\sin t \cos t}{\sqrt{1 - c \sin^2 t}} \right)'_t = \frac{c - 1}{c} \cdot \frac{1}{(1 - c \sin^2 t)^{\frac{3}{2}}} + \frac{1}{c} \cdot \sqrt{1 - c \sin^2 t}. \tag{2.50}$$

If we integrate Eq. (2.50) with respect to t over the interval $[0, \frac{\pi}{2}]$, we obtain:

$$0 = \frac{c-1}{c} \cdot \int_0^{\frac{\pi}{2}} \frac{dt}{(1-c\sin^2 t)^{\frac{3}{2}}} + \frac{1}{c} \cdot e(c)$$

and thus, returning to Eq. (2.49), we have:

$$k'(c) = \frac{1}{2cc'}\left[e(c) - c'k(c)\right], c' = 1-c. \tag{2.51}$$

Clearly, $[e(1-c)]' = -e'(1-c)$ and $[k(1-c)]' = -k'(1-c)$; using formulas (2.48) and (2.51), we obtain, for every $c \in {]}0, 1{[}$,

$$l'(c) = \frac{1}{2c} \cdot e(c) \cdot k(1-c) - \frac{1}{2c} \cdot k(c) \cdot k(1-c) - \frac{1}{2cc'} \cdot e(c) \cdot e(1-c) + \frac{1}{2c'} \cdot e(c) \cdot k(1-c) -$$

$$-\frac{1}{2c'} \cdot e(1-c) \cdot k(c) + \frac{1}{2c'} \cdot k(c) \cdot k(1-c) + \frac{1}{2cc'} \cdot e(c) \cdot e(1-c) - \frac{1}{2c} \cdot k(c) \cdot e(1-c) -$$

$$-\frac{1}{2cc'} \cdot e(c) \cdot k(1-c) + \frac{1}{2c} \cdot k(c) \cdot k(1-c) + \frac{1}{2cc'} \cdot e(1-c) \cdot k(c) - \frac{1}{2c'} \cdot k(c) \cdot k(1-c) =$$

$$= \left(\frac{1}{2c} + \frac{1}{2c'} - \frac{1}{2cc'}\right) \cdot e(c) \cdot k(1-c) + \left(-\frac{1}{2c} - \frac{1}{2c'} + \frac{1}{2cc'}\right) \cdot e(1-c) \cdot k(c) = 0.$$

Since the derivative of the function l is zero on the interval ${]}0, 1{[}$, the function is constant on this interval. Let $l_0 \in \mathbb{R}$ be such that:

$$l(c) = l_0, \text{ for every } c \in {]}0, 1{[}. \tag{2.52}$$

On the other hand

$$l(c) = \underbrace{(e(c) - k(c)) \cdot k(1-c)}_{l_1(c)} + \underbrace{e(1-c) \cdot k(c)}_{l_2(c)}.$$

$$|l_1(c)| = c \cdot \int_0^{\frac{\pi}{2}} \frac{\sin^2 t}{\sqrt{1-c\sin^2 t}} dt \cdot \int_0^{\frac{\pi}{2}} \frac{dt}{\sqrt{\cos^2 t + c\sin^2 t}} dt \leq$$

$$\leq \int_0^{\frac{\pi}{2}} \frac{dt}{\sqrt{c\cos^2 t + c\sin^2 t}} dt$$

2.2 Nonlinear Recurrences

or

$$|l_1(c)| \leqslant \sqrt{c} \cdot \frac{\pi}{2} \cdot \int_0^{\frac{\pi}{2}} \frac{\sin^2 t}{\sqrt{1 - c \sin^2 t}} dt \text{ and then } \lim_{c \to 0} l_1(c) = 0.$$

Since

$$\lim_{c \to 0} l_2(c) = \int_0^{\frac{\pi}{2}} \sqrt{\cos^2 t} \, dt \cdot \frac{\pi}{2} = \frac{\pi}{2},$$

it follows that

$$\lim_{c \to 0} l(c) = \frac{\pi}{2}.$$

But, from (2.52),

$$\lim_{c \to 0} l(c) = l_0$$

and so $l_0 = \frac{\pi}{2}$. It follows that

$$l(c) = \frac{\pi}{2}, \text{ for every } c \in]0, 1[$$

which concludes the proof. □

Remark 2.2.26 Let us note that if $x = \frac{1}{\sqrt{2}}$, then $x = x'$, and thus Legendre's formula can be written in a simplified form as follows:

$$2K\left(\frac{1}{\sqrt{2}}\right) \cdot E\left(\frac{1}{\sqrt{2}}\right) - K^2\left(\frac{1}{\sqrt{2}}\right) = \frac{\pi}{2}.$$

We will use this reduced form of Legendre's formula to prove an interesting integral formula.

Corollary 2.2.27

$$K\left(\frac{1}{\sqrt{2}}\right) = \sqrt{2} \int_0^1 \frac{1}{\sqrt{1-x^4}} dx, \quad E\left(\frac{1}{\sqrt{2}}\right) = \frac{1}{\sqrt{2}} \int_0^1 \frac{1+x^2}{\sqrt{1-x^4}} dx,$$

and thus, the relationship from the previous remark becomes:

$$\int_0^1 \frac{1}{\sqrt{1-x^4}} dx \cdot \int_0^1 \frac{x^2}{\sqrt{1-x^4}} dx = \frac{\pi}{4}.$$

Proof In integrals

$$K\left(\frac{1}{\sqrt{2}}\right) = \sqrt{2}\int_0^{\frac{\pi}{2}} \frac{dt}{\sqrt{2-\sin^2 t}} \text{ and } E\left(\frac{1}{\sqrt{2}}\right) = \frac{1}{\sqrt{2}}\int_0^{\frac{\pi}{2}} \sqrt{2-\sin^2 t}\,dt,$$

we make the change of variable $\cos t = x$ and obtain:

$$K\left(\frac{1}{\sqrt{2}}\right) = \sqrt{2}\int_0^1 \frac{1}{\sqrt{1-x^4}}dx \text{ and respectively}$$
$$E\left(\frac{1}{\sqrt{2}}\right) = \frac{1}{\sqrt{2}}\int_0^1 \frac{1+x^2}{\sqrt{1-x^4}}dx.$$

Now we substitute these values into the reduced form of Legendre's formula (see Remark 2.2.26) and obtain

$$2\int_0^1 \frac{1}{\sqrt{1-x^4}}dx \cdot \int_0^1 \frac{1+x^2}{\sqrt{1-x^4}}dx - 2\left(\int_0^1 \frac{1}{\sqrt{1-x^4}}dx\right)^2 = \frac{\pi}{2},$$

from which the desired relationship follows immediately. □

Remarks 2.2.28

(i) The formula in the previous corollary was first proved by Euler in 1782. Equivalent formulations were established earlier by Landen and Wallis.

(ii) If we make the change of variable $1-x^4 \mapsto x^2$ in the integral $\int_0^1 \frac{x^2}{\sqrt{1-x^4}}dx$, we obtain:

$$\int_0^1 \frac{x^2}{\sqrt{1-x^4}}dx = \frac{1}{2}\int_0^1 \frac{1}{\sqrt[4]{1-x^2}}dx$$

and thus the above formula reduces to:

$$\int_0^1 \frac{1}{\sqrt{1-x^4}}dx \cdot \int_0^1 \frac{1}{\sqrt[4]{1-x^2}}dx = \frac{\pi}{2}.$$

The elliptic functions and their corresponding integrals owe their names to the problem of rectifying the arc of an ellipse. However, the truly fruitful problem was the rectification of the arc of a lemniscate. In 1694, Jacob Bernoulli published an article in Acta Eruditorum about a curve "having the shape of the numeral eight". Bernoulli named this curve lemniscate (from Latin lemniscus, meaning ribbon).

The general properties of the lemniscate were studied by Giulio Fagnano (1682–1766). Fagnano demonstrated how to double the arc of a lemniscate using ruler and compass. Euler also worked on this problem, establishing a general result for adding lemniscate arcs.

2.2 Nonlinear Recurrences

After nearly half a century since Euler's discovery of the addition theorem for lemniscate arcs, Gauss reversed the problem, considering the radial coordinate r as a function of the arc length. Assuming that r changes sign when it passes through zero, the resulting function is qualitatively similar to the sine function and is called the lemniscatic sine. It is periodic with a period equal to the perimeter of the lemniscate. Gauss extended this function to the complex plane, obtaining the first elliptic function (doubly periodic and meromorphic). The real period of this function can be expressed using the complete elliptic integrals of the first kind and can be approximated using Gauss's arithmetic-geometric mean algorithm with the starting values 1 and $\sqrt{2}$. Later, Gauss showed that the arithmetic-geometric mean is generally closely related to the periods of elliptic functions.

The **lemniscate** can be defined as the geometric locus of points in the plane for which the product of the distances to two fixed points is constant. Bernoulli considered the fixed points in the plane as $F_1(-\frac{1}{\sqrt{2}}, 0)$ and $F_2(\frac{1}{\sqrt{2}}, 0)$, and the constant product as $\frac{1}{2}$. It is easy then to obtain the equation that the coordinates (x, y) of a generic point on the lemniscate must satisfy

$$(x^2 + y^2)^2 = x^2 - y^2.$$

If we switch to polar coordinates:

$$\begin{cases} x = r \cos t \\ y = r \sin t \end{cases}, r \geqslant 0, t \in [0, 2\pi[,$$

we obtain the equation in polar coordinates as

$$r^2 = \cos 2t.$$

This curve is symmetric with respect to the coordinate axes, so we will only study its behavior in the first quadrant, restricting the range of the argument t to the interval $[0, \frac{\pi}{2}]$. Thus, from the polar equation of the lemniscate, we obtain

$$\begin{cases} \cos t = \sqrt{\dfrac{1+r^2}{2}} \\ \sin t = \sqrt{\dfrac{1-r^2}{2}} \end{cases}, r \in [0, 1].$$

Returning to the equations in polar coordinates, we obtain the parametric equations of the lemniscate as:

$$\begin{cases} x = \dfrac{1}{\sqrt{2}}\sqrt{r^2 + r^4} \\ y = \dfrac{1}{\sqrt{2}}\sqrt{r^2 - r^4} \end{cases}, r \in [0, 1].$$

The functions x and y are differentiable and

$$\begin{cases} x'_r = \dfrac{1+2r^2}{\sqrt{2}\sqrt{1+r^2}} \\ y'_r = \dfrac{1-2r^2}{\sqrt{2}\sqrt{1-r^2}} \end{cases}.$$

Therefore

$$y'_x = \frac{y'_r}{x'_r} = \frac{(1-2r^2)\sqrt{1+r^2}}{(1+2r^2)\sqrt{1-r^2}}, r \in [0, 1[.$$

We present below a table of variations and a graph of the lemniscate in the first quadrant.

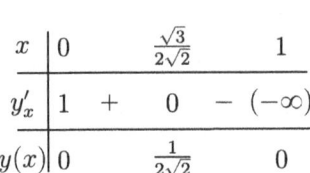
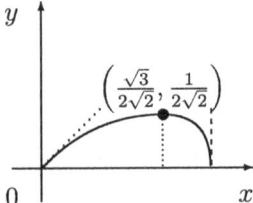

Using the symmetry of the curve with respect to the axes, we can easily obtain the complete graph of Bernoulli's lemniscate.

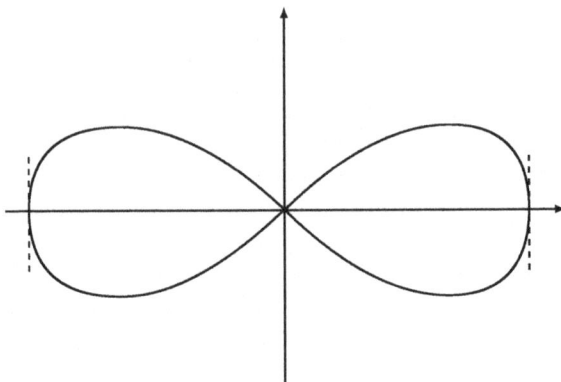

The length of this curve is:

$$L = 4 \cdot \int_0^1 \sqrt{(x'_r)^2 + (y'_r)^2}\, dr = 4 \cdot \int_0^1 \frac{1}{\sqrt{1-r^4}}\, dr.$$

2.2 Nonlinear Recurrences

We have observed in Corollary 2.2.27 that

$$\int_0^1 \frac{1}{\sqrt{1-r^4}}\, dr = \frac{1}{\sqrt{2}} \cdot K\left(\frac{1}{\sqrt{2}}\right)$$

and so

$$L = 2\sqrt{2} \cdot K\left(\frac{1}{\sqrt{2}}\right).$$

From Gauss's theorem (Theorem 2.2.21), we obtain:

$$K\left(\frac{1}{\sqrt{2}}\right) = \frac{\pi}{2 \cdot M\left(1 + \frac{1}{\sqrt{2}}, 1 - \frac{1}{\sqrt{2}}\right)} = \frac{\pi}{2 \cdot M\left(1, \frac{1}{\sqrt{2}}\right)} = \frac{\pi}{\sqrt{2} \cdot M(\sqrt{2}, 1)}$$

and thus the length of the lemniscate will be:

$$L = \frac{2\pi}{M(\sqrt{2}, 1)}.$$

Let $(a_n)_n$ and $(b_n)_n$ be the sequences of the arithmetic-geometric mean algorithm, where $a_0 = \sqrt{2}$ and $b_0 = 1$. Then $b_n \uparrow M(\sqrt{2}, 1)$ and $a_n \downarrow M(\sqrt{2}, 1)$, and thus

$$\frac{2\pi}{a_n} \uparrow L \downarrow \frac{2\pi}{b_n}.$$

We can use the arithmetic-geometric mean algorithm $(A - G)$ for the approximate calculation of the length of the lemniscate. We will provide a formula for estimating the error in this calculation:

$$L - \frac{2\pi}{a_n} < \frac{2\pi}{b_n} - \frac{2\pi}{a_n} < 2\pi(a_n - b_n)$$

and according to relationship $(*)$ in Proposition 2.2.19,

$$L - \frac{2\pi}{a_n} < 16\pi \cdot \left(\frac{\sqrt{2}-1}{8}\right)^{2^n}, \text{ for every } n \in \mathbb{N}.$$

If, for example, we want to calculate the length of the lemniscate with three decimal places of accuracy, we solve the inequality:

$$16\pi \cdot \left(\frac{\sqrt{2}-1}{8}\right)^{2^n} < 10^{-3}.$$

For $n = 2$ we obtain $16\pi \cdot \left(\frac{\sqrt{2}-1}{8}\right)^{2^2} = 0.000361\ldots < 10^{-3}$. It follows that $\frac{2\pi}{a_2}$ will give the first three decimal places of accuracy for the length of Bernoulli's lemniscate. We obtain:

$a_1 = 1.207106781186547\ldots$
$b_1 = 1.189207115002721\ldots$
$a_2 = 1.198156948094634\ldots$
$b_2 = 1.198123521493120\ldots$
$a_3 = 1.198140234793877\ldots$
$b_3 = 1.198140234677307\ldots$

and then

$\dfrac{2\pi}{a_0} = 4.442882938158366\ldots$
$\dfrac{2\pi}{b_0} = 6.283185307179586\ldots$
$\dfrac{2\pi}{a_1} = 5.205161138274292\ldots$
$\dfrac{2\pi}{b_1} = 5.283508001182123\ldots$
$\dfrac{2\pi}{a_2} = 5.244041957250595\ldots$
$\dfrac{2\pi}{b_2} = 5.244188261448521\ldots$
$\dfrac{2\pi}{a_3} = 5.244115108329133\ldots$
$\dfrac{2\pi}{b_3} = 5.244115108839346\ldots$

We observe that 5.244 is an approximation with three decimal places of accuracy for the length of the lemniscate. Moreover, it follows that for $n = 3$, an approximation of the length of the lemniscate with 9 decimal places of accuracy is obtained. This rapid increase in accuracy is due to the convergence speed of the arithmetic-geometric mean algorithm.

Since the time of Kepler and Euler, there have been attempts to find a formula for calculating the length of an ellipse. In 1602, Kepler stated that the orbit of Mars is an oval. Later, he discovered that it was actually an ellipse with the Sun at one of its foci (Kepler was the one who introduced the word "focus" in 1609). Therefore, the initial motivation for finding an approximation for the perimeter of an ellipse was the desire to accurately calculate the elliptical orbits of planets.

There is no formula for the length of an ellipse that uses elementary functions. While in the case of the lemniscate, the length is expressed using an elliptic integral of the first kind, the length of an ellipse is expressed using an elliptic integral of the second kind. An ellipse can be characterized as the geometric locus of points in a plane for which the sum of the distances to two fixed points is constant.

The implicit equation of an ellipse is:

$$\frac{x^2}{a^2} + \frac{y^2}{b^2} - 1 = 0.$$

An ellipse can also be described by parametric equations:

$$\begin{cases} x = a \cos t, \\ y = b \sin t \end{cases}, t \in [0, 2\pi],$$

2.2 Nonlinear Recurrences

and then, through elementary calculations, we obtain that the perimeter of an ellipse, L, is given by:

$$L = L(a, b) = \int_0^{2\pi} \sqrt{[x'(t)]^2 + [y'(t)]^2}\, dt = \int_0^{2\pi} \sqrt{a^2 \sin^2 t + b^2 \cos^2 t}\, dt =$$

$$= 4 \cdot \int_0^{\frac{\pi}{2}} \sqrt{a^2 \sin^2 t + b^2 \cos^2 t}\, dt = 4 \cdot \int_0^{\frac{\pi}{2}} \sqrt{a^2 \cos^2 t + b^2 \sin^2 t}\, dt =$$

$$= 4a \cdot \int_0^{\frac{\pi}{2}} \sqrt{1 - \frac{a^2 - b^2}{a^2} \sin^2 t}\, dt = 4a \cdot E(x), \text{ where } x^2 = \frac{a^2 - b^2}{a^2}.$$

In the above equality, we assumed that $a \geqslant b > 0$ and $E(x)$ represents the complete elliptic integral of the second kind (see Definition 2.2.24).

There are numerous approximations for L in the mathematical literature, which depend on the values of a and b and become more accurate as the eccentricity of the ellipse, $x = \frac{\sqrt{a^2 - b^2}}{a}$, becomes smaller.

In the following, we will present a method for approximating the perimeter of ellipses for which the semi-axes are related by the equation: $a = \sqrt{2} \cdot b$. In this case $x = \frac{1}{\sqrt{2}}$ and the perimeter is given by:

$$L = 4a \cdot E\left(\frac{1}{\sqrt{2}}\right).$$

Using the reduced form of Legendre's formula (see Remark 2.2.26), we obtain:

$$L = 2a \cdot K\left(\frac{1}{\sqrt{2}}\right) + \frac{\pi a}{K\left(\frac{1}{\sqrt{2}}\right)}.$$

Since

$$K\left(\frac{1}{\sqrt{2}}\right) = \frac{\pi}{\sqrt{2} \cdot M(\sqrt{2}, 1)},$$

$$L = \sqrt{2}a \cdot \left[\frac{\pi}{M(\sqrt{2}, 1)} + M(\sqrt{2}, 1)\right].$$

Let's assume, for simplicity, that $a = \sqrt{2}$; using once again the sequences of the arithmetic-geometric mean algorithm $(a_n)_{n \in \mathbb{N}}$ and $(b_n)_{n \in \mathbb{N}}$ for the starting points $a_0 = \sqrt{2}$ and $b_0 = 1$. Then

$$2\left(b_n + \frac{\pi}{a_n}\right) \uparrow L \downarrow 2\left(a_n + \frac{\pi}{b_n}\right)$$

and the we obtain

$$L - 2\left(b_n + \frac{\pi}{a_n}\right) < 16(1 + \pi)\left(\frac{\sqrt{2}-1}{8}\right)^{2n}, \text{ for every } n \in \mathbb{N}.$$

For $n = 2$, we obtain the perimeter of the ellipse with three decimal places exact: $L = 7.640\ldots$. For $n = 3$, we determine the first 8 decimal places exact: $L = 7.64039557\ldots$.

2.3 Exercises

(1) Find the general solution of the non-homogeneous linear recurrence

$$f(n+3) - 3f(n+2) + 3f(n+1) - f(n) = 6.$$

Indications: The characteristic equation of the associated homogeneous equation is $\lambda^3 - 3\lambda^2 + 3\lambda - 1 = 0$ and it has a triple root $\lambda_{1,2,3} = 1$. Using the notations from Theorem 2.1.16 $q = 0$ and $s = 3$, thus, we will look for a particular solution of the non-homogeneous equation in the form $f^*(n) = an^3 + bn^2 + cn + d$. In the end, we obtain $f^*(n) = n^3$. Therefore, the general solution of the non-homogeneous linear recurrence is $f(n) = n^3 + c_1 n^2 + c_2 n + c_3$.

(2) Find the general solution of the non-homogeneous linear recurrence

$$f(n+2) + f(n+1) + f(n) = n^2 - \frac{1}{3}.$$

Indications: The characteristic equation of the associated homogeneous equation is $\lambda^2 + \lambda + 1 = 0$ and it has the roots $\lambda_{1,2} = \cos\frac{2\pi}{3} \pm i \sin\frac{2\pi}{3}$. We are looking for a particular solution of the non-homogeneous equation in the form $f^*(n) = R(n)e^{0 \cdot n}$, where R is a polynomial of degree 2 (see Theorem 2.1.16). We obtain $f^*(n) = \frac{1}{3}(n^2 - 2n)$. Therefore, the general solution of the non-homogeneous linear recurrence is $f(n) = \frac{1}{3}(n^2 - 2n) + c_1 \cos\frac{2n\pi}{3} + c_2 \sin\frac{2n\pi}{3}$.

(3) Find the general solution of the non-homogeneous linear recurrence

$$f(n+3) + f(n+2) + f(n+1) + f(n) = \frac{15n - 4}{2^n}.$$

Indications: The characteristic equation of the associated homogeneous equation is $\lambda^3 + \lambda^2 + \lambda + 1 = 0$ and it has the roots $\lambda_{1,2} = \pm i$, $\lambda_3 = -1$. We are looking for a particular solution of the non-homogeneous equation in the form $f^*(n) = R(n)\frac{1}{2^n}$, where R is a polynomial of degree 1 ($\frac{1}{2^n} = e^{(-\log 2)n}$ and $e^{(-\log 2)} = \frac{1}{2}$ it is not a root of the characteristic polynomial—see Theorem 2.1.16). We obtain $f^*(n) = (8n - 8)\frac{1}{2^n}$. Therefore, the general solution of the non-homogeneous linear recurrence is $f(n) = (8n - 8)\frac{1}{2^n} + c_1(-1)^n + c_2 \cos\frac{n\pi}{2} + c_3 \sin\frac{n\pi}{2}$.

(4) Let $(F(n))_{n \in \mathbb{N}}$ be the sequence of Fibonacci numbers and let $\phi = \frac{1+\sqrt{5}}{2}$ be the golden ratio.

2.3 Exercises   125

(a) Show that, for every $n \in \mathbb{N}$, $\begin{bmatrix} F(n+1) & F(n) \\ F(n) & F(n-1) \end{bmatrix} = \begin{bmatrix} 1 & 1 \\ 1 & 0 \end{bmatrix}^n$.

(b) Prove Cassini's identity: $F(n-1)F(n+1) - F(n)^2 = (-1)^n$, $\forall n \in \mathbb{N}$.

(c) Show that ϕ has the following expression as a continued fraction:

$$\phi = 1 + \cfrac{1}{1 + \cfrac{1}{1 + \cfrac{1}{1 + \cdots}}}$$

Indications: (a) Proof by induction. The base case $n = 0$ is true. Assume the statement is true for n.
Then $\begin{bmatrix} 1 & 1 \\ 1 & 0 \end{bmatrix}^{n+1} = \begin{bmatrix} 1 & 1 \\ 1 & 0 \end{bmatrix}^n \cdot \begin{bmatrix} 1 & 1 \\ 1 & 0 \end{bmatrix} = \begin{bmatrix} F(n+1) & F(n) \\ F(n) & F(n-1) \end{bmatrix} \cdot \begin{bmatrix} 1 & 1 \\ 1 & 0 \end{bmatrix} = = \begin{bmatrix} F(n+1) + F(n) & F(n+1) \\ F(n) + F(n-1) & F(n) \end{bmatrix} = \begin{bmatrix} F(n+2) & F(n+1) \\ F(n+1) & F(n) \end{bmatrix}$.

(b) $F(n-1)F(n+1) - F(n)^2 = \begin{vmatrix} F(n+1) & F(n) \\ F(n) & F(n-1) \end{vmatrix} = \det\begin{bmatrix} F(n+1) & F(n) \\ F(n) & F(n-1) \end{bmatrix} = = \det\begin{bmatrix} 1 & 1 \\ 1 & 0 \end{bmatrix}^n = (-1)^n$.

(c) It will be observed that $\phi = 1 + \frac{1}{\phi}$.

(5) Consider the nonlinear recurrence $x_{n+1} = \frac{2}{3} \cdot \left(x_n + \frac{1}{x_n^2}\right)$, $x_0 = 1$. Show that

(a) $\sqrt[3]{2} < x_{n+1} < x_n$, for every $n \in \mathbb{N}^*$.
(b) $\lim_{n \to \infty} x_n = \sqrt[3]{2}$.
(c) $x_{n+1} - \sqrt[3]{2} \leq 2(x_n - \sqrt[3]{2})^2$, for every $n \in \mathbb{N}^*$.
(d) $x_{n+1} - \sqrt[3]{2} < 2 \cdot \left(\frac{1}{10}\right)^{2^n}$, for every $n \in \mathbb{N}^*$.

Indication: It will be taken into account that $2x^3 - 3\sqrt[3]{2}x^2 + 2 = (x - \sqrt[3]{2})^2(2x + \sqrt[3]{2})$.

(6) Show that the function $f(x) = \tan x$ has a zero in the interval $]3, 4[$. Using Newton's method, calculate x_3 starting from the initial value $x_0 = 3$. Observe that x_3 has the same first 7 decimal places as π.

Indication: Using Newton's method, we obtain the nonlinear recurrence relation $x_{n+1} = x_n - \frac{1}{2}\sin 2x_n$, $x_0 = 3$. After the first 3 iterations, we obtain: $x_1 = 3,1397077490994629364057777233059$, $x_2 = 3,1415926491252556944794381440657$, $x_3 = 3,1415926535897932384626433239544$. It is observed that $x_n \to \pi$.

(7) Consider the sequences $(x_n)_n$, $(y_n)_n \subseteq \mathbb{R}$ defined by the recurrence relations

$$\begin{cases} x_{n+1} = \dfrac{9x_n y_n^2}{(x_n + 2y_n)^2}, & x_0 = x, \\ y_{n+1} = \dfrac{x_n + 2y_n}{3}, & y_0 = y, \end{cases} \quad x, y > 0.$$

Show that

(a) $0 < x_1 \leq x_2 \leq \cdots \leq x_n \leq y_n \leq \cdots \leq y_2 \leq y_1$, for every $n \in \mathbb{N}^*$.

(b) $\lim_{n\to\infty} x_n = \lim_{n\to\infty} y_n = \sqrt[3]{xy^2}$ (it will be observed that, for every $n \in \mathbb{N}$, $x_{n+1} y_{n+1}^2 = x_n y_n^2$).

(c) If $x < y$, then $y_{n+1} - x_{n+1} < \frac{3x^2}{y} \left[\frac{y(y-x)}{3x^2} \right]^{2^{n+1}}$ (it will be observed that, for every $n \in \mathbb{N}$, $y_{n+1} - x_{n+1} < \frac{y}{3x^2}(y_n - x_n)^2$).

(d) Let $x = \frac{3}{2}, y = 2$; show that $0 < y_n - \sqrt[3]{6} < \frac{27}{8} \cdot \left(\frac{4}{27}\right)^{2^n}$.

(e) Determine the first two decimal places exactly of $\sqrt[3]{6}$.

Indications: (a)+(b) Let $f(x, y) = \frac{9xy^2}{(x+2y)^2}$, $g(x, y) = \frac{x+2y}{3}$. The recurrence relations are written as follows

$$\begin{cases} x_{n+1} = f(x_n, y_n), \\ y_{n+1} = g(x_n, y_n). \end{cases}$$ Then $f, g \in \mathcal{M}$ and $f \prec g$. According to Theorem 2.2.15, $0 < x_1 \leq x_2 \leq \cdots \leq x_n \leq y_n \leq \cdots \leq y_2 \leq y_1$ and there is $\varphi \in \mathcal{M}$ such that $\lim_n x_n = \lim_n y_n = \varphi(x, y)$. We remark that $\sqrt[3]{x_{n+1} y_{n+1}^2} = \sqrt[3]{x_n y_n^2}$ and then $\sqrt[3]{x_1 y_1^2} = \sqrt[3]{xy^2}$. Since $\psi(x, y) = \sqrt[3]{xy^2}$ is a mean which satisfies the property of invariance relative to f and g, it follows that $\psi = \varphi$ (see Remark 2.2.17). Therefore $\lim_n x_n = \lim_n y_n = \sqrt[3]{xy^2}$. (c) $y_n - x_n = \frac{(y_{n-1} - x_{n-1})^2 (x_{n-1} + 8y_{n-1})}{3(x_{n-1} + 2y_{n-1})^2} < \frac{9y}{27x^2}(y_{n-1} - x_{n-1})^2$ and then, iteratively, $y_n - x_n < \left(\frac{y}{3x^2}\right)^{1+2+\cdots+2^{n-1}} (y-x)^{2^n}$. (d) $0 < y_n - \sqrt[3]{6} < y_n - x_n < \frac{27}{8}\left(\frac{4}{27}\right)^{2^n}$. (e) For $n = 2$, $0 < y_2 - \sqrt[3]{6} < \frac{1}{100}$. Then, the first two decimal places of $y_2 = \frac{1979}{1083} = 1, 81\ldots$ are also the first two decimal places of $\sqrt[3]{6}$.

References

1. Almkvist, G., & Berndt, B. (1988). Gauss, Landen, Ramanujan, the arithmetic-geometric mean, ellipses and the Ladies Diary. *The American Mathematical Monthly, 95*, 585–608.
2. Borwein, J. M., & Borwein, P. B. (1984). The arithmetic-geometric mean and fast computation of elementary functions. *SIAM Review, 26*(3), 351–366.
3. Carlson, B. C. (1971). Algorithms involving arithmetic and geometric means. *The American Mathematical Monthly, 78*, 496–505.
4. Carlson, B. C. (1998). Elliptic integrals: symmetry and symbolic integration, Tricomi's Ideas and Contemporary Applied Mathematics (Rome/Turin, 1997). *Atti dei Convegni Lincei, 147*, 161–181.
5. Dunlap, R. A. (1997). *The golden ratio and Fibonacci numbers.* New Jersey: World Scientific Publishing Co.
6. Elaydi, S. (2005). *An introduction to difference equations, Undergraduate texts in mathematics.* New York: Springer Science + Business Media, Inc.
7. Fowler, D., & Robson, E. (1998). Square root approximations in Old Babylonian mathematics: YBC 7289 in context. *Historia Mathematica, 25*, 366–378.
8. Ghyka, M. (2016). *The golden number: Pythagorean rites and rhythms in the development of western civilization.* Inner Traditions/Bear.
9. Kelley, W. G., & Peterson, A. C. (2001). *Difference equations. An introduction with applications.* San Diego: Academic Press.
10. Neuman, E., & Sándor, J. (2003). On the Schwab-Borchardt mean. *Mathematica Pannonica, 14*(2), 253–266.
11. Newman, D. J. (1982). Rational approximation versus fast computer method. In *Lecture on approximation and value distribution, Sém. Math. Sup.* (Vol. 79, pp. 149–174).
12. Schoenberg, I. J. (1982). *Mathematical time exposures.* The Mathematical Association of America.
13. Singh, P. (1985). The so-called Fibonacci numbers in ancient and medieval India. *Historia Mathematica, 12*(3), 229–244.

Chapter 3
Elements of Asymptotic Analysis

Some mathematical objects cannot be explicitly expressed. For example, there are functions defined through integrals or as solutions to equations that cannot be expressed using elementary functions. However, we can still compare the behavior at the limit of such functions with that of known functions. Asymptotic analysis provides a wide range of methods for such comparisons.

One of these methods, initiated by Laplace, uses the fact that studying the behavior at infinity of a function of the form

$$f(t) = \int_\alpha^\beta \Phi(t, x) dx$$

can be accomplished by examining the maximum $\max_x \Phi(t, x)$ and then integrating the function Φ only over a neighborhood of the point where the maximum is attained. This integration can be done by approximating the function Φ with simpler functions. The result is usually an asymptotic expansion of the function f. Although the method uses this general technique, its practical application varies from case to case.

The first paragraph mentions the definitions of the Bachmann–Landau symbols O and o, providing several examples to facilitate a correct understanding of working with them.

According to Poincaré, asymptotic sequences and series are defined. Multiple examples of asymptotic series are given, which either do not converge to the function from which the expansion originates or represent expansions of different functions. The paragraph concludes by studying operations with asymptotic series.

The second paragraph is dedicated to the study of asymptotic approximations of various parameter integrals. Laplace's method of expansion is presented, emphasizing the importance of finding divergent asymptotic expansions. Integrals of the form $\int_\alpha^\beta e^{th(x)} dx$ are evaluated, which are useful in subsequent asymptotic expansions. Watson's lemma is introduced, and examples of asymptotic behavior are provided.

© The Author(s), under exclusive license to Springer Nature Switzerland AG 2024
L. C. Florescu, *Selected Topics in Mathematical Analysis*, Birkhäuser Advanced Texts Basler Lehrbücher, https://doi.org/10.1007/978-3-031-67784-7_3

One of the examples, reviewed in various ways and particular cases, aims to obtain conditions for the asymptotic expansion of integrals of the form $\int_\alpha^\beta e^{-tx^2} h(x)dx$. In the last subsection of this paragraph, the previous results are applied for studying the asymptotic behavior at infinity of the Γ function (the Stirling's formula). The concepts and results presented in the first two paragraphs of this chapter provide the reader with only an introduction to asymptotic analysis. For additional information, we refer to the already classic monographs in the field [1] and [2].

The last paragraph focuses on presenting asymptotic methods in the study of the behavior of solutions of difference equations. Poincaré's theorem concerning the limiting behavior of solutions of homogeneous recurrences is demonstrated, and consequences of this theorem on the asymptotic form of solutions are provided.

3.1 Asymptotic Sequences and Series

3.1.1 The Bachmann–Landau Symbols

Definition 3.1.1 Let $A \subseteq \mathbb{R}$, let $x_0 \in \bar{\mathbb{R}}$ be an cluster point of A and let $f, g : A \to \mathbb{R}$ be two functions. We say that f **is of order big** O **of** g **as** $x \to x_0$ if there exist $M > 0$ and a neighborhood $V \subseteq \bar{\mathbb{R}}$ of x_0 such that $|f(x)| \leqslant M|g(x)|$, for every $x \in V \cap A \setminus \{x_0\}$.

We denote this by $f(x) = O(g(x))(x \to x_0)$. If $x_0 \in \mathbb{R}$, then $f(x) = O(g(x))(x \to x_0)$ means that there exist $M > 0$ and $\delta > 0$ such that $|f(x)| \leqslant M|g(x)|$, for every $x \in A$ with $0 < |x - x_0| < \delta$.

If $x_0 = +\infty(-\infty)$, then $f(x) = O(g(x))(x \to +\infty)((x \to -\infty))$ if and only if there exist $M > 0$ and $a \in A$ such that $|f(x)| \leqslant M|g(x)|$, for every $x \in A$, $x > a (x < a)$.

If $f = (a_n)_n$, $g = (b_n)_n$ are sequences, then $x_0 = +\infty$ and we write $a_n = O(b_n)(n \to +\infty)$.

Examples 3.1.2 The following examples are easy to verify, and we leave them as exercises for the reader.

(1) $x^n = O(e^x)(x \to \infty)$, $\forall n \in \mathbb{N}$, (5) $\sin x = O(1)(x \to \infty)$,
(2) $x = O(\sin x)(x \to 0)$, (6) $\log x = O(x)(x \to \infty)$
(3) $x^2 = O(x)(x \to 0)$, (7) $x = O(x^2)(x \to \infty)$,
(4) $\sin x = O(x)(x \to 0)$, (8) $\sin x = O(x)(x \to \infty)$.

Remarks 3.1.3

(i) Let us observe that the formula $f(x) = O(1)(x \to x_0)$ actually states that f is bounded in a neighborhood of x_0. Indeed, from this formula, it follows that there exist $M > 0$ and a neighborhood V of the point x_0 such that $|f(x)| \leqslant M$, for every $x \in V \cap A \setminus \{x_0\}$.
(ii) Some formulas involving the symbol O require a few explanations.

3.1 Asymptotic Sequences and Series

(a) The formula $O(x) + O(x^2) = O(x)(x \to 0)$ is interpreted as follows: For any two functions f and g such that $f(x) = O(x)(x \to 0)$ and $g(x) = O(x^2)(x \to 0)$, $f(x) + g(x) = O(x)(x \to 0)$.

(b) In a similar manner, the following formulas are interpreted: $O(x) + O(x^2) = O(x^2)(x \to \infty)$, $O(x) \cdot O(x^2) = O(x^3)(x \to 0)$, $e^{O(1)} = O(1)(x \to x_0)$, for every $x_0 \in \mathbb{R}$, $f(x) \cdot O(g(x)) = O(f(x)g(x))(x \to x_0)$.

(c) The formula $e^{-x} = 1 - x + O(x^2)(x \to 0)$ implies the existence of a function f such that $f(x) = O(x^2)(x \to 0)$ and $e^{-x} = 1 - x + f(x)$.

(d) The formula $x^{-1}O(1) = O(1) + O(x^{-2})\ (x \to 0)$ is interpreted as follows: for every function f with $f(x) = O(1)(x \to 0)$ there exist two functions g and h, such that $g(x) = O(1)(x \to 0)$ and $h(x) = O(x^{-2})(x \to 0)$ and $x^{-1} f(x) = g(x) + h(x)$.

(iii) It is evident that the equal sign (=) is not appropriate for relationships involving the symbol O because it suggests symmetry, and these formulas are not symmetric.

For example, the formulas $n = O(n^2)(n \to \infty)$ and $n^2 = O(n^2)(n \to \infty)$ are correct. If the equality here were symmetric, we would reach the absurd conclusion $n = n^2$.

Donald Knuth refers to such statements as "one-way equalities".

However, as can easily be shown, the relationship involving the symbol O is reflexive and transitive.

Definition 3.1.4 Let $A \subseteq \mathbb{R}$, let $x_0 \in \bar{\mathbb{R}}$ be an cluster point of A and let $f, g : A \to \mathbb{R}$ be two functions. We say that f **is of order little o of g as** $x \to x_0$ if for every $\varepsilon > 0$ there exists a neighborhood $V \subseteq \bar{\mathbb{R}}$ of x_0 such that $|f(x)| < \varepsilon |g(x)|$, for every $x \in V \cap A \setminus \{x_0\}$.

We denote this by $f(x) = o(g(x))(x \to x_0)$.

If $x_0 \in \mathbb{R}$, then $f(x) = o(g(x))(x \to x_0)$ means that for every $\varepsilon > 0$ there exists $\delta > 0$ such that $|f(x)| < \varepsilon |g(x)|$, for every $x \in A$ with $0 < |x - x_0| < \delta$.

If $x_0 = +\infty (-\infty)$, then $f(x) = o(g(x))(x \to +\infty)((x \to -\infty))$ if and only if for every $\varepsilon > 0$ there exists $a \in A$ such that $|f(x)| < \varepsilon |g(x)|$, for every $x \in A, x > a (x < a)$.

If $f = (a_n)_n, g = (b_n)_n$ are sequences, then $x_0 = +\infty$ and we write $a_n = o(b_n)(n \to +\infty)$.

We remark that, if $g(x) \neq 0$, then $f(x) = o(g(x))(x \to x_0)$ if and only if $\lim_{x \to x_0} \frac{f(x)}{g(x)} = 0$.

Examples 3.1.5 The following examples are easy to verify, and we leave them as exercises for the reader.

(1) $\cos x = 1 + o(x)(x \to 0)$,
(2) $o(f(x) \cdot g(x)) = o(f(x)) \cdot O(g(x))(x \to 0)$,
(3) $o(f(x) \cdot g(x)) = f(x) \cdot o(g(x))(x \to 0)$,
(4) $e^{o(x)} = 1 + o(x)(x \to 0)$,
(5) $\log n = o(n)(n \to \infty)$,
(6) $n = o(e^n)(n \to \infty)$,
(7) $e^n = o(n!)(n \to \infty)$,
(8) $n! = o(n^n)(n \to \infty)$.

Numerous operations involving the symbols O and o can be performed using formulas. As an example, let's assume $f_n(x) = O(g_n(x))(x \to x_0)$ for $n = 1, 2, 3, \cdots, N$. Then $\sum_{n=1}^{N} a_n \cdot f_n(x) = O\left(\sum_{n=1}^{N} |a_n| \cdot |g_n(x)|\right)(x \to x_0)$, where $a_1, a_2, \cdots a_N \in \mathbb{R}$. Indeed, since $f_n(x) = O(g_n(x))(x \to x_0)$, there exist $M_1, \cdots M_N > 0$ and $V_1, \cdots, V_N$ neighborhoods of x_0 such that $|f_n(x)| \leq M_n \cdot |g_n(x)|$, for every $x \in A \cap V \setminus \{x_0\}$. Let $M = \max\{M_1, \cdots M_N\}$ and $V = \cap_{n=1}^{N} V_n$. Then

$$\left|\sum_{n=1}^{N} a_n \cdot f_n(x)\right| \leq \sum_{n=1}^{N} |a_n| \cdot |f_n(x)| \leq M \cdot \sum_{n=1}^{N} |a_n| \cdot |g_n(x)|, \forall x \in A \cap V \setminus \{x_0\}.$$

Here are some other results whose justification we propose:

1. If $f(x) = O(g(x))(x \to x_0)$, then $|f(x)|^\alpha = O(|g(x)|^\alpha)(x \to x_0)$, for $\alpha > 0$.
2. If, for every $i = 1, \cdots n$, $a_i \in \mathbb{R}$, $f_i(x) = O(g_i(x))(x \to x_0)$, and $|g_i(x)| \leq |g(x)|$, then $\sum_{i=1}^{n} a_i \cdot f_i(x) = O(g(x))(x \to x_0)$.
3. If, for every $i = 1, \cdots n$, $f_i(x) = O(g_i(x))(x \to x_0)$, then $\prod_{i=1}^{n} f_i(x) = O(\prod_{i=1}^{n} g_i(x))(x \to x_0)$.

If in the above formulas, the symbol O is replaced with the symbol o, the formulas remain true.

Definition 3.1.6 We say that a function f is **asymptotically equivalent** or **asymptotically equal** to a function g, when $x \to x_0$, and we write $f(x) \sim g(x)$ $(x \to x_0)$, if $\lim_{x \to x_0} \dfrac{f(x)}{g(x)} = 1$.

The relationship $f(x) \sim g(x)(x \to x_0)$ is equivalent to $f(x) = g(x)(1 + o(1))(x \to x_0)$, $f(x) = g(x) + o(g(x))(x \to x_0)$, or $f(x) = e^{o(1)}g(x)(x \to x_0)$.

Asymptotic equality is an equivalence relation (reflexive, symmetric, and transitive).

Examples 3.1.7

(1) $\sin x \sim x$ $(x \to 0)$,
(2) $\log(1+x) \sim x$ $(x \to 0)$,
(3) $e^x - 1 \sim x$ $(x \to 0)$,
(4) $n! \sim e^{-n} n^n \sqrt{2\pi n}$ $(n \to \infty)$.

To determine the "asymptotic behavior" of a given function f as x approaches x_0 means to find a simple function g that is asymptotically equivalent to f. Here, "simple" means that its explicit evaluation is not difficult when x is close to x_0.

3.1 Asymptotic Sequences and Series

3.1.2 Asymptotic Series

A function $f : A \to \mathbb{R}$ is said to be expandable in a power series about the point $x_0 \in A$ if there exists $r > 0$ such that, for every $x \in A$ with $|x - x_0| < r$, $f(x) = \sum_0^\infty a_n (x - x_0)^n$.

f is expandable to infinity if there exists $R > 0$ such that, for every $x \in A$ with $|x| > R$, $f(x) = \sum_0^\infty \frac{a_n}{x^n}$.

Let $x \in A$ and let $(S_n(x))_n$ be the sequence of partial sums of one of the series above: $S_n(x) = \sum_{m=0}^n a_m (x - x_0)^m$ or $S_n(x) = \sum_{m=0}^n \frac{a_m}{x^m}$; then $\lim_n |f(x) - S_n(x)| = 0$. Thus, as n increases, $S_n(x)$ represents an approximation of f with increasing accuracy.

If f is not expandable into a power series about the given point, we can still describe its behavior using the partial sums $S_n(x)$ of a different approximate series called an asymptotic expansion. Such series are usually not convergent, but nonetheless, $|f(x) - S_n(x)| \to 0$ for a fixed n as x approaches x_0 or ∞.

Definition 3.1.8 For every $n \in \mathbb{N}^*$ let $f_n : A \to \mathbb{R}$ be a function and let x_0 be a cluster point of A. The sequence $(f_n)_{n \in \mathbb{N}^*}$, is an **asymptotic sequence** as $x \to x_0$, if the following conditions are fulfilled:

1. $\lim_{x \to x_0} f_n(x) = 0$, for every $n \in \mathbb{N}^*$.
2. For every $n \in \mathbb{N}^*$, there exists V a neighborhood of x_0 such that $f_n(x) \neq 0$, for every $x \in V \cap A \setminus \{x_0\}$.
3. $f_{n+1}(x) = o(f_n(x))(x \to x_0)$, for every $n \in \mathbb{N}^*$.

So, $(f_n)_n$ is an asymptotic sequence as $x \to x_0$ if and only if, for all $n \in \mathbb{N}^*$, $\lim_{x \to x_0} \frac{f_{n+1}(x)}{f_n(x)} = 0$. Here, it is understood that $f_n(x)$ does not vanish in a neighborhood of x_0 (except possibly at x_0), and $\lim_{x \to x_0} f_n(x) = 0$, for every $n \in \mathbb{N}^*$.

Remark 3.1.9 If (f_n) is an asymptotic sequence as $x \to x_0$, then, for every $n \in \mathbb{N}^*$, $O(f_{n+1}(x)) = o(f_n(x))(x \to x_0)$.

Examples 3.1.10 Below are some examples of asymptotic sequences:
(1) $\{(x - x_0)^n\}$ as $x \to x_0$.
(2) $\{(\log x)^{-n}\}$ as $x \to \infty$.
(3) $\{e^{-x} x^{-a_n}\}$ as $x \to \infty$, $((a_n)_n \subseteq \mathbb{R}$ is a strictly decreasing sequence).
(4) $f_n(x) = e^{-nx}$ as $x \to \infty$.

Theorem 3.1.11 *Let x_0 be a cluster point of A, let $(f_n)_{n \geqslant 1}$ be an asymptotic sequence as $x \to x_0$, and let $(a_n)_{n \geqslant 1}$ be a sequence of real numbers with the first term, $a_1 \neq 0$; the following statements are equivalent:*

(i) $f(x) = \sum_{n=1}^{N} a_n f_n(x) + o(f_N(x))(x \to x_0)$, for any $N \in \mathbb{N}^*$.
(ii) $f(x) = \sum_{n=1}^{N} a_n f_n(x) + O(f_{N+1}(x))(x \to x_0)$, for any $N \in \mathbb{N}^*$.
Any of these equivalent conditions implies:
(iii) $f(x) \sim \sum_{n=1}^{N} a_n f_n(x)(x \to x_0)$, for any $N \in \mathbb{N}^*$.

Proof (i) $\Rightarrow$ (ii):
Let an arbitrary $N \in \mathbb{N}^*$; from (i) $f(x) = \sum_{n=1}^{N-1} a_n f_n(x) + a_N f_N(x) + o(f_N(x))$. There exists a function $g : A \to \mathbb{R}$ such that $g(x) = o(f_N(x))(x \to x_0)$ and $f(x) = \sum_{n=1}^{N-1} a_n \cdot f_n(x) + a_N \cdot f_N(x) + g(x)$. Since $g(x) = o(f_N(x))(x \to x_0)$, there exists V a neighborhood of x_0 such that, for every $x \in V \cap A \setminus \{x_0\}$, $|g(x)| < |f_N(x)|$. Let now $M = |a_N| + 1$; for every $x \in V \cap A \setminus \{x_0\}$, $|f(x) - \sum_{n=1}^{N-1} a_n f_n(x)| \leq |a_N| \cdot |f_N(x)| + |f_N(x)| = M \cdot |f_N(x)|$. Therefore, for every $N \geq 2$, $f(x) - \sum_{n=1}^{N-1} a_n f_n(x) = O(f_N(x))(x \to x_0)$, and so, we have (ii).

(ii) $\Rightarrow$ (i):
Let an arbitrary $N \in \mathbb{N}^*$; from (ii), there exist $M > 0$ and a neighborhood V of x_0 such that, for every $x \in V \cap A \setminus \{x_0\}$,

$$\left| f(x) - \sum_{n=1}^{N} a_n f_n(x) \right| \leq M \cdot |f_{N+1}(x)|.$$

Since $\lim_{x \to x_0} \frac{f_{N+1}(x)}{f_N(x)} = 0$, for every $\varepsilon > 0$, there exists another neighborhood of x_0, $W \subseteq V$, such that, for every $x \in W \cap A \setminus \{x_0\}$, $\left|\frac{f_{N+1}(x)}{f_N(x)}\right| < \frac{\varepsilon}{M}$. Then, for every $x \in W \cap A \setminus \{x_0\} \subseteq V \cap A \setminus \{x_0\}$,

$$\left| f(x) - \sum_{n=1}^{N} a_n f_n(x) \right| \leq M \cdot \left|\frac{f_{N+1}(x)}{f_N(x)}\right| \cdot |f_N(x)| < \varepsilon \cdot |f_N(x)|,$$

and so $f(x) = \sum_{n=1}^{N} a_n f_n(x) + o(f_N(x))(x \to x_0)$, for every $N \in \mathbb{N}^*$.

Let's now demonstrate that (ii) $\Rightarrow$ (iii):
For every $N \in \mathbb{N}^*$, there exist $M > 0$ and V a neighborhood of x_0 such that, for every $x \in V \cap A \setminus \{x_0\}$,

$$\left| f(x) - \sum_{n=1}^{N} a_n f_n(x) \right| \leq M \cdot |f_{N+1}(x)|. \tag{3.1}$$

Let $\varepsilon > 0$ be so that $\varepsilon \cdot \sum_{n=2}^{N} |a_n| < M$ and let

$$\varepsilon_1 = \frac{\varepsilon |a_1|}{2M} > 0. \tag{3.2}$$

3.1 Asymptotic Sequences and Series

Since $\lim_{x \to x_0} \dfrac{f_n(x)}{f_1(x)} = \lim_{x \to x_0} \dfrac{f_n(x)}{f_{n-1}(x)} \cdot \dfrac{f_{n-1}(x)}{f_{n-2}(x)} \cdots \dfrac{f_2(x)}{f_1(x)} = 0$, for every $n = 2, \ldots, N+1$, there exist another neighborhood of x_0, $W \subseteq V$, such that, for every $x \in W \cap A \setminus \{x_0\}$,

$$\left| \frac{f_n(x)}{f_1(x)} \right| < \varepsilon_1, \text{ for every } n = 2, \ldots, N+1. \tag{3.3}$$

From (3.1), (3.2), and (3.3), for every $x \in W \cap A \setminus \{x_0\}$,

$$\left| \frac{f(x)}{\sum_{n=1}^{N} a_n f_n(x)} - 1 \right| = \frac{|f(x) - \sum_{n=1}^{N} a_n f_n(x)|}{|f_1(x)| \cdot \left| a_1 + \sum_{n=2}^{N} a_n \frac{f_n(x)}{f_1(x)} \right|} \leqslant$$

$$\leqslant \frac{M \cdot |f_{N+1}(x)|}{|f_1(x)| \cdot \left(|a_1| - \sum_{n=2}^{N} |a_n| \left| \frac{f_n(x)}{f_1(x)} \right| \right)} \leqslant M \cdot \frac{\left| \frac{f_{N+1}(x)}{f_1(x)} \right|}{|a_1| - \varepsilon_1 \sum_{n=2}^{N} |a_n|} <$$

$$< M \cdot \frac{\varepsilon_1}{|a_1| - \frac{\varepsilon |a_1|}{2M} \cdot \sum_{n=2}^{N} |a_n|} < M \cdot \frac{\varepsilon_1}{|a_1| - \frac{|a_1|}{2}} = \frac{2M}{|a_1|} \cdot \varepsilon_1 = \varepsilon,$$

what means that

$$\lim_{x \to x_0} \frac{f(x)}{\sum_{n=1}^{N} a_n f_n(x)} = 1$$

and then, for every $N \in \mathbb{N}^*$,

$$f(x) \sim \sum_{n=1}^{N} a_n f_n(x)(x \to x_0).$$

$\square$

Definition 3.1.12 Let $(f_n)_n$ be an asymptotic sequence as $x \to x_0$; we say that the series $\sum_{n=1}^{\infty} a_n f_n(x)$ is an **asymptotic expansion** or **asymptotic approximation** of $f(x)$ as $x \to x_0$ if one of the equivalent conditions (i) or (ii) from the previous theorem is satisfied.

This definition was given by Poincaré in 1886.

Furthermore, we will use the notation $f(x) \approx \sum_{n=1}^{\infty} a_n f_n(x)(x \to x_0)$ to represent the situation where the series $\sum_{n=1}^{\infty} a_n f_n(x)$ is an asymptotic expansion of $f(x)$ as $x \to x_0$.

Remarks 3.1.13

(i) If $f(x) \approx \sum_{n=1}^{\infty} a_n f_n(x)(x \to x_0)$, then $f(x) = a_1 f_1(x) + o(f_1(x))$ and then there exists $\lim_{x \to x_0} f(x) = 0$.

In the case where $\lim_{x \to x_0} f(x) = a_0 \in \mathbb{R}$ and $f(x) - a_0 \approx \sum_{n=1}^{\infty} a_n f_n(x)$ $(x \to x_0)$, we will agree to say that $f(x) \approx \sum_{n=0}^{\infty} a_n f_n(x)(x \to x_0)$ where, for every $x \in A$, we denote $f_0(x) = 1$.

(ii) If there exists an asymptotic expansion for the function f with the given asymptotic sequence $(f_n(x))$, then it is unique, and the coefficients a_n are uniquely determined by the following relationships.

$a_0 = \lim_{x \to x_0} f(x)$

$a_1 = \lim_{x \to x_0} \dfrac{f(x)}{f_1(x)}$

$a_2 = \lim_{x \to x_0} \dfrac{f(x) - a_1 f_1(x)}{f_2(x)}$

$\dotsb$

$a_N = \lim_{x \to x_0} \left\{ \dfrac{f(x) - \sum_{n=1}^{N-1} a_n f_n(x)}{f_N(x)} \right\}$

$\dotsb$

Indeed, from (i) of Theorem 3.1.11, for every $N \in \mathbb{N}^*$ and every $\varepsilon > 0$, there exists V a neighborhood of x_0, such that either

$$\left| f(x) - \sum_{1}^{N} a_n f_n(x) \right| < \varepsilon \cdot f_N(x), \text{ for every } x \in V \cap A \setminus \{x_0\}$$

holds, or equivalently,

$$\left| \dfrac{f(x) - \sum_{1}^{N-1} a_n f_n(x)}{f_N(x)} - a_N \right| < \varepsilon$$

holds. This result shows that a function admits a unique asymptotic expansion with respect to a given asymptotic sequence.

(iii) The first non-zero term in the asymptotic expansion $\sum_{n=1}^{\infty} a_n f_n(x)$ $(x \to x_0)$ is called the **dominant term** of the expansion. For example, if $a_1 \neq 0$, then $f(x) \sim a_0 + a_1 f_1(x)(x \to x_0)$.

(iv) The relationship $f(x) \approx \sum_{n=0}^{\infty} a_n f_n(x)(x \to x_0)$ does not imply that the series $\sum_{n=1}^{\infty} a_n f_n(x)$ is convergent. An asymptotic expansion can, of course, be convergent; if it is so, it is usually less useful than if it were divergent because in the case of a divergent series, only a few terms are needed for each x to provide a good approximation of the function.

The class of convergent power series is the simplest class of asymptotic series. Suppose that f is the sum of a power series: $f(x) = a_0 + a_1 x + a_2 x^2 + \cdots$ when $|x| \leqslant p$, where p is an arbitrary positive number less than the radius of convergence of the series. Then

$$f(x) \approx a_0 + a_1 x + a_2 x^2 + \cdots (x \to 0).$$

3.1 Asymptotic Sequences and Series

The proof is simple: for every $n \in \mathbb{N}$,

$$\left| f(x) - \sum_{k=0}^{n} a_k x^k \right| = \left| \sum_{k=n+1}^{\infty} a_k x^k \right| \leqslant |x|^{n+1} \cdot \underbrace{\sum_{k=n+1}^{\infty} |a_k| \cdot |x|^{k-n-1}}_{g(x)}.$$

Since $\lim_{x \to 0} g(x) = |a_{n+1}|$, there exists $\delta > 0$ such that, for every $x \in \,]-\delta, \delta[$, $g(x) < |a_{n+1}| + 1 = M$. Then

$$\left| f(x) - \sum_{k=0}^{n} a_k x^k \right| \leqslant M \cdot |x|^{n+1}$$

and then $f(x) = \sum_{k=1}^{n} a_k x^k + O(x^{n+1})$, for every $n \in \mathbb{N}$. According to (ii) of Theorem 3.1.11 this means that

$$f(x) \approx \sum_{n=0}^{\infty} a_n x^n \quad (x \to 0).$$

If a function admits a Taylor expansion on a neighborhood of x_0, then the Taylor series is a convergent asymptotic expansion.

The following example presents a more interesting asymptotic expansion.

Example 3.1.14 Let $E :]0, +\infty[\to \mathbb{R}$ be the function defined by

$$E(t) = \int_{t}^{\infty} x^{-1} e^{-x} dx.$$

Let's find an asymptotic expansion of $E(t)$ as $t \to \infty$.

Firstly, we remark that $E(t) \leqslant \frac{1}{t} \cdot \int_{t}^{\infty} e^{-x} dx = \frac{1}{te^t}$, and then $\lim_{t \to \infty} E(t) = 0$; according to (i) and (ii) of Remark 3.1.13, $a_0 = 0$.

Integrating by parts, we obtain:

$$E(t) = \left[-\frac{e^{-x}}{x} \right]_{t}^{\infty} - \int_{t}^{\infty} e^{-x} x^{-2} dx = \frac{e^{-t}}{t} + \left[\frac{e^{-x}}{x^2} \right]_{t}^{\infty} + 2 \int_{t}^{\infty} e^{-x} x^{-3} dx$$

and repeating the process, we obtain

$$E(t) = e^{-t} \left\{ \frac{1}{t} - \frac{1}{t^2} + \frac{2!}{t^3} - \frac{3!}{t^4} + \cdots + \frac{(-1)^{n-1}(n-1)!}{t^n} \right\} +$$

$$+ (-1)^n n! \int_{t}^{\infty} e^{-x} x^{-(n+1)} dx = s_n(t) + r_n(t)$$

where the partial sums s_n and the remainder r_n are:

$$s_n(t) = e^{-t}\left\{\frac{1}{t} - \frac{1}{t^2} + \frac{2!}{t^3} - \frac{3!}{t^4} + \cdots + \frac{(-1)^{n-1}(n-1)!}{t^n}\right\},$$

$$r_n(t) = (-1)^n n! \int_t^\infty e^{-x} x^{-(n+1)} dx.$$

The series for which $(s_n(t))_n$ is the sequence of partial sums is divergent for every fixed t because the absolute value of the general term, $\frac{(n-1)!}{t^n}$, tends to ∞ as $n \to \infty$.

If we consider n fixed and t sufficiently large, then:

$$|r_n(t)| = n! \int_t^\infty e^{-x} x^{-(n+1)} dx \leqslant \frac{n!}{t^{n+1}} \int_t^\infty e^{-x} dx = \frac{n!}{t^{n+1}} e^{-t} \xrightarrow[t\to\infty]{} 0.$$

Moreover, for a fixed value of n, the ratio between $r_n(t)$ and the last term in $s_n(t)$ is:

$$\left|\frac{r_n(t)}{(n-1)! t^{-n} e^{-t}}\right| < \frac{n}{t} \to 0 \text{ when } t \to \infty,$$

hence, if we denote by $f_n(t) = \frac{(-1)^n(n-1)!}{t^n}$,

$$\lim_{t\to+\infty} \frac{E(t) - s_n(t)}{f_n(t)} = 0$$

which is equivalent to: $E(t) - s_n(t) = o(f_n(t))(t \to \infty)$ and then

$$E(t) = s_n(t) + o(f_n(t))(t \to \infty), \text{ for every } n \in \mathbb{N}^*.$$

The last relationship shows that we have the following asymptotic expansion for the function E:

$$E(t) \approx \sum_{n=1}^\infty \frac{(-1)^{n-1}(n-1)!}{t^n} \cdot e^{-t} (t \to \infty).$$

Using (iii) of Theorem 3.1.11, for every $n \in \mathbb{N}^*$,

$$E(t) \sim e^{-t}\left\{\frac{1}{t} - \frac{1}{t^2} + \frac{2!}{t^3} - \frac{3!}{t^4} + \cdots + \frac{(-1)^{n-1}(n-1)!}{t^n}\right\} (t \to \infty).$$

3.1.3 Operations with Asymptotic Series

Proposition 3.1.15 *Let $f, g : A \to \mathbb{R}$, let x_0 be a cluster point of A and let $(f_n)_{n \in \mathbb{N}}$ be an asymptotic sequence as $x \to x_0$ (from (i) of Remark 3.1.13, $f_0(x) = 1$). If we suppose that $f(x) \approx \sum_{n=0}^{\infty} a_n f_n(x) (x \to x_0)$ and $g(x) \approx \sum_{n=0}^{\infty} b_n f_n(x) (x \to x_0)$, then, for every $\alpha, \beta \in \mathbb{R}$,*

$$\alpha f(x) + \beta g(x) \approx \sum_{n=0}^{\infty} (\alpha a_n + \beta b_n) f_n(x) \, (x \to x_0).$$

Proof For every $N \in \mathbb{N}$,

$$f(x) = \sum_{0}^{N} a_n f_n(x) + O(f_{N+1}(x))(x \to x_0),$$

$$g(x) = \sum_{0}^{N} b_n f_n(x) + O(f_{N+1}(x))(x \to x_0),$$

and therefore

$$\alpha f(x) + \beta g(x) = \sum_{0}^{N} (\alpha a_n + \beta b_n) f_n(x) + O(f_{N+1}(x))(x \to x_0).$$

□

Remark 3.1.16 The Cauchy product of two asymptotic series is not necessarily an asymptotic series. Indeed, by formally multiplying the two series, terms of the form $a_n b_m f_n(x) f_m(x)$ are obtained, and the sequence $(f_n f_m)_{n,m}$ cannot, in general, be arranged as an asymptotic sequence. In the particular case of the asymptotic sequences $(x^n)_{n \in \mathbb{N}}$ as $(x \to 0)$ or $(x^{-n})_{n \in \mathbb{N}}$ as $(x \to \infty)$, the aforementioned products can be arranged as asymptotic sequences.

Proposition 3.1.17 *Let $(a_n)_n, (b_n)_n \subseteq \mathbb{R}$ be two arbitrary sequences, let $(c_n)_n \subseteq \mathbb{R}$ be the Cauchy product sequence ($c_n = a_0 b_n + a_1 b_{n-1} + \cdots a_n b_0$, for every $n \in \mathbb{N}$) and let $f, g : A \to \mathbb{R}$.*

(1) If 0 is a cluster point of A, $f(x) \approx \sum_{n=0}^{\infty} a_n x^n (x \to 0)$ and $g(x) \approx \sum_{n=0}^{\infty} b_n x^n (x \to 0)$, then $f(x) \cdot g(x) \approx \sum_{n=0}^{\infty} c_n x^n (x \to 0)$.

(2) If $+\infty$ is a cluster point of A, $f(x) \approx \sum_{n=0}^{\infty} a_n x^{-n} (x \to \infty)$ and $g(x) \approx \sum_{n=0}^{\infty} b_n x^{-n} (x \to \infty)$, then $f(x) \cdot g(x) \approx \sum_{n=0}^{\infty} c_n x^{-n} (x \to \infty)$.

Proof

(1) From the assumption, for every $N \in \mathbb{N}$, there exist $f_1, g_1 : A \to \mathbb{R}$, $f_1(x) = O(x^{N+1})(x \to 0)$, $g_1(x) = O(x^{N+1})(x \to 0)$ such that

$$f(x) = \sum_0^N a_n x^n + f_1(x), \quad g(x) = \sum_0^N b_n x^n + g_1(x).$$

Multiplying the two aforementioned relationships, we obtain

$$f(x) \cdot g(x) = \left(\sum_0^N a_n x^n\right) \cdot \left(\sum_0^N b_n x^n\right) +$$

$$+ \left(\sum_0^N a_n x^n\right) \cdot g_1(x) + \left(\sum_0^N b_n x^n\right) \cdot f_1(x) + f_1(x) \cdot g_1(x).$$

It is immediately noticeable that

$$\left(\sum_0^N a_n x^n\right) \cdot \left(\sum_0^N b_n x^n\right) = \sum_0^N c_n x^n + O(x^{N+1})(x \to 0)$$

and since $\sum_0^N a_n x^n$ and $\sum_0^N b_n x^n$ are bounded in a neighborhood of 0,

$$\left(\sum_0^N a_n x^n\right) \cdot g_1(x) = O(x^{N+1})(x \to 0),$$

$$\left(\sum_0^N b_n x^n\right) \cdot f_1(x) = O(x^{N+1})(x \to 0).$$

It is evident that $f_1(x) \cdot g_1(x) = O(x^{N+1})(x \to 0)$ and thus

$$f(x) \cdot g(x) = \sum_0^N c_n x^n + O(x^{N+1})(x \to 0), \text{ for every } N \in \mathbb{N}$$

which, according to Theorem 3.1.11, leads us to

$$f(x) \cdot g(x) \approx \sum_0^\infty c_n x^n (x \to 0).$$

The proof of point (2) is similar. □

3.1 Asymptotic Sequences and Series

We will now address the integration and differentiation of asymptotic expansions. For easily understandable reasons, we will only consider the cases of asymptotic sequences $(x^n)_{n\in\mathbb{N}}$ as $(x \to 0)$ and $(x^{-n})_{n\in\mathbb{N}}$ as $(x \to \infty)$.

Theorem 3.1.18 *Let A be a bounded or unbounded interval, and let $f : A \to \mathbb{R}$ have the property that $f \in \mathfrak{R}_{[c,d]}$ for every $[c, d] \subset A$ (meaning f is Riemann integrable on any closed subinterval of A).*

(1) If 0 is a cluster point of A and if $f(x) \approx \sum_0^\infty a_n x^n (x \to 0)$, then there exists a neighborhood V of 0 such that f is Riemann integrable on $]0, x]$, for every $x \in V \cap A \setminus \{0\}$ and

$$\int_0^x f(t)dt \approx \sum_0^\infty \frac{a_n}{n+1} \cdot x^{n+1} \; (x \to 0).$$

(2) If ∞ is a cluster point of A and if $f(x) \approx \sum_0^\infty a_n \cdot x^{-n} (x \to \infty)$, then there exists a neighborhood V of $+\infty$ such that the function $g : A \to \mathbb{R}$, $g(t) = f(t) - a_0 - \dfrac{a_1}{t}$, is Riemann integrable on $[x, +\infty[$, for every $x \in V \cap A$ and

$$\int_x^\infty \left[f(t) - a_0 - \frac{a_1}{t} \right] dt \approx \sum_{n=1}^\infty \frac{a_{n+1}}{n} \cdot x^{-n} \; (x \to \infty).$$

Proof In the above statement, $\mathfrak{R}_{]0,x]}$ (respectively $\mathfrak{R}_{[x,\infty[}$) denotes the set of all functions Riemann integrable in the generalized sense over $]0, x]$ (respectively on $[x, \infty[$), that is, those functions for which the limit $\lim_{y \downarrow 0} \int_y^x f(t)dt \in \mathbb{R}$ (respectively the limit $\lim_{y \uparrow \infty} \int_x^y f(t)dt \in \mathbb{R}$) exists.

(1) Since $f(x) \approx \sum_0^\infty a_n \cdot x^n (x \to 0)$, $f(x) = a_0 + a_1 x + O(x^2)(x \to 0)$ therefore there exist $\delta > 0$, $M > 0$ such that $|f(x) - a_0 - a_1 x| \leq M \cdot x^2$, for every $x \in [-\delta, \delta] \cap A \setminus \{0\}$. Since $\int_0^\delta M \cdot x^2 dx$ converges, $f(x) - a_0 - a_1 x \in \mathfrak{R}_{]0,\delta]}$ from where $f \in \mathfrak{R}_{]0,\delta]}$ and then $f \in \mathfrak{R}_{]0,x]}$, for every $x \in [-\delta, \delta] \cap A$.

Using the hypothesis again, for every $N \in \mathbb{N}^*$, there exist $g(x) = O(x^{N+1})(x \to 0)$ and a neighborhood V of 0 such that

$$f(x) = a_0 + a_1 x + \cdots + a_N x^N + g(x), \text{ for every } x \in V \cap A \setminus \{0\}.$$

From here, it follows that $g \in \mathfrak{R}_{]0,x]}$, for every $x \in V \cap A \setminus \{0\}$, and if we integrate the relationship above

$$\int_0^x f(t)dt = a_0 x + \frac{a_1}{2}x^2 \cdots + \frac{a_N}{N+1}x^{N+1} + \int_0^x g(t)dt.$$

Since $|g(x)| \leq Mx^{N+1}$, for every $x \in V \cap A \setminus \{0\}$,

$$\left| \int_0^x g(t)dt \right| \leq \frac{M}{N+2} x^{N+2}$$

and so $\int_0^x g(t)dt = O(x^{N+2})(x \to 0)$. It follows that, for every $N \in \mathbb{N}$,

$$\int_0^x f(t)dt = \sum_{k=0}^N \frac{a_k}{k+1} x^{k+1} + O(x^{N+2})(x \to 0)$$

what means that

$$\int_0^x f(t)dt \approx \sum_{n=0}^\infty \frac{a_n}{n+1} x^{n+1} (x \to 0).$$

(2) We suppose now that $+\infty$ is a cluster point of A and $f(x) \approx \sum_0^\infty a_n x^{-n} (x \to \infty)$; then $f(x) = a_0 + \frac{a_1}{x} + O(x^{-2})(x \to \infty)$ and then there exist $M > 0, \delta > 0$ such that

$$\left| f(x) - a_0 - \frac{a_1}{x} \right| \leq \frac{M}{x^2}, \text{ for every } x \in [\delta, +\infty[$$

Since $\int_\delta^\infty \frac{M}{x^2} dx$ is convergent it follows that $f(x) - a_0 - \frac{a_1}{x} \in \mathfrak{R}_{[\delta, +\infty)}$ and then $f(t) - a_0 - \frac{a_1}{t} \in \mathfrak{R}_{[x, +\infty[}$, for every $x \in [\delta, +\infty[$.

Now, for every $N \in \mathbb{N}^*$, there exists $g(x) = O(x^{-(N+1)})(x \to \infty)$ such that $f(x) - a_0 - \frac{a_1}{x} = \frac{a_2}{x^2} + \cdots + \frac{a_N}{x^N} + g(x)$.

It follows that $g \in \mathfrak{R}_{[x, +\infty)}$, for every $x \in [\delta, +\infty[$; if we integrate the relationship above

$$\int_x^\infty \left[f(t) - a_0 - \frac{a_1}{t} \right] dt = \frac{a_2}{x} + \cdots + \frac{a_N}{N-1} \cdot \frac{1}{x^{N-1}} + \int_x^\infty g(x)dx.$$

Since $g(x) = O(x^{-N-1})(x \to \infty)$, there exist $M > 0, \delta_1 > \delta$ such that

$$|g(x)| \leq M \cdot x^{-N-1}, \text{ for every } x \in [\delta_1, +\infty[,$$

from where

$$\left| \int_x^\infty g(t)dt \right| \leq \frac{M}{N} \cdot x^{-N}, \text{ hence}$$

$$\int_x^\infty \left[f(t) - a_0 - \frac{a_1}{t} \right] dt = \sum_{k=2}^N \frac{a_k}{k-1} \cdot x^{-k+1} + O(x^{-N})(x \to \infty),$$

3.1 Asymptotic Sequences and Series

for every $N \geqslant 2$, what means that

$$\int_x^\infty \left[f(t) - a_0 - \frac{a_1}{t} \right] dt \approx \sum_{n=2}^\infty \frac{a_n}{n-1} \cdot x^{-n+1} (x \to \infty) =$$

$$= \sum_{n=1}^\infty \frac{a_{n+1}}{n} \cdot x^{-n} (x \to \infty).$$

$\square$

Theorem 3.1.19 *Let A be a bounded or unbounded interval, and let $f : A \to \mathbb{R}$ be a differentiable function such that $f' \in \mathfrak{R}_{[c,d]}$, for every $[c, d] \subseteq A$.*

(1) If 0 is a cluster point of A, $f'(x) \approx \sum_{n=0}^\infty b_n x^n (x \to 0)$ and if there exists $\lim_{x \to 0} f(x) = a_0 \in \mathbb{R}$, then

$$f(x) \approx a_0 + \sum_{n=1}^\infty \frac{b_{n-1}}{n} \cdot x^n (x \to 0).$$

(2) If ∞ is a cluster point of A, $f'(x) \approx \sum_{n=0}^\infty b_n x^{-n} (x \to \infty)$ and if there exists $\lim_{x \to \infty} f(x) = a_0 \in \mathbb{R}$, then $b_0 = b_1 = 0$ and

$$f(x) \approx a_0 - \sum_{n=1}^\infty \frac{b_{n+1}}{n} \cdot x^{-n} (x \to \infty).$$

Proof

(1) We can apply point (1) of the preceding theorem to the function f'. As a result, there exists V, a neighborhood of 0, such that $f' \in \mathfrak{R}_{[0,x]}$, for every $x \in V \cap A \setminus \{0\}$ and

$$\int_0^x f'(t) dt \approx \sum_0^\infty \frac{b_n}{n+1} \cdot x^{n+1} (x \to 0).$$

But $\int_0^x f'(t) dt = f(x) - \lim_{y \to 0} f(y) = f(x) - a_0$ and then

$$f(x) \approx a_0 + \sum_0^\infty \frac{b_n}{n+1} \cdot x^{n+1} (x \to 0).$$

(2) We apply the point (2) of the preceding theorem to the function f'; there exists a neighborhood V of $+\infty$ such that the function $g : A \to \mathbb{R}, g(t) = f'(t) - b_0 - \frac{b_1}{t}$, is integrable over $[x, +\infty[$, for every $x \in V \cap A$ and

$$\int_x^\infty \left[f' - b_0 - \frac{b_1}{t} \right] dt \approx \sum_{n=1}^\infty \frac{b_{n+1}}{n} \cdot x^{-n} \quad (x \to \infty).$$

However, the integral $\int_x^\infty f'(t) dt = \lim_{t \to \infty} f(t) - f(x) = a_0 - f(x)$ is convergent, and since $\int_x^\infty g(t) dt$ also converges, it must be the case that $b_0 = b_1 = 0$. It follows that

$$f(x) \approx a_0 - \sum_{n=1}^\infty \frac{b_{n+1}}{n} \cdot x^{-n} \quad (x \to \infty).$$

□

3.2 Laplace's Method

We aim to study the asymptotic behavior of integrals of the form

$$\int_\alpha^\beta f(t, x) dx \quad (t \to \infty),$$

where the interval of integration can be bounded or unbounded.

The general idea will be to restrict the integration interval $]\alpha, \beta[$ to a small interval $]\varepsilon, \eta[$ where the function f has maximum values, and where the asymptotic behavior, as $t \to \infty$, is the same as on $]\alpha, \beta[$. The advantage of this restriction to a small interval is that in such a situation, we can approximate the function f with simpler functions whose integral exhibits an asymptotic behavior that is relatively easy to study.

First, we will recall some properties of the Euler's **Gamma function**, a function defined by

$$\Gamma(a) = \int_{0+0}^{+\infty} x^{a-1} \cdot e^{-x} dx.$$

The generalized (or improper) mixed integral $\int_{0+0}^\infty$ is convergent if both integrals $\int_{0+0}^1$ and $\int_1^\infty$ are convergent.

3.2 Laplace's Method

Since $x^{a-1} \cdot e^{-1} \leqslant x^{a-1} \cdot e^{-x} \leqslant x^{a-1}$, for every $x \in]0, 1]$, the integral $\int_{0+0}^{1} x^{a-1} \cdot e^{-x} dx$ converges if and only if $\int_{0+0}^{1} x^{a-1} dx$ converges. But

$$\int_{0+0}^{1} x^{a-1} dx = \lim_{u \downarrow 0} \int_{u}^{1} x^{a-1} dx = \begin{cases} \lim_{u \downarrow 0} \left(\frac{1}{a} \cdot x^{a} \big|_{u}^{1} \right), & a \neq 0, \\ \lim_{u \downarrow 0} \left(\log x \big|_{u}^{1} \right), & a = 0 \end{cases} =$$

$$= \begin{cases} \frac{1}{a}, & a > 0, \\ +\infty, & a \leqslant 0. \end{cases}$$

It follows that $\int_{0+0}^{1} x^{a-1} \cdot e^{-x} dx$ converges if and only if $a > 0$.

For the second integral, we observe that $x^{a-1} \cdot e^{-x} = O\left(e^{-\frac{x}{2}}\right) (x \to \infty)$ and since $\int_{1}^{+\infty} e^{-\frac{x}{2}} dx$ is convergent, it follows that $\int_{1}^{+\infty} x^{a-1} \cdot e^{-x} dx$ converges, for every $a \in \mathbb{R}$.

This implies that $\Gamma :]0, +\infty[\to \mathbb{R}$.

The following proposition recalls some of the properties of the function Γ.

Proposition 3.2.1

(1) $\Gamma(a+1) = a \cdot \Gamma(a)$, for every $a > 0$.
(2) $\Gamma(n+1) = n!$, for every $n \in \mathbb{N}$.
(3) $\Gamma\left(\frac{1}{2}\right) = \int_{-\infty}^{+\infty} e^{-x^2} dx = \sqrt{\pi}$.
(4) $\Gamma\left(n + \frac{1}{2}\right) = \frac{(2n-1)!!}{2^n} \cdot \sqrt{\pi}$, for any $n \in \mathbb{N}$ $((2n-1)!! = 1 \cdot 3 \cdot 5 \cdots (2n-1))$.

Proof

(1) $\Gamma(a+1) = \int_{0+0}^{+\infty} x^{a} \cdot e^{-x} dx = -\int_{0+0}^{+\infty} x^{a} \cdot (e^{-x})' dx = -x^{a} \cdot e^{-x} \Big|_{0+0}^{+\infty} +$
$+ a \cdot \int_{0+0}^{+\infty} x^{a-1} \cdot e^{-x} dx = a \cdot \Gamma(a)$.

(2) $\Gamma(1) = \int_{0}^{+\infty} e^{-x} dx = 1 = 0!$. From (1), $\Gamma(2) = 1 \cdot \Gamma(1) = 1, \Gamma(3) = 2 \cdot \Gamma(2), \cdots, \Gamma(n+1) = n \cdot \Gamma(n)$.

By multiplying the above relationships, we obtain the result from (2).

(3) $\Gamma\left(\frac{1}{2}\right) = \int_{0+0}^{+\infty} \frac{1}{\sqrt{x}} \cdot e^{-x} dx$; in the previous integral, we perform the change of variable $x = y^2$ and we obtain $\Gamma\left(\frac{1}{2}\right) = \int_{0+0}^{+\infty} \frac{1}{y} \cdot e^{-y^2} \cdot 2y dy = 2 \cdot \int_{0}^{+\infty} e^{-y^2} dy = \int_{-\infty}^{+\infty} e^{-y^2} dy = \sqrt{\pi}$.

The final equality can be proven in various ways; here, we propose a demonstration based on the change of variables formula for double integrals.
$$I^2 = \int_{-\infty}^{+\infty} e^{-x^2} dx \cdot \int_{-\infty}^{+\infty} e^{-y^2} dy = \iint_{\mathbb{R}^2} e^{-(x^2+y^2)} dx dy.$$
In the double integral, we switch to polar coordinates:
$$\begin{cases} x = u \cos v, \\ y = u \sin v \end{cases}, u \geq 0, v \in [0, 2\pi]. \text{ The Jacobian of this transformation is}$$

$$\frac{D(x, y)}{D(u, v)} = \begin{vmatrix} x'_u & x'_v \\ y'_u & y'_v \end{vmatrix} = u$$

and then

$$I^2 = \iint_{[0,+\infty[\times [0,2\pi]} e^{-u^2} \cdot \frac{D(x, y)}{D(u, v)} du dv = \int_0^{+\infty} e^{-u^2} \cdot u du \cdot \int_0^{2\pi} dv =$$

$$= \left(-\frac{1}{2} \cdot e^{-u^2} \Big|_0^{+\infty} \right) \cdot 2\pi = \pi.$$

(4) $\Gamma \left(n + \frac{1}{2} \right) = \Gamma \left(\frac{2n+1}{2} \right) = \frac{2n-1}{2} \cdot \Gamma \left(\frac{2n-1}{2} \right) =$
$= \frac{2n-1}{2} \cdot \frac{2n-3}{2} \cdot \Gamma \left(\frac{2n-3}{2} \right) = \cdots = \frac{(2n-1)(2n-3) \cdots 3 \cdot 1}{2^n} \cdot \Gamma \left(\frac{1}{2} \right) =$
$= \frac{(2n-1)!!}{2^n} \cdot \sqrt{\pi}.$

□

The properties of the function Γ presented in the previous proposition allow for the calculation of the values of certain integrals that will frequently arise in what follows.

Proposition 3.2.2

(1) $\int_0^{+\infty} e^{-tx^2} \cdot x^{2n} dx = \frac{\sqrt{\pi}}{2} \cdot \frac{(2n-1)!!}{2^n} \cdot t^{-\frac{2n+1}{2}}$, *for every* $n \in \mathbb{N}$, *and every* $t > 0$ $((2n-1)!! = 1 \cdot 3 \cdot 5 \cdots (2n-1)$ *and* $(-1)!! = 1)$.

(2) $\int_0^{+\infty} e^{-tx^2} \cdot x^{2n+1} dx = \frac{n!}{2} \cdot t^{-n-1}$, *for every* $n \in \mathbb{N}$, *and every* $t > 0$.

(3) $\int_0^{+\infty} e^{-tx} \cdot x^n dx = n! \cdot t^{-n-1}$, *for every* $n \in \mathbb{N}$, *and every* $t > 0$.

3.2 Laplace's Method

Proof

(1) In the integral $\int_0^{+\infty} e^{-tx^2} \cdot x^{2n} dx$ we make the change of variable $tx^2 = y$ and we obtain

$$\int_0^{+\infty} e^{-tx^2} \cdot x^{2n} dx = \int_0^{+\infty} e^{-y} \cdot y^n \cdot t^{-n} \cdot \frac{1}{\sqrt{t}} \cdot \frac{1}{2\sqrt{y}} dy = \frac{1}{2} \cdot t^{-\frac{2n+1}{2}} \cdot \Gamma\left(n + \frac{1}{2}\right) =$$

$$= \frac{1}{2} \cdot t^{-\frac{2n+1}{2}} \cdot \frac{(2n-1)!!}{2^n} \cdot \sqrt{\pi}.$$

(2) $\int_0^{+\infty} e^{-tx^2} \cdot x^{2n+1} dx = \int_0^{+\infty} e^{-y} \cdot y^{\frac{2n+1}{2}} \cdot t^{-\frac{2n+1}{2}} \cdot \frac{1}{\sqrt{t}} \cdot \frac{1}{2\sqrt{y}} dy =$

$$= \frac{1}{2} \cdot t^{-n-1} \cdot \Gamma(n+1) = \frac{n!}{2} \cdot t^{-n-1}.$$

(3) With the change of variable $x = y^2$ we obtain $\int_0^{+\infty} e^{-tx} \cdot x^n dx =$

$$= 2 \int_0^{+\infty} e^{-ty^2} \cdot y^{2n+1} dy = n! t^{-n-1}.$$

□

We will present an example to illustrate Laplace's method.

Example 3.2.3 Let's study the asymptotic behavior as $t \to +\infty$ of the function defined by the integral:

$$f(t) = \int_{-\infty}^{+\infty} e^{-tx^2} \cdot \log(1 + x + x^2) dx.$$

Let's observe initially that, for every $t \in]-1, 1[$,

$$\frac{1}{1+t} = 1 - t + t^2 - t^3 + \cdots + (-1)^{n+1} t^n + \cdots$$

We integrate the previous relationship over $[0, x]$; so, for every $x \in]-1, 1[$:

$$\log(1+x) = x - \frac{x^2}{2} + \frac{x^3}{3} - \frac{x^4}{4} + \cdots + (-1)^{n+1} \frac{x^n}{n} + \cdots. \quad (3.4)$$

We use formula (3.4) to expand $\log(1+x+x^2)$; in order to apply it, we need $x+x^2$ to be in the interval $]-1, 1[$, which is equivalent to $x \in \left]\frac{-1-\sqrt{5}}{2}, \frac{-1+\sqrt{5}}{2}\right[$.

So, for all $x \in \left[-\frac{1}{2}, \frac{1}{2}\right] \subseteq \left]\frac{-1-\sqrt{5}}{2}, \frac{-1+\sqrt{5}}{2}\right[$,

$$\log(1+x+x^2) = (x+x^2) - \frac{1}{2}(x+x^2)^2 + \cdots + \frac{(-1)^{n+1}}{n}(x+x^2)^n + \cdots \quad (3.5)$$

Grouping the similar terms, for every $x \in \left[-\frac{1}{2}, \frac{1}{2}\right]$,

$$\log(1 + x + x^2) = x + \frac{1}{2}x^2 - \frac{2}{3}x^3 + \frac{1}{4}x^4 + \frac{1}{5}x^5 - \frac{2}{6}x^6 + \cdots \quad (3.6)$$

$$\cdots + \frac{1}{3n-2}x^{3n-2} + \frac{1}{3n-1}x^{3n-1} - \frac{2}{3n}x^{3n} + \cdots.$$

Let $(a_n)_{n \geq 1}$ be the sequence defined by

$$a_{3n-2} = \frac{1}{3n-2}, \; a_{3n-1} = \frac{1}{3n-1}, \; a_{3n} = -\frac{2}{3n}, \text{ for every } n \geq 1.$$

Then (3.6) can be rewritten as

$$\log(1 + x + x^2) = \sum_{n=1}^{\infty} a_n x^n, \text{ for every } x \in \left[-\frac{1}{2}, \frac{1}{2}\right]. \quad (3.7)$$

For every $N \in \mathbb{N}$, we will define the function $R_N : \mathbb{R} \to \mathbb{R}$, by $R_N(x) = \log(1 + x + x^2) - \sum_{n=1}^{2N-1} a_n x^n$, for every $x \in \mathbb{R}$. We apply Lagrange's theorem to the function R_N on the interval $[0, x]$; let $c \in]0, x[$ such that $R_N(x) - R_N(0) = R'_N(c) \cdot x$, or

$$R_N(x) = R'_N(c) \cdot x. \quad (3.8)$$

But $R'_N(x) = \frac{2x+1}{x^2+x+1} - 1 - x + 2x^2 - x^3 - x^4 + 2x^5 + \cdots - (2N-1)a_{2N-1}x^{2N-2}$. Bringing to a common denominator and simplifying the terms, we obtain:

$$R'_N(x) = -\frac{1}{x^2 + x + 1}\left[(2N - 2)a_{2N-2}x^{2N-1} + (2N - 1)a_{2N-1}x^{2N} + \right.$$

$$\left. + (2N - 1)a_{2N-1}x^{2N-1}\right].$$

Taking into account that the pair of numbers $((2N - 2)a_{2N-2}, (2N - 1)a_{2N-1})$ can only take the values $(1, 1), (1, -2), (-2, 1)$, it follows that $|R'_N(x)| < 4|x|^{2N-1}$ and then, from (3.8),

$$|R_N(x)| < 4|x|^{2N}. \quad (3.9)$$

3.2 Laplace's Method

It follows that

$$f(t) = \int_{-\infty}^{\infty} e^{-tx^2} \log(1+x+x^2)dx = \int_{-\infty}^{\infty} \left[\sum_{n=1}^{2N-1} a_n x^n + R_N(x)\right] e^{-tx^2} dx =$$

$$= \sum_{n=1}^{2N-1} a_n \int_{-\infty}^{\infty} e^{-tx^2} x^n dx + \int_{-\infty}^{\infty} e^{-tx^2} R_N(x) dx.$$

We evaluate the last term in the above sum using Proposition 3.2.2:

$$\left|\int_{-\infty}^{\infty} e^{-tx^2} R_N(x)dx\right| \leq 4 \int_{-\infty}^{\infty} e^{-tx^2} x^{2N} dx = 8 \int_0^{\infty} e^{-tx^2} x^{2N} dx =$$

$$= 4\sqrt{\pi} \frac{(2N-1)!!}{2^N} \cdot t^{-\frac{2N+1}{2}} = O\left(t^{-(N+\frac{1}{2})}\right).$$

We observe that $\int_{-\infty}^{\infty} e^{-tx^2} x^n dx = 0$ for any odd n, and thus, by using Proposition 3.2.2 again,

$$f(t) = \sum_{n=1}^{N-1} \sqrt{\pi} \frac{(2n-1)!!}{2^n} a_{2n} t^{-(n+\frac{1}{2})} + O\left(t^{-(N+\frac{1}{2})}\right), \text{ for every } N \in \mathbb{N}.$$

Point (ii) of Theorem 3.1.11 shows us that

$$f(t) \approx \sum_{n=1}^{\infty} \sqrt{\pi} \frac{(2n-1)!!}{2^n} a_{2n} t^{-(n+\frac{1}{2})} \; (t \to \infty).$$

In what follows, we will analyze the asymptotic behavior of several classes of integrals.

3.2.1 Integrals of the Type $\int_\alpha^\beta e^{th(x)} dx$

Theorem 3.2.4 *Let $\alpha, \beta \in \bar{\mathbb{R}}$ such that $-\infty \leq \alpha < 0 < \beta \leq +\infty$ and let $h :]\alpha, \beta[\to \mathbb{R}$ be a function with the properties:*

(1) *h is continuous on $]\alpha, \beta[$.*
(2) *$h(x) < h(0) = 0$, for every $x \in]\alpha, \beta[\setminus \{0\}$.*
(3) *There exist $b, c > 0$ such that, $h(x) \leq -b$, for every $x \in]\alpha, \beta[, |x| \geq c$.*
(4) *h is twice differentiable at the origin, and $h''(0) < 0$,*
(5) *$\int_\alpha^\beta e^{h(x)} dx$ is convergent.*

Then $\int_\alpha^\beta e^{th(x)}dx$ converges, for every $t \geq 1$ and

$$\int_\alpha^\beta e^{th(x)}dx \sim \sqrt{\frac{2\pi}{-th''(0)}} \quad (t \to \infty).$$

Proof We remark that $h(x) \leq 0$, for every $x \in]\alpha, \beta[$ and then $e^{th(x)} \leq e^{h(x)}$, for every $x \in]\alpha, \beta[$ and every $t \geq 1$. From (5), $\int_\alpha^\beta e^{th(x)}dx$ converges, for every $t \geq 1$.

Condition (4) ensures that h is differentiable in a neighborhood of the origin $V \subseteq]\alpha, \beta[$; since (2) tells us that h has a maximum point at the origin, then, by Fermat's theorem, $h'(0) = 0$.

Therefore $h''(0) = \lim_{x \to 0} \frac{h'(x) - h'(0)}{x} = \lim_{x \to 0} \frac{h'(x)}{x}$.

We define the function $\varphi :]\alpha, \beta[\to \mathbb{R}$ by $\varphi(x) = h(x) - \frac{1}{2}x^2 \cdot h''(0)$.

φ is differentiable in V and $\varphi'(x) = h'(x) - xh''(0)$, for every $x \in V$. Therefore, $\varphi(0) = \varphi'(0) = 0$ and $\varphi''(0) = \lim_{x \to 0} \frac{\varphi'(x)}{x} = \lim_{x \to 0} \left[\frac{h'(x)}{x} - h''(0)\right] = 0$.

Let $\varepsilon > 0$ such that $2\varepsilon < -h''(0)$; there exists $\delta > 0$ such that φ is differentiable on $[-\delta, \delta] \subseteq V \subseteq]\alpha, \beta[$ and

$$|\varphi'(x)| < \varepsilon \cdot |x|, \text{ for every } x \in]\alpha, \beta[\text{ with } |x| \leq \delta. \tag{3.10}$$

We can choose $\delta < c$, and since h is continuous on the compact set $C = [-c, -\delta] \cup [\delta, c]$, there exists $x_0 \in C$ such that $h(x) \leq h(x_0)$, for every $x \in C$. From (2), $h(x_0) < 0$; we denote by $M = \max\{b, -h(x_0) > 0$. From (3) it follows that

$$h(x) \leq -M, \text{ for every } x \in]\alpha, \beta[\setminus] - \delta, \delta[. \tag{3.11}$$

Let $x \in [-\delta, \delta]$; we apply Lagrange's theorem on the interval $[0, x]$. Therefore, there exists a point c between 0 and x such that $\varphi(x) - \varphi(0) = \varphi'(c) \cdot x$, from where, using (3.10),

$$|\varphi(x)| < \varepsilon \cdot x^2, \text{ for every } x \in [-\delta, \delta]. \tag{3.12}$$

From (3.12) we obtain

$$\frac{1}{2} \cdot x^2 \cdot [h''(0) - 2\varepsilon] < h(x) < \frac{1}{2} \cdot x^2 \cdot [h''(0) + 2\varepsilon], \text{ for every } x \in [-\delta, \delta]. \tag{3.13}$$

We multiply inequalities (3.13) by $t \geq 1$, exponentiate them, integrate over the interval $[-\delta, \delta]$, and obtain:

$$\underbrace{\int_{-\delta}^{\delta} e^{\frac{1}{2}tx^2[h''(0)-2\varepsilon]}dx}_{I_1} < \int_{-\delta}^{\delta} e^{th(x)}dx < \underbrace{\int_{-\delta}^{\delta} e^{\frac{1}{2}tx^2[h''(0)+2\varepsilon]}dx}_{I_2}. \tag{3.14}$$

3.2 Laplace's Method

According to Proposition 3.2.2 and to the inequality $2\varepsilon < -h''(0)$ it follows:

$$I_2 \leqslant \int_{-\infty}^{\infty} e^{\frac{1}{2}tx^2[h''(0)+2\varepsilon]}dx = 2\int_{0}^{\infty} e^{-\frac{1}{2}tx^2[-h''(0)-2\varepsilon]}dx = \tag{3.15}$$

$$= \sqrt{\pi}\frac{1}{\sqrt{\frac{1}{2}t[-h''(0)-2\varepsilon]}} = \sqrt{\frac{2\pi}{-t[h''(0)+2\varepsilon]}}.$$

Using (3.11), (3.14), and (3.15) we obtain

$$\int_{\alpha}^{\beta} e^{th(x)}dx = \int_{]\alpha,\beta[\setminus]-\delta,\delta[} e^{(t-1)h(x)} \cdot e^{h(x)}dx + \int_{-\delta}^{\delta} e^{th(x)}dx <$$

$$< e^{-(t-1)M} \cdot \int_{]\alpha,\beta[\setminus]-\delta,\delta[} e^{h(x)}dx + I_2 \leqslant$$

$$\leqslant e^{-tM} \cdot e^M \cdot \int_{\alpha}^{\beta} e^{h(x)}dx + \sqrt{\frac{2\pi}{-t[h''(0)+2\varepsilon]}}.$$

Since $\int_{-\infty}^{+\infty} e^{h(x)}dx$ converges, we denote by $K = e^M \int_{-\infty}^{+\infty} e^{h(x)}dx \in \mathbb{R}$ and, from the above relationship, we obtain:

$$\int_{\alpha}^{\beta} e^{th(x)}dx < K \cdot e^{-tM} + \sqrt{\frac{2\pi}{-t[h''(0)+2\varepsilon]}}, \text{ for every } t \geqslant 1. \tag{3.16}$$

Dividing Eq. (3.16) by $\sqrt{\frac{2\pi}{-th''(0)}}$ and taking the upper limit as $t \to +\infty$, we obtain:

$$\limsup_{t \to +\infty} \frac{\int_{\alpha}^{\beta} e^{th(x)}dx}{\sqrt{\frac{2\pi}{-th''(0)}}} \leqslant \sqrt{\frac{h''(0)}{h''(0)+\varepsilon}}, \text{ for every } \varepsilon \in \left]0, \frac{-h''(0)}{2}\right[. \tag{3.17}$$

If $\varepsilon \to 0$, then

$$\limsup_{t \to +\infty} \frac{\int_{\alpha}^{\beta} e^{th(x)}dx}{\sqrt{\frac{2\pi}{-th''(0)}}} \leqslant 1. \tag{3.18}$$

On the other hand, using again Eqs. (3.14) and Proposition 3.2.2,

$$\int_\alpha^\beta e^{th(x)}dx \geq \int_{-\delta}^\delta e^{th(x)}dx > I_1 =$$

$$= \int_\alpha^\beta e^{\frac{1}{2}tx^2[h''(0)-2\varepsilon]}dx - \int_{]\alpha,-\delta[\cup]\delta,\beta[} e^{\frac{1}{2}tx^2[h''(0)-2\varepsilon]}dx =$$

$$= \sqrt{\frac{2\pi}{-t[h''(0)-2\varepsilon]}} - \int_{]\alpha,-\delta[\cup]\delta,\beta[} e^{\frac{1}{2}(t-1)x^2[h''(0)-2\varepsilon]} \cdot e^{\frac{1}{2}x^2[h''(0)-2\varepsilon]}dx.$$

We remark that $\frac{1}{2}x^2[h''(0)-2\varepsilon] \leq \frac{\delta^2}{2}[h''(0)-2\varepsilon] \equiv -M(\varepsilon) < 0$, for every $x \in]\alpha, \beta[$ with $|x| \geq \delta$ and then, from the previous inequalities we obtain

$$\int_\alpha^\beta e^{th(x)}dx > \sqrt{\frac{2\pi}{-t[h''(0)-2\varepsilon]}} - e^{-\frac{\delta^2}{2}[2\varepsilon-h''(0)](t-1)} \cdot \int_\alpha^\beta e^{-\frac{1}{2}x^2[2\varepsilon-h''(0)]}dx =$$

$$= \sqrt{\frac{2\pi}{-t[h''(0)-2\varepsilon]}} - e^{-tM(\varepsilon)} \cdot e^{M(\varepsilon)} \cdot \sqrt{\frac{2\pi}{2\varepsilon-h''(0)}}.$$

We denote by $K(\varepsilon) = e^{M(\varepsilon)} \cdot \sqrt{\frac{2\pi}{2\varepsilon-h''(0)}} \in \mathbb{R}$ and then

$$\int_\alpha^\beta e^{th(x)}dx > \sqrt{\frac{2\pi}{-t[h''(0)-2\varepsilon]}} - K(\varepsilon) \cdot e^{-tM(\varepsilon)}. \tag{3.19}$$

We divide Eq. (3.19) by $\sqrt{\frac{2\pi}{-th''(0)}}$ and take the lower limit as $t \to +\infty$:

$$\liminf_{t \to +\infty} \frac{\int_\alpha^\beta e^{th(x)}dx}{\sqrt{\frac{2\pi}{-th''(0)}}} \geq \sqrt{\frac{h''(0)}{h''(0)-2\varepsilon}}, \text{ for every } \varepsilon \in \left]0, \frac{-h''(0)}{2}\right[. \tag{3.20}$$

In (3.20), we take the limit as $\varepsilon \to 0$ and obtain:

$$\liminf_{t \to +\infty} \frac{\int_\alpha^\beta e^{th(x)}dx}{\sqrt{\frac{2\pi}{-th''(0)}}} \geq 1. \tag{3.21}$$

3.2 Laplace's Method

From (3.18) and (3.21) it follows that there exists

$$\lim_{t \to +\infty} \frac{\int_\alpha^\beta e^{th(x)} dx}{\sqrt{\dfrac{2\pi}{-h''(0)}}} = 1.$$

□

Remark 3.2.5 A sufficient condition for conditions (2) and (3) to be satisfied is that $h(0) = 0$ and $h'(x) > 0 > h'(y)$ for every $x < 0 < y$.

Corollary 3.2.6 *Let $h :]\alpha, \beta[\to \mathbb{R}$ be a function with the properties:*

(1) h is continuous on $]\alpha, \beta[$.
(2) $h(x) < h(0)$, for every $x \in]\alpha, \beta[\setminus \{0\}$.
(3) There exist $b, c > 0$ such that, $h(x) - h(0) \leq -b$, for every $x \in]\alpha, \beta[, |x| \geq c$.
(4) h is twice differentiable at the origin, and $h''(0) < 0$,
(5) $\int_\alpha^\beta e^{h(x)} dx$ is convergent.

Then

$$\int_\alpha^\beta e^{th(x)} dx \sim e^{t \cdot h(0)} \sqrt{\frac{2\pi}{-th''(0)}} \quad (t \to +\infty).$$

Proof The function $h_1 :]\alpha, \beta[\to \mathbb{R}, h_1(x) = h(x) - h(0)$, for every $x \in]\alpha, \beta[$, check the conditions of the previous theorem. □

Remark 3.2.7 We can observe a change in the behavior of the integral based on the value that the function h takes at the origin. Therefore

If $h(0) > 0$, then $\int_\alpha^\beta e^{th(x)} dx \to +\infty$ at the speed of $e^{th(0)}$;
If $h(0) = 0$, then $\int_\alpha^\beta e^{th(x)} dx \to 0$ at the speed of $\frac{1}{\sqrt{t}}$;
If $h(0) < 0$, then $\int_\alpha^\beta e^{th(x)} dx \to 0$ at the speed of $e^{th(0)}$.

Examples 3.2.8

(1) $\displaystyle\int_{-\infty}^{+\infty} e^{-tx^2} \cdot \left(\frac{x^2+2}{2}\right)^t dx \sim \sqrt{\frac{2\pi}{t}} \quad (t \to +\infty)$.

Indeed, $\displaystyle\int_{-\infty}^{+\infty} e^{-tx^2} \cdot \left(\frac{x^2+2}{2}\right)^t dx = \int_{-\infty}^{+\infty} e^{t\left[\log\left(\frac{x^2+2}{2}\right) - x^2\right]} dx$. If we denote by $h(x) = \log(x^2+2) - x^2 - \log 2$, then h is continuous on $\mathbb{R}$, $h(0) = 0$, $h'(x) = -2 \cdot \frac{x^3+x}{x^2+2}$, $h''(0) = -1$. Thus, conditions (1)–(5) of Theorem 3.2.4 are satisfied (for the verification of conditions 2 and 3, see Remark 3.2.5).

(2) Calculate $\lim_{t\to+\infty} \sqrt{t} \cdot \int_{-\infty}^{+\infty} e^{-tx^2(x^2+1)} dx$.

If $h(x) = -x^4 - x^2$, then $h(0) = 0$, $h'(x) = -2x(2x^2+1)$, and $g''(0) = -2$. According to Theorem 3.2.4, $\int_{-\infty}^{+\infty} e^{-tx^2(x^2+1)} dx \sim \sqrt{\frac{\pi}{t}}$ ($t \to +\infty$), and then $\lim_{t\to+\infty} \sqrt{t} \cdot \int_{-\infty}^{+\infty} e^{-tx^2(x^2+1)} dx = \sqrt{\pi}$.

(3) Evaluate the asymptotic behavior of integral $\int_{-\infty}^{+\infty} e^{-tx^2}(x^2+2)^t dx$ as t approaches $+\infty$.

If $h(x) = \log(x^2+2) - x^2$, then $h(0) = \log 2$, $h'(x) = -2\frac{x^3+x}{x^2+2}$, and $h''(0) = -1$. According to Corollary 3.2.6, $\int_{-\infty}^{+\infty} e^{-tx^2}(x^2+2)^t dx \sim e^{t\log 2} \cdot \sqrt{\frac{2\pi}{t}} = 2^t \cdot \sqrt{\frac{2\pi}{t}}$.

3.2.2 Integrals of the Type $\int_0^{+\infty} e^{-tx} x^\lambda g(x) dx$

Theorem 3.2.9 (Watson's Lemma) *Let $g : [0, +\infty[\to \mathbb{R}$ be a function with the properties:*

(1) g is continuous on $[0, +\infty[$.
(2) There exist $r > 0$, and $(a_n)_n \subseteq \mathbb{R}$ such that $g(x) = \sum_{n=0}^{\infty} a_n x^n$, for every $x \in [0, r]$.
(3) There exist $K > 0$, and $c \in \mathbb{R}$ such that $|g(x)| \leq K \cdot e^{cx}$, for every $x \in [0, +\infty[$.

Then, for every $\lambda > -1$,

$$\int_0^{+\infty} e^{-tx} x^\lambda g(x) dx \approx \sum_{n=0}^{+\infty} a_n \Gamma(\lambda+n+1) t^{-(\lambda+n+1)} \quad (t \to \infty).$$

Proof Let $\beta \in \mathbb{R}$ such that $-\lambda < \beta < 1$; then $\lim_{x\to 0} x^\beta \left[e^{-tx} x^\lambda |g(x)|\right] = 0 < +\infty$ and therefore $\int_0^\infty e^{-tx} x^\lambda g(x) dx$ is absolutely convergent at 0, for every $t \geq 0$.

$$\left|e^{-tx} x^\lambda g(x)\right| \leq K e^{(-t+c)x} x^\lambda \text{ and so}$$

$$\lim_{x\to\infty} x^2 \left[e^{-tx} x^\lambda |g(x)|\right] \leq \lim_{x\to\infty} K e^{(-t+c)x} x^{\lambda+2} = 0, \text{ for } t > c.$$

It follows that $\int_0^\infty e^{-tx} x^\lambda g(x) dx$ is absolutely convergent at ∞, for $t > c$.

Therefore $\int_0^\infty e^{-tx} x^\lambda g(x) dx$ it is absolutely convergent when t is in a neighborhood of $+\infty$.

$$\int_0^\infty e^{-tx} x^\lambda g(x) dx = \underbrace{\int_0^r e^{-tx} x^\lambda g(x) dx}_{I_1(t)} + \underbrace{\int_r^\infty e^{-tx} x^\lambda g(x) dx}_{I_2(t)}.$$

3.2 Laplace's Method

For $t > c + 1$ we have

$$|I_2(t)| \leqslant K \int_r^\infty e^{(-t+c)x} x^\lambda dx = K \int_r^\infty e^{(-t+c+1)x} e^{-x} x^\lambda dx \leqslant \qquad (3.22)$$

$$\leqslant K e^{(-t+c+1)r} \int_r^\infty e^{-x} x^\lambda dx = K_1 e^{-tr}.$$

According to (3.22)

$$I_2(t) = O(e^{-tr}) \quad (t \to +\infty).$$

For every $N \in \mathbb{N}$ let us denote by $r_N(x) = g(x) - \sum_{n=0}^N a_n x^n$.

For every $x \in [0, r]$, $r_N(x) = \sum_{n=N+1}^\infty a_n x^n = x^{N+1} \cdot (a_{N+1} + a_{N+2}x + \cdots) = x^{N+1} \cdot g_1(x)$. The function g_1 is continuous on $[0, r]$ and then, there exists $L > 0$ such that

$$|r_N(x)| \leqslant L \cdot x^{N+1}, \text{ for every } x \in [0, r]. \qquad (3.23)$$

$$I_1(t) = \int_0^r e^{-tx} x^\lambda g(x) dx = \sum_{n=0}^N a_n \cdot \int_0^r e^{-tx} x^{\lambda+n} dx + \int_0^r e^{-tx} x^\lambda r_N(x) dx. \qquad (3.24)$$

From (3.23) it follows that

$$\left| \int_0^r e^{-tx} x^\lambda r_N(x) dx \right| \leqslant L \cdot \int_0^r e^{-tx} x^{\lambda+N+1} dx. \qquad (3.25)$$

Let's evaluate the integrals $I_\alpha(t) = \int_0^r e^{-tx} x^\alpha dx$; first, we perform the change of variable $tx = y$ and we obtain

$$I_\alpha(t) = \int_0^{tr} e^{-y} t^{-(\alpha+1)} y^\alpha dy = t^{-(\alpha+1)} \left[\int_0^\infty e^{-y} y^\alpha dy - \underbrace{\int_{tr}^\infty e^{-y} y^\alpha dy}_{J(t)} \right] = \qquad (3.26)$$

$$= t^{-(\alpha+1)} \cdot \Gamma(\alpha+1) - t^{-(\alpha+1)} \cdot J(t).$$

In $J(t)$ we make the change of variable $y = tr(1 + u)$ and we obtain

$$J(t) = \int_0^\infty e^{-tr} e^{-tru} (tr)^{\alpha+1} (1+u)^\alpha du.$$

Since $1+u \leqslant e^u$, for every $u \in \mathbb{R}$,

$$J(t) \leqslant e^{-tr}(tr)^{\alpha+1} \int_0^\infty e^{(-tr+\alpha)u} du = e^{-tr}(tr)^{\alpha+1} \frac{1}{tr-\alpha} \sim$$

$$\sim e^{-tr}(tr)^\alpha \quad (t \to +\infty)$$

and then

$$t^{-(\alpha+1)} \cdot J(t) = O\left(e^{-tr}\right) \quad (t \to +\infty). \tag{3.27}$$

From (3.25) and (3.26) we obtain

$$\left| \int_0^r e^{-tx} x^\lambda r_N(x) dx \right| \leqslant L \cdot I_{\lambda+N+1}(t) =$$

$$= L \cdot \Gamma(\lambda+N+2) \cdot t^{-(\lambda+N+2)} + t^{-(\lambda+N+2)} \cdot J(t).$$

But $O\left(e^{-tr}\right) = O\left(t^{-(\lambda+N+2)}\right) (t \to +\infty)$, and then, according to (3.27),

$$\int_0^r e^{-tx} x^\lambda r_N(x) dx = O\left(t^{-(\lambda+N+2)}\right) (t \to +\infty). \tag{3.28}$$

From (3.24), (3.26), (3.27), and (3.28) it follows that

$$I_1(t) = \sum_{n=0}^N a_n I_{\lambda+n}(t) + O\left(t^{-(\lambda+N+2)}\right) = \sum_{n=0}^N a_n \Gamma(\lambda+n+1) t^{-(\lambda+n+1)} +$$

$$+ O\left(e^{-tr}\right) + O\left(t^{(\lambda+N+2)}\right) (t \to +\infty),$$

or

$$I_1(t) = \sum_{n=0}^N a_n \Gamma(\lambda+n+1) t^{-(\lambda+n+1)} + O\left(t^{-(\lambda+N+2)}\right) (t \to +\infty) \tag{3.29}$$

and then, for every $N \in \mathbb{N}$,

$$\int_0^\infty e^{-tx} x^\lambda g(x) dx = I_1(t) + O\left(e^{-tr}\right) =$$

$$= \sum_{n=0}^N a_n \Gamma(\lambda+n+1) t^{-(\lambda+n+1)} + O\left(t^{-(\lambda+N+2)}\right) (t \to +\infty),$$

3.2 Laplace's Method

which is equivalent to

$$\int_0^\infty e^{-tx} x^\lambda g(x)dx \approx \sum_{n=0}^\infty a_n \Gamma(\lambda+n+1) t^{-(\lambda+n+1)} \quad (t \to \infty).$$

□

Remark 3.2.10 The result from the previous theorem remains valid if the half-right interval $[0, +\infty[$ is replaced with the interval $[0, T[$, where $T \in \mathbb{R}_+$.

Examples 3.2.11

(1) Let be the function $f(t) = \int_0^{+\infty} e^{-tx} \log(1+x+x^2)dx$; determine the asymptotic expansion for as $t \to +\infty$.

In Example 3.2.3 we have obtained the following expansion:

$$\log(1+x+x^2) = \sum_{n=1}^\infty a_n x^n, \text{ for every } x \in \left[-\frac{1}{2}, \frac{1}{2}\right], \text{ where}$$

$a_{3n-2} = \frac{1}{3n-2}, a_{3n-1} = \frac{1}{3n-1}, a_{3n} = -\frac{2}{3n},$ for every $n \in \mathbb{N}^*$.

It follows that the function $g : [0, +\infty[\to \mathbb{R}$, defined as $g(x) = \log(1+x+x^2)$ for $x \in [0, +\infty[$, is continuous and can be expanded into a power series in a neighborhood of the origin. Condition (3) from Watson's lemma is also satisfied (for every $x \in [0, +\infty[, \log(1+x+x^2) \leq 1+x \leq e^x)$, and therefore

$$\int_0^\infty e^{-tx} \log(1+x+x^2)dx \approx \sum_{n=1}^\infty a_n \Gamma(n+1) t^{-(n+1)} \quad (t \to \infty) \text{ or}$$

$$\int_0^\infty e^{-tx} \log(1+x+x^2)dx \approx \sum_{n=1}^\infty \frac{a_n \cdot n!}{t^{n+1}} \quad (t \to \infty).$$

(2) In Example 3.2.3, we have shown that

$$\int_{-\infty}^{+\infty} e^{-tx^2} \cdot \log(1+x+x^2)dx \approx \sum_{n=1}^\infty \sqrt{\pi} \frac{(2n-1)!!}{2^n} a_{2n} t^{-(n+\frac{1}{2})} \quad (t \to \infty),$$

where $(a_n)_n$ is the sequence of coefficients of the series expansion of the function $g : [-\frac{1}{2}, \frac{1}{2}] \to \mathbb{R}, g(x) = \log(1+x+x^2)$. We will find this result as an application of Watson's lemma.

First, we will transform the integral

$$\int_{-\infty}^{+\infty} e^{-tx^2} \cdot \log(1+x+x^2)dx = \int_{-\infty}^{0} e^{-tx^2} \cdot \log(1+x+x^2)dx +$$

$$+ \int_{0}^{+\infty} e^{-tx^2} \cdot \log(1+x+x^2)dx =$$

$$= \int_{0}^{+\infty} e^{-tx^2} \cdot \left[\log(1-x+x^2) + \log(1+x+x^2)\right]dx =$$

$$= \int_{0}^{+\infty} e^{-tx^2} \cdot \log(1+x^2+x^4)dx.$$

If we perform the change of variable in the last integral $x \mapsto \sqrt{x}$ we obtain

$$f(t) = \frac{1}{2} \int_{0}^{\infty} e^{-tx} x^{-\frac{1}{2}} \log(1+x+x^2)dx.$$

We can apply Watson's lemma to this last integral, and we obtain

$$f(t) \approx \frac{1}{2} \sum_{n=1}^{\infty} a_n \Gamma\left(n + \frac{1}{2}\right) \cdot t^{-(n+\frac{1}{2})} \quad (t \to \infty)$$

or

$$f(t) \approx \sqrt{\pi} \sum_{n=1}^{\infty} \frac{(2n-1)!!}{2^n} \cdot \frac{a_n}{2} \cdot t^{-(n+\frac{1}{2})} \quad (t \to \infty).$$

It can be easily shown that the sequence $(a_n)_n$ satisfies the recurrence relationship $a_n = 2a_{2n}$ for $n \in \mathbb{N}^*$, and thus we recover the announced result.

Another application of Watson's lemma is presented in the next subsection.

3.2.3 Integrals of the Type $\int_{\alpha}^{\beta} e^{-tx^2} h(x) dx$

Theorem 3.2.12 Let $\alpha, \beta \in \bar{\mathbb{R}}$ such that $-\infty \leq \alpha < 0 < \beta \leq +\infty$ and let $h :]\alpha, \beta[\to \mathbb{R}$ be a continuous function that satisfies the following properties:

(1) There exist $0 < r < \min\{-\alpha, \beta\}$ and a sequence $(a_n)_n \subseteq \mathbb{R}$ such that $h(x) = \sum_{n=0}^{\infty} a_n x^n$, for every $x \in \mathbb{R}$ with $|x| \leq r$.
(2) There exist $K, c > 0$ such that $|h(x)| \leq K e^{c|x|}$, for every $x \in]\alpha, \beta[$.

3.2 Laplace's Method

Then $\int_\alpha^\beta e^{-tx^2} h(x) dx$ is convergent, for every $t > 0$, and

$$\int_\alpha^\beta e^{-tx^2} h(x) dx \approx \sum_{n=0}^\infty a_{2n} \Gamma\left(n + \frac{1}{2}\right) \cdot t^{-(n+\frac{1}{2})} \quad (t \to \infty).$$

Proof We remark that, for every $t > 0$ and every $x \in]\alpha, \beta[$, $\left|e^{-tx^2} h(x)\right| \leqslant K e^{-tx^2 + c|x|}$ and that the integral $\int_\alpha^\beta e^{-tx^2 + c|x|} dx$ is convergent; it follows that the integral $\int_\alpha^\beta e^{-tx^2} h(x) dx$ is absolutely convergent, for every $t > 0$.

We perform a change of variable in the integral $x = -\sqrt{y}$ on the interval $]\alpha, 0]$ and $x = \sqrt{y}$ on $[0, \beta[$ and we obtain

$$F(t) = \int_\alpha^\beta e^{-tx^2} h(x) dx = \int_\alpha^0 e^{-tx^2} h(x) dx + \int_0^\beta e^{-tx^2} h(x) dx =$$

$$= \frac{1}{2} \int_0^{\alpha^2} e^{-ty} y^{-\frac{1}{2}} h(-\sqrt{y}) dy + \frac{1}{2} \int_0^{\beta^2} e^{-ty} y^{-\frac{1}{2}} h(\sqrt{y}) dy =$$

$$= \frac{1}{2} \int_0^{r^2} e^{-ty} y^{-\frac{1}{2}} \left[h(-\sqrt{y}) + h(\sqrt{y})\right] dy +$$

$$+ \frac{1}{2} \int_{r^2}^{\alpha^2} e^{-ty} y^{-\frac{1}{2}} h(-\sqrt{y}) dy + \frac{1}{2} \int_{r^2}^{\beta^2} e^{-ty} y^{-\frac{1}{2}} h(\sqrt{y}) dy.$$

We remark that

$$\left|\frac{1}{2} \int_{r^2}^{\alpha^2} e^{-ty} y^{-\frac{1}{2}} h(-\sqrt{y}) dy\right| \leqslant \frac{K}{2} \int_{r^2}^{\alpha^2} e^{-ty} y^{-\frac{1}{2}} e^{-c\sqrt{y}} dy \leqslant$$

$$\leqslant \frac{K}{2} \cdot e^{-tr^2} \int_0^\infty y^{-\frac{1}{2}} e^{-c\sqrt{y}} dy = K_1 \cdot e^{-tr^2},$$

and then, for every $N \in \mathbb{N}$,

$$\frac{1}{2} \int_{r^2}^{\alpha^2} e^{-ty} y^{-\frac{1}{2}} h(-\sqrt{y}) dy = O\left(e^{-tr^2}\right) = O\left(t^{-(N+\frac{3}{2})}\right) \quad (t \to +\infty). \qquad (3.30)$$

Similarly, we obtain the inequality

$$\left|\frac{1}{2} \int_{r^2}^{\beta^2} e^{-ty} y^{-\frac{1}{2}} h(\sqrt{y}) dy\right| \leqslant K_2 \cdot e^{-tr^2}$$

and therefore, for every $N \in \mathbb{N}$,

$$\frac{1}{2}\int_{r^2}^{\beta^2} e^{-ty} y^{-\frac{1}{2}} h(\sqrt{y}) dy = O\left(e^{-tr^2}\right) = O\left(t^{-(N+\frac{3}{2})}\right) (t \to +\infty). \quad (3.31)$$

Let now the function $g : [0, r^2] \to \mathbb{R}$, defined by $g(y) = h(-\sqrt{y}) + h(\sqrt{y})$. Then $g(y) = 2 \cdot \sum_{n=0}^{+\infty} a_{2n} y^n$, for every $y \in [0, r^2]$.

Applying Watson's lemma and Remark 3.2.10, it follows that

$$\frac{1}{2} \cdot \int_0^{r^2} e^{-ty} y^{-\frac{1}{2}} g(y) dy \approx \frac{1}{2} \cdot \sum_{n=0}^{\infty} 2a_{2n} \Gamma\left(n + \frac{1}{2}\right) \cdot t^{-\left(n+\frac{1}{2}\right)} \quad (t \to \infty),$$

from where, for every $N \in \mathbb{N}$,

$$\frac{1}{2} \cdot \int_0^{r^2} e^{-ty} y^{-\frac{1}{2}} g(y) dy = \sum_{n=0}^{N} a_{2n} \Gamma\left(n + \frac{1}{2}\right) t^{-\left(n+\frac{1}{2}\right)} + O\left(t^{-\left(N+\frac{3}{2}\right)}\right). \quad (3.32)$$

From (3.30), (3.31), and (3.32), for every $N \in \mathbb{N}$,

$$F(t) = \sum_{n=0}^{N} a_{2n} \Gamma\left(n + \frac{1}{2}\right) t^{-\left(n+\frac{1}{2}\right)} + O\left(t^{-\left(N+\frac{3}{2}\right)}\right),$$

which is equivalent to

$$\int_\alpha^\beta e^{-tx^2} h(x) dx \approx \sum_{n=0}^{\infty} a_{2n} \Gamma\left(n + \frac{1}{2}\right) \cdot t^{-(n+\frac{1}{2})} \quad (t \to \infty).$$

$\square$

3.2.4 Stirling's Formula

In this short paragraph, we will use the results from the preceding sections to derive various forms of Stirling's formula.

Let $\Gamma :\,]-1, +\infty[\, \to \mathbb{R}$ be the function defined by $\Gamma(t+1) = \int_0^\infty e^{-x} x^t dx$, for every $t > -1$.

Making the variable change $x = t + y$, we obtain:

$$\Gamma(t+1) = e^{-t} \int_{-t}^{\infty} e^{-y} \cdot (t+y)^t dy = e^{-t} \cdot t^t \int_{-t}^{\infty} e^{-y} \cdot \left(1 + \frac{y}{t}\right)^t dy$$

3.2 Laplace's Method

formula which, after the new variable change $y = tx$, becomes:

$$\Gamma(t+1) = e^{-t} \cdot t^{t+1} \int_{-1}^{\infty} e^{-tx} \cdot (1+x)^t dx \text{ or}$$

$$\Gamma(t+1) = e^{-t} \cdot t^{t+1} \int_{-1}^{\infty} e^{t[-x+\log(1+x)]} dx. \tag{3.33}$$

If we denote by $h(x) = -x + \log(1 + x)$, the function $h :]-1, +\infty[\to \mathbb{R}$ is continuous and $h(x) < h(0) = 0$, for every $x \in]-1, +\infty[\setminus \{0\}$.

We aim to apply to the above integral Theorem 3.2.4 ($\alpha = -1, \beta = +\infty$); conditions (1) and (2) are satisfied. Taking into account Remark 3.2.5, condition (3) is also verified.

$h''(0) = -1 < 0$ and $\int_{-1}^{\infty} e^{h(x)} dx = e \cdot \Gamma(2) = e$ is convergent. Then, from the quoted theorem, we obtain:

$$\int_{-1}^{\infty} e^{th(x)} dx \sim \sqrt{\frac{2\pi}{t}} \quad (t \to \infty) \text{ and therefore}$$

$$\Gamma(t+1) \sim \sqrt{2\pi t} \cdot \left(\frac{t}{e}\right)^t \quad (t \to \infty).$$

This is a first form of Stirling's formula. If we particularly consider the sequence $(t_n)_{n \in \mathbb{N}}$, defined by $t_n = n$, for every $n \in \mathbb{N}$, we obtain

$$\Gamma(n+1) = n! \sim \sqrt{2\pi n} \left(\frac{n}{e}\right)^n \quad (n \to +\infty).$$

We emphasize that the relationship mentioned above states that

$$n! = \sqrt{2\pi n} \left(\frac{n}{e}\right)^n (1 + \varepsilon_n), \text{ where } \varepsilon_n \to 0.$$

We can attempt to refine Stirling's formula as follows.

In the integral $\int_{-1}^{+\infty} e^{th(x)} dx$, we make the change of variable

$$y^2 = -2h(x) = 2x - 2\log(1+x), \tag{*}$$

and we obtain

$$\int_{-1}^{\infty} e^{th(x)} dx = \int_{-1}^{0} e^{th(x)} dx + \int_{0}^{\infty} e^{th(x)} dx =$$

$$= \int_{-\infty}^{0} e^{-\frac{ty^2}{2}} x'(y) dy + \int_{0}^{\infty} e^{-\frac{ty^2}{2}} x'(y) dy = \int_{-\infty}^{\infty} e^{-\frac{ty^2}{2}} x'(y) dy,$$

so that

$$\Gamma(t+1) = e^{-t} \cdot t^{t+1}\sqrt{2} \cdot \int_{-\infty}^{\infty} e^{-tz^2} x'(\sqrt{2}z)dz. \quad (3.34)$$

We intend to apply Theorem 3.2.12 to the integral from (3.34). For this purpose, we need to make some clarifications about the function x'. First, let's observe that the substitution $(*)$ allows us to define the function $y :]-1, +\infty[\to \mathbb{R}$ by

$$y(x) = \begin{cases} -\sqrt{2x - 2\log(1+x)}, & x \in]-1, 0], \\ \sqrt{2x - 2\log(1+x)}, & x \in]0, +\infty[\end{cases}.$$

The function y is strictly increasing and surjective. Therefore, we can define its inverse function $x : \mathbb{R} \to]-1, +\infty[$. We can also assume that the function x is analytic in a neighborhood of the origin. Thus, let's assume that there exists $r > 0$ such that:

$$x(y) = x(0) + \frac{x'(0)}{1!} \cdot y + \frac{x''(0)}{2!} \cdot y^2 + \cdots, \text{ for every } y \in [-r, r].$$

We will determine the first three terms of the aforementioned series. By assuming increasing calculation difficulties, the procedure can be continued to determine terms of higher rank than three.

We differentiate the relationship $(*)$ with respect to y and obtain:

$$x'(y) = (1+x) \cdot \frac{y}{x},$$

$$x''(y) = (1+x) \cdot \frac{x^2 - y^2}{x^3},$$

$$x'''(y) = (1+x) \cdot \frac{y}{x} \cdot \frac{3y^2 + 2xy^2 - 3x^2}{x^4}.$$

$x(0) = 0$ and since $\lim_{y \to 0} \frac{y}{x(y)} = 1$ and $\lim_{y \to 0} \frac{x^2(y) - y^2}{x^3(y)} = \frac{2}{3}$, we obtain $x'(0) = 1, x''(0) = \frac{2}{3}, x'''(0) = \frac{1}{6}$. Therefore

$$x(y) = y + \frac{1}{3} \cdot y^2 + \frac{1}{36} \cdot y^3 + \cdots,$$

from where

$$x'(y) = 1 + \frac{2}{3} \cdot y + \frac{1}{12} \cdot y^2 + \cdots \text{ for every } y \in [-r, r].$$

3.2 Laplace's Method

It can be easily shown that, for every $x \in [0, +\infty[$, $2x - 2\log(1+x) \leqslant x^2$, from where $y \leqslant x(y)$, for every $y \geqslant 0$ or

$$|x'(y)| = \frac{y}{x} + y \leqslant 1 + y \leqslant e^y, \text{ for every } y \geqslant 0. \tag{3.35}$$

On the other hand, since $\lim_{y \uparrow 0} \frac{y}{x(y)} = 1$, there exists $y_0 < 0$ such that

$$\frac{1}{2} < \frac{y}{x(y)} < \frac{3}{2}, \text{ for every } y \in]y_0, 0[. \tag{3.36}$$

From (3.36) it follows that

$$|x'(y)| = y + \frac{y}{x} < y + \frac{3}{2} < \frac{3}{2} \cdot e^{-y}, \text{ for every } y \in]y_0, 0[. \tag{3.37}$$

Since x is an increasing function, for every $y \leqslant y_0$, $x(y) \leqslant x(y_0) = x_0$ and then

$$|x'(y)| = y + \frac{y}{x} \leqslant y + \frac{y}{x_0} = -y \cdot \frac{1+x_0}{-x_0} < -\frac{1+x_0}{x_0} e^{-y}, \text{ for every } y \leqslant y_0. \tag{3.38}$$

Let us denote $K = \max\left\{\frac{3}{2}, -\frac{1+x_0}{x_0}\right\} > 0$. From (3.37) and (3.38)

$$|x'(y)| < Ke^{-y} = Ke^{|y|}, \text{ for every } y < 0. \tag{3.39}$$

Finally, from (3.35) and (3.39), we obtain

$$|x'(y)| \leqslant Ke^{|y|}, \text{ for every } y \in \mathbb{R}. \tag{*}$$

It follows that

$$|x'(\sqrt{2}z)| \leqslant Ke^{\sqrt{2}|z|}, \text{ for every } z \in \mathbb{R}, \text{ and} \tag{3.40}$$

$$x'(\sqrt{2}z) = 1 + \frac{2\sqrt{2}}{3}z + \frac{1}{6}z^2 + \cdots, \text{ for every } z \in \left[-\frac{r}{\sqrt{2}}, \frac{r}{\sqrt{2}}\right]. \tag{3.41}$$

Equations (3.40) and (3.41) correspond exactly to conditions (1) and (2) from Theorem 3.2.12; the conclusion of this theorem leads us to

$$\int_{-\infty}^{+\infty} e^{tz^2} x'(\sqrt{2}z) dz \approx \sum_{n=0}^{\infty} a_{2n} \Gamma\left(n + \frac{1}{2}\right) \cdot t^{-(n+\frac{1}{2})} (t \to +\infty),$$

where $(a_{2n})_n$ is the sequence of even coefficients from the expansion (3.41), so that $a_0 = 1, a_2 = \frac{1}{6}$. It follows from here and from relationship (3.34) above that

$$\Gamma(t+1) \approx e^{-t} t^{t+1} \sqrt{2} \left[\Gamma\left(\frac{1}{2}\right) t^{-\frac{1}{2}} + \frac{1}{6} \cdot \Gamma\left(\frac{3}{2}\right) t^{-\frac{3}{2}} + \cdots \right] (t \to +\infty).$$

According to Proposition 3.2.1,

$$\Gamma(t+1) \approx \sqrt{2\pi t} \left(\frac{t}{e}\right)^t \cdot \left(1 + \frac{1}{12t} + \cdots\right) (t \to +\infty).$$

In the particular case where $t_n = n$, for every $n \in \mathbb{N}$, we obtain

$$\Gamma(n+1) = n! \sim \sqrt{2\pi n} \left(\frac{n}{e}\right)^n \left(1 + \frac{1}{12n}\right) (n \to +\infty).$$

3.3 Poincaré's Theorem

Results of asymptotic analysis are successfully applied in studying the behavior of solutions to difference equations. We will briefly address this subject in the last paragraph. An important result is Poincaré's theorem regarding the behavior of solutions in a class of homogeneous difference equations - Poincaré-type equations. We present a general demonstration of this theorem, which will incorporate some of the ideas suggested by [4] for the case when $p = 2$, as well as those of [3, 5]. This theorem allows us to obtain information about the asymptotic behavior of solutions to these equations. The paragraph concludes with counterexamples that illustrate the limits of the presented results.

Definition 3.3.1 A homogeneous linear recurrence

$$f(n+p) + a_1(n) f(n+p-1) + \cdots + a_p(n) f(n) = 0, \text{ for every } n \in \mathbb{N},$$

is of **Poincaré type** if there exist

$$\lim_{n \to +\infty} a_i(n) = a_i \in \mathbb{R}, \text{ for every } i = 1, 2, \cdots, p.$$

This condition would suggest that the solutions of such equations will exhibit asymptotic behavior at infinity similar to the solutions of the homogeneous linear equation with constant coefficients

$$f(n+p) + a_1 f(n+p-1) + \cdots + a_p f(n) = 0, \text{ for every } n \in \mathbb{N}.$$

3.3 Poincaré's Theorem

For the convenience of calculations, we will denote the coefficients of a Poincaré equation as $a_i(n) = a_i + \alpha_i(n)$, for every $n \in \mathbb{N}$.

Thus, a Poincaré-type equation takes the form

$$f(n+p)+[a_1+\alpha_1(n)]f(n+p-1)+\cdots+[a_p+\alpha_p(n)]f(n) = 0, \forall n \in \mathbb{N}, \quad (3.42)$$

where

$$\lim_{n \to +\infty} \alpha_i(n) = 0, \text{ for every } i = 1, 2, \cdots, p. \quad (3.43)$$

Theorem 3.3.2 (Poincaré's Theorem) *Consider a Poincaré-type equation (3.42). We assume that the roots $\lambda_1, \lambda_2, \cdots, \lambda_p$ of the characteristic equation*

$$\lambda^p + a_1 \lambda^{p-1} + \cdots + a_p = 0, \quad (3.44)$$

possess the property that $|\lambda_i| \neq |\lambda_j|$, for every $i \neq j, i, j = 1, 2, \cdots, p$.

Then for every nonzero solution f of Eq. (3.42) there exists a root λ_i such that

$$\lim_{n \to +\infty} \frac{f(n+1)}{f(n)} = \lambda_i.$$

Proof Let f be a nonzero solution of recurrence (3.42). We consider the linear system of equations

$$\begin{cases} x_1(n) + x_2(n) + \cdots + x_p(n) &= f(n) \\ \lambda_1 x_1(n) + \lambda_2 x_2(n) + \cdots + \lambda_p x_p(n) &= f(n+1) \\ \cdots & \cdots \\ \lambda_1^{p-1} x_1(n) + \lambda_2^{p-1} x_2(n) + \cdots + \lambda_p^{p-1} x_p(n) &= f(n+p-1). \end{cases}, n \in \mathbb{N}.$$

(3.45)

The determinant of this system is a Vandermonde determinant:

$$D = \begin{vmatrix} 1 & 1 & \cdots & 1 \\ \lambda_1 & \lambda_2 & \cdots & \lambda_p \\ \cdots & \cdots & \cdots & \cdots \\ \lambda_1^{p-1} & \lambda_2^{p-1} & \cdots & \lambda_p^{p-1} \end{vmatrix} = \prod_{1 \leq i < j \leq p} (\lambda_j - \lambda_i) \neq 0.$$

It follows that, for any $n \in \mathbb{N}$, the system (3.45) has a unique solution, $x_1(n), x_2(n), \cdots, x_p(n)$, where, for every $k = 1, \cdots, p$, and every $n \in \mathbb{N}$,

$$x_k(n) = \frac{1}{D} \begin{vmatrix} 1 & \cdots & 1 & f(n) & 1 & \cdots & 1 \\ \lambda_1 & \cdots & \lambda_{k-1} & f(n+1) & \lambda_{k+1} & \cdots & \lambda_p \\ \cdots & & & & & & \cdots \\ \lambda_1^{p-1} & \cdots & \lambda_{k-1}^{p-1} & f(n+p-1) & \lambda_{k+1}^{p-1} & \cdots & \lambda_p^{p-1} \end{vmatrix}.$$

Using the above determinant and the relationships (3.45) and (3.42), we obtain:
$D \cdot x_k(n+1) =$

$$= \begin{vmatrix} 1 & \ldots & 1 & \sum_{j=1}^{p} \lambda_j x_j(n) & 1 & \ldots & 1 \\ \lambda_1 & \ldots & \lambda_{k-1} & \sum_{j=1}^{p} \lambda_j^2 x_j(n) & \lambda_{k+1} & \ldots & \lambda_p \\ \vdots & & & \vdots & & & \vdots \\ \lambda_1^{p-1} & \ldots & \lambda_{k-1}^{p-1} & -\sum_{i=1}^{p}\sum_{j=1}^{p}(a_i+\alpha_i(n))\lambda_j^{p-i}x_j(n) & \lambda_{k+1}^{p-1} & \ldots & \lambda_p^{p-1} \end{vmatrix}.$$

Considering (3.44), it follows that $\sum_{i=1}^{p} a_i \lambda_j^{p-i} = -\lambda_j^p$ and therefore

$$-\sum_{i=1}^{p}\sum_{j=1}^{p}(a_i+\alpha_i(n))\lambda_j^{p-i}x_j(n) = -\sum_{j=1}^{p}\left(\sum_{i=1}^{p}a_i\lambda_j^{p-i}\right)x_j(n)-$$

$$-\sum_{j=1}^{p}\sum_{i=1}^{p}\alpha_i(n)\lambda_j^{p-i}x_j(n) = \sum_{j=1}^{p}\lambda_j^p x_j(n) - \sum_{j=1}^{p}\sum_{i=1}^{p}\alpha_i(n)\lambda_j^{p-i}x_j(n).$$

We will use the above relationship to express $x_k(n+1)$ as a difference of two determinants:

$$x_k(n+1) = \frac{1}{D}\begin{vmatrix} 1 & \ldots & 1 & \sum_{j=1}^{p}\lambda_j x_j(n) & 1 & \ldots & 1 \\ \lambda_1 & \ldots & \lambda_{k-1} & \sum_{j=1}^{p}\lambda_j^2 x_j(n) & \lambda_{k+1} & \ldots & \lambda_p \\ \vdots & & & \vdots & & & \vdots \\ \lambda_1^{p-1} & \ldots & \lambda_{k-1}^{p-1} & \sum_{j=1}^{p}\lambda_j^p x_j(n) & \lambda_{k+1}^{p-1} & \ldots & \lambda_p^{p-1} \end{vmatrix} -$$

$$-\frac{1}{D}\begin{vmatrix} 1 & \ldots & 1 & 0 & 1 & \ldots & 1 \\ \lambda_1 & \ldots & \lambda_{k-1} & 0 & \lambda_{k+1} & \ldots & \lambda_p \\ \vdots & & & \vdots & & & \vdots \\ \lambda_1^{p-1} & \ldots & \lambda_{k-1}^{p-1} & \sum_{i=1}^{p}\sum_{j=1}^{p}\alpha_i(n)\lambda_j^{p-i}x_j(n) & \lambda_{k+1}^{p-1} & \ldots & \lambda_p^{p-1} \end{vmatrix} =$$

$$= \frac{1}{D}\sum_{j=1}^{p}\lambda_j x_j(n)\begin{vmatrix} 1 & \ldots & 1 & 1 & 1 & \ldots & 1 \\ \lambda_1 & \ldots & \lambda_{k-1} & \lambda_j & \lambda_{k+1} & \ldots & \lambda_p \\ \vdots & & & & & & \vdots \\ \lambda_1^{p-1} & \ldots & \lambda_{k-1}^{p-1} & \lambda_j^{p-1} & \lambda_{k+1}^{p-1} & \ldots & \lambda_p^{p-1} \end{vmatrix} +$$

$$+(-1)^{p+k-1}\frac{1}{D}\sum_{i,j=1}^{p}\alpha_i(n)\lambda_j^{p-i}x_j(n)\begin{vmatrix} 1 & \ldots & 1 & 1 & \ldots & 1 \\ \lambda_1 & \ldots & \lambda_{k-1} & \lambda_{k+1} & \ldots & \lambda_p \\ \vdots & & & & & \vdots \\ \lambda_1^{p-2} & \ldots & \lambda_{k-1}^{p-2} & \lambda_{k+1}^{p-2} & \ldots & \lambda_p^{p-2} \end{vmatrix} =$$

3.3 Poincaré's Theorem

$$= \lambda_k x_k(n) + (-1)^{p+k-1} \frac{1}{D} \sum_{i,j=1}^{p} \alpha_i(n) \lambda_j^{p-i} x_j(n) \cdot \prod_{\substack{1 \leq s < t \leq p \\ s,t \neq k}} (\lambda_t - \lambda_s) =$$

$$= \lambda_k x_k(n) + (-1)^{p+k-1} \frac{1}{D_k} \sum_{i,j=1}^{p} \alpha_i(n) \lambda_j^{p-i} x_j(n),$$

where $D_k = \prod_{1 \leq t < k}(\lambda_k - \lambda_t) \cdot \prod_{k < t \leq p}(\lambda_t - \lambda_k)$. Following the calculations above, we have obtained that, for every $k = 1, \cdots, p$ and every $n \in \mathbb{N}$,

$$x_k(n+1) = \lambda_k x_k(n) + \frac{(-1)^{p+k-1}}{D_k} \sum_{i,j=1}^{p} \alpha_i(n) \lambda_j^{p-i} x_j(n). \quad (3.46)$$

Let us suppose that $|\lambda_1| < |\lambda_2| < \cdots < |\lambda_p|$. Then there exists $\varepsilon_0 > 0$ such that

$$\frac{|\lambda_i| + \varepsilon_0}{|\lambda_j| - \varepsilon_0} < 1, \text{ for every } 1 \leq i < j \leq p. \quad (3.47)$$

According to (3.43), for every $\varepsilon \in]0, \varepsilon_0[$, there exists $n_0 \in \mathbb{N}$ such that

$$\left| \frac{(-1)^{p+k-1}}{D_k} \sum_{i=1}^{p} \alpha_i(n) \lambda_j^{p-i} \right| < \frac{\varepsilon}{p}, \text{ for every } n \geq n_0, k, j = 1, \cdots, p. \quad (3.48)$$

For every $n \in \mathbb{N}$, let r_n be the smallest number from the set $\{1, 2, \cdots, p\}$ for which $|x_{r_n}|(n) = \max\{|x_1(n)|, \cdots, |x_p(n)|\}$. Then, from (3.46) and (3.48), we obtain, for every $k \in \{1, \cdots, p\}$, and every $n \geq n_0$:

$$\begin{cases} |x_k(n+1)| < |\lambda_k||x_k(n)| + \frac{\varepsilon}{p} \sum_{j=1}^{p} |x_j(n)| \leq |\lambda_k||x_k(n)| + \varepsilon|x_{r_n}(n)|, \\ |x_k(n+1)| > |\lambda_k||x_k(n)| - \frac{\varepsilon}{p} \sum_{j=1}^{p} |x_j(n)| \geq |\lambda_k||x_k(n)| - \varepsilon|x_{r_n}(n)|. \end{cases} \quad (3.49)$$

Let's now show that, for every $n \geq n_0$, $r_n \leq r_{n+1}$. If we assume that there exists $n \geq n_0$ such that $r_{n+1} < r_n$, then, according to (3.49) and (3.47),

$$1 \leq \left| \frac{x_{r_{n+1}}(n+1)}{x_{r_n}(n+1)} \right| < \frac{|\lambda_{r_{n+1}}||x_{r_{n+1}}(n)| + \varepsilon|x_{r_n}(n)|}{|\lambda_{r_n}||x_{r_n}(n)| - \varepsilon|x_{r_n}(n)|} =$$

$$= \frac{|\lambda_{r_{n+1}}| \left| \frac{x_{r_{n+1}}(n)}{x_{r_n}(n)} \right| + \varepsilon}{|\lambda_{r_n}| - \varepsilon} \leq \frac{|\lambda_{r_{n+1}}| + \varepsilon}{|\lambda_{r_n}| - \varepsilon} < 1,$$

which is absurd.

Now, since $(r_n)_n \subseteq \{1, \cdots, p\}$ is an increasing sequence, there exist $n_1 \geq n_0$ and $k_0 \in \{1, \cdots, p\}$ such that $r_n = k_0$, for every $n \geq n_1$; therefore

$$|x_{k_0}(n)| = \max\{|x_1(n)|, \cdots, |x_p(n)|\}, \text{ for every } n \geq n_1. \quad (3.50)$$

Taking into consideration the significance of k_0, from (3.49) it follows that, for every $k \in \{1, 2, \cdots, p\}$, every $n \geq n_1$, and every $\varepsilon \in]0, \varepsilon_0[$,

$$|\lambda_k||x_k(n)| - \varepsilon|x_{k_0}(n)| < |x_k(n+1)| < |\lambda_k||x_k(n)| + \varepsilon|x_{k_0}(n)|. \quad (3.51)$$

Let's now demonstrate that:

$$\lim_{n \to +\infty} \frac{x_k(n)}{x_{k_0}(n)} = 0, \text{ for every } k \in \{1, 2, \cdots, p\} \setminus \{k_0\}. \quad (3.52)$$

We denote by $a = \limsup_{n \to +\infty} \left|\frac{x_k(n)}{x_{k_0}(n)}\right|$; obviously $0 \leq a \leq 1$.

Let's first consider the case where $k > k_0$. Taking into consideration the significance of a, there exists a sequence of natural number $m_n \uparrow +\infty$ such that $\lim_{n \to +\infty} \left|\frac{x_k(m_n)}{x_{k_0}(m_n)}\right| = a$. There exists $n_2 \geq n_1$ such that

$$\left|\frac{x_k(m_n)}{x_{k_0}(m_n)}\right| < a + \varepsilon, \text{ for every } n \geq n_2. \quad (3.53)$$

According to (3.51) and (3.53),

$$a + \varepsilon > \left|\frac{x_k(m_n+1)}{x_{k_0}(m_n+1)}\right| \geq \frac{|\lambda_k||x_k(m_n)| - \varepsilon|x_{k_0}(m_n)|}{|\lambda_{k_0}||x_{k_0}(m_n)| + \varepsilon|x_{k_0}(m_n)|} = \frac{|\lambda_k|\left|\frac{x_k(m_n)}{x_{k_0}(m_n)}\right| - \varepsilon}{|\lambda_{k_0}| + \varepsilon}.$$

Taking the limit as n approaches infinity in the previous relationship, we obtain

$$a + \varepsilon \geq \frac{|\lambda_k|a - \varepsilon}{|\lambda_{k_0}| + \varepsilon}, \text{ for every } \varepsilon \in]0, \varepsilon_0[,$$

so that $|\lambda_{k_0}|a \geq |\lambda_k|a$ and, since $|\lambda_{k_0}| < |\lambda_k|$ it follows that $a = 0$. Therefore

$$\lim_{n \to +\infty} \frac{x_k(n)}{x_{k_0}(n)} = 0, \text{ for every } k > k_0. \quad (3.54)$$

If $k < k_0$, then $|\lambda_k| < |\lambda_{k_0}|$; then there exists $\varepsilon_1 < \varepsilon_0$ such that $|\lambda_k| < |\lambda_{k_0}| - \varepsilon$, for every $\varepsilon \in]0, \varepsilon_1[$. Let now $\varepsilon \in]0, \varepsilon_1[$ and $n \geq n_1$ be arbitrary; from (3.51) we obtain:

$$\left|\frac{x_k(n+1)}{x_{k_0}(n+1)}\right| \leq \frac{|\lambda_k||x_k(n)| + \varepsilon|x_{k_0}(n)|}{|\lambda_{k_0}||x_{k_0}(n)| - \varepsilon|x_{k_0}(n)|} = \frac{|\lambda_k|}{|\lambda_{k_0}| - \varepsilon} \cdot \left|\frac{x_k(n)}{x_{k_0}(n)}\right| + \frac{\varepsilon}{|\lambda_{k_0}| - \varepsilon}.$$

3.3 Poincaré's Theorem

Using the above inequality for n taking values from n_1 to $n-1$, we obtain

$$\left|\frac{x_k(n)}{x_{k_0}(n)}\right| \leq \left(\frac{|\lambda_k|}{|\lambda_{k_0}|-\varepsilon}\right)^{n-n_1} \cdot \left|\frac{x_k(n_1)}{x_{k_0}(n_1)}\right| + \frac{1-\left(\frac{|\lambda_k|}{|\lambda_{k_0}|-\varepsilon}\right)^{n-n_1}}{1-\frac{|\lambda_k|}{|\lambda_{k_0}|-\varepsilon}} \cdot \frac{\varepsilon}{|\lambda_{k_0}|-\varepsilon}, \forall n \geq n_1.$$

We take the limit as $n \to +\infty$ in the above inequality:

$$a = \limsup_{n \to +\infty} \left|\frac{x_k(n)}{x_{k_0}(n)}\right| \leq \frac{\varepsilon}{|\lambda_{k_0}|-|\lambda_k|-\varepsilon}, \text{ for every } \varepsilon \in]0, \varepsilon_0[.$$

Finally, if we let $\varepsilon \downarrow 0$ in this last inequality, we obtain: $a \leq 0$, which implies $a = 0$. Therefore

$$\lim_{n \to +\infty} \frac{x_k(n)}{x_{k_0}(n)} = 0, \text{ for every } k < k_0. \tag{3.55}$$

Equations (3.54) and (3.55) show that (3.52) is satisfied.
From the first two equations in (3.45), we obtain, for every $n \geq n_1$,

$$\frac{x_1(n)}{x_{k_0}(n)} + \cdots + \frac{x_{k_0-1}(n)}{x_{k_0}(n)} + 1 + \frac{x_{k_0+1}(n)}{x_{k_0}(n)} + \cdots + \frac{x_p(n)}{x_{k_0}(n)} = \frac{f(n)}{x_{k_0}(n)},$$

$$\lambda_1 \frac{x_1(n)}{x_{k_0}(n)} + \cdots + \lambda_{k_0-1} \frac{x_{k_0-1}(n)}{x_{k_0}(n)} + \lambda_{k_0} + \lambda_{k_0+1} \frac{x_{k_0+1}(n)}{x_{k_0}(n)} + \cdots + \lambda_p \frac{x_p(n)}{x_{k_0}(n)} =$$

$$= \frac{f(n+1)}{x_{k_0}(n)}.$$

Taking into account (3.52) and taking the limit in the above relationships, it follows that:

$$\lim_{n \to \infty} \frac{f(n)}{x_{k_0}(n)} = 1, \text{ and } \lim_{n \to \infty} \frac{f(n+1)}{x_{k_0}(n)} = \lambda_{k_0}, \text{ from where}$$

$$\lim_{n \to +\infty} \frac{f(n+1)}{f(n)} = \lambda_{k_0}.$$

□

Remark 3.3.3 A natural question that arises in connection with the previous theorem is whether any root of the characteristic equation is the limit of a sequence of the form $\left(\frac{f(n+1)}{f(n)}\right)_{n \in \mathbb{N}}$, where f is a non-zero solution of the Poincaré-type equation. The answer is provided by a theorem of Oskar Perron: If to the statement of Poincaré's theorem we add the condition that $a_p + \alpha_p(n) \neq 0$, for every $n \in \mathbb{N}$, then

Eq. (3.42) has a linearly independent system of solutions $f_1(n), f_2(n), \cdots, f_p(n)$ with the property that $\lim_{n \to +\infty} \frac{f_k(n+1)}{f_k(n)} = \lambda_k$, for every $k = 1, 2, \cdots, p$.

Poincaré's theorem allows us to study the asymptotic behavior of solutions of certain homogeneous recurrences.

Theorem 3.3.4 *Consider the recurrence*

$$f(n+p) + [a_1 + \alpha_1(n)] f(n+p-1) + \cdots + [a_p + \alpha_p(n)] f(n) = 0, \forall n \in \mathbb{N}, \quad (3.56)$$

where

$$\lim_{n \to +\infty} \alpha_i(n) = 0, \text{ for every } i = 1, 2, \cdots, p,$$

and let $\lambda_1, \lambda_2, \cdots, \lambda_p$ *be the roots of the characteristic equation*

$$\lambda^p + a_1 \lambda^{p-1} + \cdots + a_p = 0. \quad (3.57)$$

If we suppose that $|\lambda_i| \neq |\lambda_j|$, *for every* $i \neq j$, $j = 1, 2, \cdots, p$, *then, according to Poincaré's theorem, for every nonzero solution* f *of Eq. (3.56) there exists a root of (3.57)*, λ, *such that* $\lim_{n \to +\infty} \frac{f(n+1)}{f(n)} = \lambda$. *Then*

$f(n) = \pm \lambda^n e^{o(n)} (n \to \infty)$ *if* $\lambda \neq 0$, *and*
$|f(n)| = e^{-g(n)}$, *where* $n = o(g(n))(n \to \infty)$ *if* $\lambda = 0$.

Proof If $\lambda > 0$, then there exists n_0 such that $\frac{f(n+1)}{f(n)} > 0$, for every $n \geq n_0$, and thus f has a constant sign on $\mathbb{N}$. Since f is a solution of a homogeneous equation, we can assume that $f(n) > 0$ for any $n \geq n_0$. Let us denote by $g(n) = \frac{f(n)}{\lambda^n}$. Then $\frac{g(n+1)}{g(n)} \to 1$ and, if $h(n) = \log g(n)$, then $h(n+1) - h(n) \to 0$. Therefore, for every $\varepsilon > 0$, there exists $n_1 \geq n_0$ such that

$$|h(n+1) - h(n)| < \varepsilon, \text{ for every } n \geq n_1.$$

Using the above inequality for n taking values from n_1 to $n-1$, we obtain

$$|h(n) - h(n_1)| \leq \sum_{k=n_1+1}^{n} |h(k) - h(k-1)| < \varepsilon(n - n_1),$$

from where

$$\left| \frac{h(n)}{n} - \frac{h(n_1)}{n} \right| < \varepsilon \left(1 - \frac{n_1}{n} \right). \quad (3.58)$$

3.3 Poincaré's Theorem

From (3.58)

$$\left|\frac{h(n)}{n}\right| \leqslant \left|\frac{h(n)}{n} - \frac{h(n_1)}{n}\right| + \left|\frac{h(n_1)}{n}\right| < \varepsilon\left(1 - \frac{n_1}{n}\right) + \frac{|h(n_1)|}{n}.$$

Then we remark that $\lim_{n\to\infty} \frac{h(n)}{n} = 0$, meaning that $h(n) = o(n)(n \to \infty)$. Therefore

$$f(n) = \lambda^n g(n) = \lambda^n e^{h(n)} = \lambda^n e^{o(n)} (n \to \infty).$$

If $f(n) < 0$, for every $n \geqslant n_0$, then $f(n) = -\lambda^n e^{o(n)} (n \to \infty)$.

In the case where $\lambda < 0$, $f(n)$ has an oscillatory sign, thus $f(n) = (-1)^n g(n)$ with $g(n) > 0$. In addition $\frac{g(n+1)}{g(n)} = -\frac{f(n+1)}{f(n)} \to -\lambda$. With a similar reasoning, we obtain that $g(n) = (-\lambda)^n e^{o(n)} (n \to \infty)$ or $f(n) = \pm \lambda^n e^{o(n)} (n \to \infty)$.

If $\lambda = 0$, then $\lim_{n\to\infty} \frac{f(n+1)}{f(n)} = 0$ and so $\lim_{n\to\infty} \log\left|\frac{f(n+1)}{f(n)}\right| \to -\infty$. For every $M > 0$ there exists $n_0 \in \mathbb{N}$ such that

$$\log\left|\frac{f(n+1)}{f(n)}\right| < -M, \text{ for every } n \geqslant n_0. \tag{3.59}$$

From (3.59), for n taking values from n_0 to $n - 1$, we obtain

$$\log\left|\frac{f(n)}{f(n_0)}\right| = \sum_{k=n_0+1}^{n} \log\left|\frac{f(k)}{f(k-1)}\right| < -(n - n_0) \cdot M, \text{ for every } n \geqslant n_0 + 1.$$

Therefore

$$\frac{\log|f(n)|}{n} < \left(-1 + \frac{n_0}{n}\right) \cdot M + \frac{1}{n} \log|f(n_0)|, \text{ for every } n \geqslant n_0 + 1,$$

and so

$$\limsup_{n\to\infty} \frac{\log|f(n)|}{n} \leqslant -M, \text{ for every } M > 0.$$

It follows that $\lim_{n\to\infty} \frac{\log|f(n)|}{n} = -\infty$, or $\lim_{n\to\infty} \frac{n}{-\log|f(n)|} = 0$.

This means that $n = o(g(n))(n \to +\infty)$, where $g(n) = -\log|f(n)|$ and so $|f(n)| = e^{-g(n)}$. □

Examples 3.3.5

(1) If the roots of the characteristic equation have equal absolute values, then it is possible for the conclusion of Poincaré's theorem not to be satisfied. Indeed, consider the second-order homogeneous Poincaré-type equation $f(n+2) - \frac{n-(-1)^n}{n} f(n) = 0$, for every $n \in \mathbb{N}$. The characteristic equation is $\lambda^2 - 1 = 0$

and its roots have the same absolute value. The sequence $f : \mathbb{N} \to \mathbb{R}$, defined by

$$f(n) = \begin{cases} \dfrac{(2k)!!}{(2k-1)!!}, & n = 2k+1, \\ \dfrac{(2k-1)!!}{(2k)!!}, & n = 2k+2, \end{cases} \quad \text{for every } n \in \mathbb{N},$$

is a nonzero solution of the equation, but the sequence $\left(\dfrac{f(n+1)}{f(n)}\right)_{n \in \mathbb{N}}$ is a divergent one.

Indeed, from the Stirling's formula: $n! = \sqrt{2\pi n}\left(\dfrac{n}{e}\right)^n (1 + \varepsilon_n)$, $\varepsilon_n \to 0$, we obtain that

$$(2k)!! = \sqrt{2\pi k}\left(\dfrac{2k}{e}\right)^k (1 + \varepsilon_k),$$
$$(2k-1)!! = \sqrt{2}\left(\dfrac{2k}{e}\right)^k \dfrac{1 + \varepsilon_{2k}}{1 + \varepsilon_k}.$$

It follows that

$$\dfrac{f(n+1)}{f(n)} = \begin{cases} \left[\dfrac{(2k-1)!!}{(2k)!!}\right]^2 = \dfrac{1}{\pi k}\dfrac{(1+\varepsilon_{2k})^2}{(1+\varepsilon_k)^4}, & n = 2k+1, \\ \dfrac{2k-1}{2k}\left[\dfrac{(2k)!!}{(2k-1)!!}\right]^2 = \dfrac{2k-1}{2k} \cdot \pi k \cdot \dfrac{(1+\varepsilon_k)^4}{(1+\varepsilon_{2k})^2}, & n = 2k. \end{cases}$$

Therefore $\dfrac{f(2k+2)}{f(2k+1)} \xrightarrow[k\to\infty]{} 0$ and $\dfrac{f(2k+1)}{f(2k)} \xrightarrow[k\to\infty]{} +\infty$.

(2) If $a_p + \alpha_p(n) = 0$ for an infinity of terms $n \in \mathbb{N}$, then there exists at least one root λ of the characteristic equation such that $\lambda \neq \lim_{n\to\infty} \dfrac{f(n+1)}{f(n)}$, for every nonzero solution of the Poincaré equation.

Let the Poincaré-type equation

$$f(n+2) - f(n+1) + \dfrac{1-(-1)^n}{n} f(n) = 0, \text{ for every } n \in \mathbb{N}^*.$$

For this equation $a_p + \alpha_p(n) = \dfrac{1-(-1)^n}{n} = 0$ for all even values of n.

$$\begin{cases} f(2n+2) - f(2n+1) = 0, \\ f(2n+1) - f(2n) + \dfrac{2}{2n-1} f(2n-1) = 0, \end{cases} \text{ for every } n \in \mathbb{N}^*. \quad (*)$$

From the first equation of $(*)$, $f(2n) = f(2n-1)$, for every $n \in \mathbb{N}^*$, and then from the second equation of $(*)$, $f(2n+1) = \dfrac{2n-3}{2n-1} f(2n-1)$, for every

$n \geqslant 2$. It follows that the general solution of the equation is $f(2n+1) = \frac{1}{2n-1}f(3) = f(2n+2)$, for every $n \in \mathbb{N}^*$.

Then, for every nonzero solution of the equation f, $\frac{f(n+1)}{f(n)} \to 1$.

The characteristic equation is $\lambda^2 - \lambda = 0$; we can remark that $0 \neq \lim_n \frac{f(n+1)}{f(n)}$.

3.4 Exercises

(1) Calculate $\lim_{t \to +\infty} t \int_{-\infty}^{+\infty} e^{-x^2} \log(\frac{1}{t}x^2 + \frac{1}{\sqrt{t}}x + 1) dx$.

   Indications: With the change of variable $x = \sqrt{t}y$ we obtain: $t \int_{-\infty}^{+\infty} e^{-x^2} \log(\frac{1}{t}x^2 + \frac{1}{\sqrt{t}}x + 1) dx = t\sqrt{t} \int_{-\infty}^{+\infty} e^{-ty^2} \log(y^2 + y + 1) dy \equiv t\sqrt{t} f(t)$. According to Example 3.2.3, $f(t) \approx \sum_{n=1}^{\infty} \sqrt{\pi} \frac{(2n-1)!!}{2^n} a_{2n} t^{-(n+\frac{1}{2})}$ $(t \to \infty)$, and, from (iii) of Theorem 3.1.11, $f(t) \sim \frac{\sqrt{\pi}}{2} a_2 t^{-\frac{3}{2}}$ $(t \to +\infty)$, where $a_2 = \frac{1}{2}$. Thus, it follows that $\lim_{t \to +\infty} t\sqrt{t} f(t) = \frac{\sqrt{\pi}}{4}$.

(2) Let $h: \mathbb{R} \to \mathbb{R}$, $h(x) = -x^4 + 2x^3 - 2x^2 - 1$; show that $\int_{-\infty}^{+\infty} e^{th(x)} dx \sim \sqrt{\frac{\pi}{2t}} e^{-t}$ $(t \to +\infty)$.

   Indications: According with Corollary 3.2.6, h is continuous on $\mathbb{R}$, $h(x) = -x^2(x^2 - 2x + 2) - 1 < -1 = h(0)$, for every $x \in \mathbb{R} \setminus \{0\}$. Moreover, $h(x) - h(0) \leqslant -1$, for every $x \in \mathbb{R}$ with $|x| \geqslant 1$. $h''(0) = -4 < 0$ and $\int_{-\infty}^{+\infty} e^{h(x)} dx \leqslant e^{h(0)} \int_{-\infty}^{+\infty} e^{-x^2} dx = \sqrt{\pi} e^{h(0)}$. It follows from the above corollary that $\int_{-\infty}^{+\infty} e^{th(x)} dx \sim \sqrt{\frac{\pi}{2t}} e^{-t}$ $(t \to +\infty)$.

(3) Show that $\int_0^{+\infty} e^{-tx}(1+x^2)^{-\frac{1}{2}} dx \approx \sum_{n=0}^{+\infty} (-1)^n \frac{[(2n-1)!!]^2}{t^{2n+1}}$ $(t \to +\infty)$ $(0!! = 0! = 1)$. Calculate $\lim_{t \to +\infty} t \cdot \int_0^{+\infty} e^{-tx}(1+x^2)^{-\frac{1}{2}} dx$.

   Indications: We apply Watson's lemma (Theorem 3.2.9) with $\lambda = 0$ and $g(x) = (1+x^2)^{-\frac{1}{2}}$. Using the binomial series, we get, for every $x \in [0, 1]$, $g(x) = \sum_{n=0}^{+\infty} (-1)^n \frac{(2n-1)!!}{2^n \cdot n!} x^{2n}$ $((-1)!! = 1)$. Therefore $a_{2n} = (-1)^n \frac{(2n-1)!!}{2^n \cdot n!}$ and $a_{2n+1} = 0$. Furthermore $|g(x)| \leqslant e^x$, for every $x \in [0, +\infty[$; therefore, $f(t) = \int_0^{+\infty} e^{-tx}(1+x^2)^{-\frac{1}{2}} dx \approx \sum_{n=0}^{+\infty} (-1)^n \frac{[(2n-1)!!]^2}{t^{2n+1}}$ $(t \to +\infty)$. According to (iii) of Theorem 3.1.11, $f(t) \sim \frac{1}{t}$ $(t \to +\infty)$, so that $\lim_{t \to +\infty} t f(t) = 1$.

(4) Evaluate the asymptotic behavior of integral $f(t) = \int_{-\infty}^{+\infty} e^{-tx^2} x \log(1+x) dx$ as $t \to +\infty$; calculate $\lim_{t \to +\infty} t\sqrt{t} f(t)$.

   Indications: Let's apply Theorem 3.2.12 to the function $h:]-\infty, +\infty[\to \mathbb{R}$, $h(x) = x \log(1+x)$. We notice that $h(x) = x^2 - \frac{x^3}{2} + \frac{x^4}{3} + \cdots + (-1)^n \frac{x^{n+2}}{n+1} + \cdots$, for any $x \in]-1, 1[$. Thus, $a_0 = 0$ and $a_{2n} = \frac{1}{2n-1}$, for any $n \in \mathbb{N}^*$. It follows from the aforementioned theorem that $f(t) \approx \sqrt{\pi} \sum_{n=1}^{+\infty} \frac{(2n-3)!!}{2^n} t^{-(n+\frac{1}{2})}$ $(t \to +\infty)$. According to (iii) of Theorem 3.1.11, $f(t) \sim \frac{\sqrt{\pi}}{2} t^{-\frac{3}{2}}$ $(t \to +\infty)$, so that $\lim_{t \to +\infty} t\sqrt{t} f(t) = \frac{\sqrt{\pi}}{2}$.

(5) Let's consider the Poincaré-type linear recurrence

$$f(n+p) + a_1(n)f(n+p-1) + \cdots + a_p(n)f(n) = 0, \text{ for every } n \in \mathbb{N}$$

($\lim_{n\to+\infty} a_i(n) = a_i \in \mathbb{R}$, for every $i = 1, 2, \cdots, p$). Assume that f is a non-zero solution of the recurrence for which exists $\lim_{n\to+\infty} \frac{f(n+1)}{f(n)} = \alpha \in \mathbb{R}$. Then α is a solution of the characteristic equation $\lambda^p + a_1\lambda^{p-1} + \cdots + a_p = 0$.

Indications: For every $k = 1, \cdots, p$, $\frac{f(n+k)}{f(n)} = \frac{f(n+k)}{f(n+k-1)} \cdot \frac{f(n+k-1)}{f(n+k-2)} \cdots \frac{f(n+1)}{f(n)} \to \alpha^k$. The result is obtained by taking the limit as $n \to +\infty$ in the relation $\frac{f(n+p)}{f(n)} + a_1(n)\frac{f(n+p-1)}{f(n)} + \cdots + a_{p-1}(n)\frac{f(n+1)}{f(n)} + a_p(n) = 0$.

(6) Consider the Poincaré-type recurrence

$$f(n+2) - \frac{3n+3}{n+2}f(n+1) + \frac{2n+2}{n+2}f(n) = 0, \forall n \in \mathbb{N}.$$

Investigate the asymptotic behavior of the solutions of the recurrence. Using the substitution $g(n) = nf(n)$, find the solutions of the recurrence and compare them with their asymptotic form.

Indications: The characteristic equation associated with the recurrence is $\lambda^2 - 3\lambda + 2 = 0$ and has the roots $\lambda_1 = 1, \lambda_2 = 2$. According to Perron's theorem (see Remark 3.3.3), the recurrence has a linearly independent system of solutions $\{f_1, f_2\}$ such that $\lim_{n\to+\infty} \frac{f_k(n+1)}{f_k(n)} = \lambda_k, k = 1, 2$. In this case, Theorem 3.3.4 presents the following asymptotic behavior of the solutions: $f_1(n) = e^{n\alpha(n)}$, $f_2(n) = 2^n e^{n\beta(n)}$, where $\alpha(n) \to 0$, $\beta(n) \to 0$. If we make the substitution $g(n) = nf(n)$, we obtain the linear homogeneous equation with constant coefficients $g(n+2) - 3g(n+1) + 2g(n) = 0$, which admits a linearly independent system of solutions $\{g_1, g_2\}$, where $g_1(n) = 1, g_2(n) = 2^n$. Then the solutions of our recurrence are $f_1(n) = \frac{1}{n}, f_2(n) = \frac{2^n}{n}$. We note that we can express these solutions in the form $f_1(n) = e^{n\left(-\frac{\log n}{n}\right)}$, $f_2(n) = 2^n e^{n\left(-\frac{\log n}{n}\right)}$, from which we obtain $\alpha(n) = \beta(n) = -\frac{\log n}{n} \to 0$.

References

1. Copson, E. T. (1965). *Asymptotic expansions*. Cambridge: Cambridge University Press.
2. De Bruijn, N. G. (1958). *Asymptotic methods in analisys*. Amsterdam: North-Holland Publishing Co.
3. Elaydi, S. (2005). *An introduction to difference equations. Undergraduate texts in mathematics* (3rd ed.). Springer.
4. Kelley, W. G., & Peterson, A. C. (2001). *Difference equations. An introduction with applications* (2nd ed.). San Diego: Academic Press.
5. Trench, W. F. (1993). Asymptotic behavior of solutions of solutions of Poincaré difference equations. *Proceedings of the American Mathematical Society, 119*(2), 431–438.

Chapter 4
Integration in Finite Terms

It is often said, in jest or in earnest, that "differentiation is mechanical while integration is an art". This saying suggests that the process of differentiation involves the application of a well-defined set of rules and procedures that can be strictly followed to obtain results. In contrast, when we calculate integrals, it often requires a more creative and intuitive approach because there are numerous methods and techniques to tackle different types of integrals. Integration can be more subjective and may require "artistic" or intuitive thinking to find solutions, as there isn't always a rigid set of rules that leads directly to an answer.

The antiderivative of a function $f : I \to \mathbb{R}$ is a differentiable function $g : I \to \mathbb{R}$ such that $g' = f$. If the function f is continuous, then it has antiderivatives; one such antiderivative is defined by

$$g(x) = \int_a^x f(t)dt, \text{ where } a \in I.$$

However, such a solution does not satisfy us. If the function f is elementary, we want the antiderivative to be expressed as a finite combination of elementary functions; in other words, in finite terms or closed form. What do we mean by "elementary function" here? Of course, this class of functions should include polynomial functions and, more generally, rational functions, algebraic functions, exponential and logarithmic functions, trigonometric and inverse trigonometric functions, hyperbolic functions and their inverses, and should allow common operations with them: addition, multiplication, composition. Certainly, this is not a definition. One of the objectives of this chapter is to precisely specify what we mean by elementary functions.

The first to work on expressing indefinite integrals was Pierre-Simon de Laplace, and he also completely developed the calculation algorithm for the antiderivatives of rational functions.

However, the first rigorous result regarding integration in finite terms belongs to Joseph Liouville. In a series of works published between 1833 and 1841, Liouville provides the representation of functions that admit an antiderivative expressible through elementary functions. On this occasion, he confirms a long-standing intuitive fact, namely that elliptic functions do not admit primitives representable by finite expressions of elementary functions. Major contributions to Liouville's work were made by Dmitry Dmitrievich Morduhai-Boltovskoi and Joseph Fels Ritt.

In 1946, Alexander Markowich Ostrowski extended Liouville's principle to a broader class of holomorphic functions, except for a countable set of isolated singularities (see [2]). He considered differential fields of such functions (Liouville fields) and constructed algebraic, logarithmic, and exponential extensions of these fields inductively. An elementary function is an element of such an extension. It should be noted that the concept of a differential field belongs to J. F. Ritt, who published a book in 1948 that would become the classic presentation of integration in finite terms (see [4]). Ritt's work was further developed by Maxwell Alexander Rosenlicht and Robert Henry Risch. In 1968, Rosenlicht formalized the notions of transcendental logarithmic and transcendental exponential functions and reformulated Laplace's principle as follows:

If f belongs to the differential field F of characteristic 0, and if the equation $g' = f$ admits a solution in an elementary extension of F, an extension that has the same field of constants K as F, then there exist $c_1, \cdots, c_m \in K$, $u_0, u_1, \cdots, u_m \in F$ such that $f = u_0' + \sum_{k=1}^{m} c_k \cdot \dfrac{u_k'}{u_k}$ (see [5]).

As an application, Rosenlicht establishes the non-elementary nature of integrals $\int e^{z^2} dz$, $\int \dfrac{e^z}{z} dz$ (see [6]).

R. H. Risch developed a theoretical algorithm to translate Liouville's principle into the language of computer algebra (see [3]). Transforming this algorithm into an executable computer program is a complex task that has been carried out in multiple stages. However, as of today, a complete implementation of Risch's algorithm is not known. Thus, MAPLE 2021, which utilizes this program for computing antiderivatives, does not seem to fully employ this algorithm.

In this chapter, our aim is to state and demonstrate Liouville's principle in a less formalized framework, specifically in the context of holomorphic functions:

If f belongs to a Liouville field L of holomorphic functions, then it admits an antiderivative elementary with respect to L iff there exist $m \in \mathbb{N}^*$, $u_0, u_1, \cdots, u_m \in L$, $c_1, \cdots, c_m \in \mathbb{C}$ such that $f = u_0' + \sum_{k=1}^{m} c_k \cdot \dfrac{u_k'}{u_k}$.

The simplest Liouville field of functions is that of rational functions $R(\mathbb{C})$. Any function from this field admits an elementary antiderivative (see Corollary 4.2.4). In short, by elementary function we mean a function that belongs to some repeated but finite extensions of $R(\mathbb{C})$ with algebraic, logarithmic, or exponential functions. All known functions: sin, cos, tan, sinh, cosh, arcsin, arccos, arctan, arsinh, arcosh are elementary functions. We will illustrate with a brief example how Liouville's principle works.

4 Integration in Finite Terms

Let $f \in R(\mathbb{C})$ defined by $f(z) = \dfrac{1}{z^2+1}$. For f to admit an elementary antiderivative with respect to $R(\mathbb{C})$ it is necessary and sufficient that there exists $m \in \mathbb{N}^*, u_0, u_1, \cdots, u_m \in R(\mathbb{C}), c_1, \cdots, c_m \in \mathbb{C}$ such that

$$\frac{1}{z^2+1} = u_0'(z) + \sum_{k=1}^{m} c_k \cdot \frac{u_k'(z)}{u_k(z)}.$$

The left-hand side of the above equality has two simple poles: $+i, -i$. It is observed from here that u_0 must be a polynomial in z (otherwise it would have multiple poles), $m = 2, u_1(z) = z - i, u_2(z) = z + i$. Thus, the equality can be written as: $\dfrac{1}{z^2+1} = u_0'(z) + \dfrac{c_1}{z-i} + \dfrac{c_2}{z+i}$. This equality is verified if u_0 is constant and $c_1 = -\dfrac{i}{2}, c_2 = \dfrac{i}{2}$. It follows that

$$\int f(z)dz = \frac{i}{2} \cdot \log \frac{z+i}{z-i},$$

is an elementary antiderivative of f. To dispel any potential confusion, let us note that

$$\frac{i}{2} \cdot \log \frac{z+i}{z-i} = -\frac{i}{2} \cdot \log \frac{z-i}{z+i} = \frac{1}{2i}\left[\log(-1) + \log \frac{i-z}{i+z}\right] =$$

$$= \frac{1}{2i} \cdot \log \frac{i-z}{i+z} + \frac{\pi}{2} = \arctan z + \frac{\pi}{2}.$$

To rigorously define elementary functions, it is necessary to present some elements of the theory of field extensions, as well as Liouville's fields and their algebraic, logarithmic, and exponential extensions. These will be the subject of Sect. 4.1 of this chapter. Section 4.2 is dedicated to proving Liouville's principle.

At the end of the chapter, applications of this principle are provided for finding conditions under which expressions of the form $F(w(z), z)$ have primitives expressible in elementary functions of both w and z, where $w(z) = \int p(z)dz$. This problem is fully addressed in specific cases such as $F(w(z), z) = A_0(z) \cdot w(z) + A_1(z)$ where A_0 and A_1 are rational functions, as well as in the case where F is a polynomial in w. The integrability conditions obtained can be expressed in an algorithmic form, allowing for the writing of programs that can effectively compute the antiderivatives.

4.1 Field Extensions

In this paragraph, by field we will understand a commutative field.

Definition 4.1.1 Let k be a subfield of the field K; in this case, K is also referred to as an **extension** of the field k.

K is called a **finite extension** of k if there exist $x_1, \cdots, x_n \in K$ such that, for any $x \in K$, x can be uniquely written as $x = a_1 x_1 + \cdots + a_n x_n$ with $a_1, \cdots, a_n \in k$. The set $\{x_1, \cdots, x_n\}$ is called the **basis** of the extension K over the field k.

We will denote by $k[X]$ the set of polynomials with coefficients from the field k. Thus, an element of $k[X]$ is a **polynomial** of the form $p = a_0 + a_1 X + \cdots + a_n X^n$, where $a_0, a_1, \cdots, a_n \in k$, and $a_n \neq 0$. If $n = 0$, then $p = a_0$ is a **constant polynomial**. If $n = 0$ and $a_0 = 0$, then $p = \underline{0}$ is the **zero polynomial**. The set of polynomials $k[X]$ is organized as a commutative ring with unity with respect to the usual operations of addition and multiplication. $\underline{0}$ is the additive identity, and the constant polynomial $\underline{1}$ is the multiplicative identity.

Let $p = a_0 + a_1 X + \cdots + a_n X^n$ be a polynomial, with $a_n \neq 0$; we say that the **degree** of the polynomial p is n, and we denote this as $\deg(p) = n$. If $n = 0$ and $a_0 \neq 0$, then $\deg(p) = 0$. If $n = 0$ and $a_0 = 0$ (meaning p is the zero polynomial), then $\deg(p) = -\infty$.

Definition 4.1.2 Let k be a subfield of K and let $\theta \in K$; then there is a smallest subfield of K that contains k and θ. It is the intersection of all subfields of K that contain k and θ, and is denoted by $k(\theta)$. One says that $k(\theta)$ is the field **generated by θ over** k. Obviously, $k \subseteq k[\theta] \subseteq k(\theta) \subseteq K$. Here, $k[\theta] = \{p(\theta) : p \in k[X]\}$ denote the set of values in θ for all polynomials $p \in k[X]$.

Definition 4.1.3 An element $\theta \in K$ is called **algebraic over** k if there exists a polynomial $p \in k[X]$ such that $p(\theta) = 0$. Additionally, since any field is a Euclidean domain, there exists a polynomial $p \in k[X]$ of minimal degree such that $p(\theta) = 0$; certainly, we can choose p in such a way that the coefficient of the highest-degree term is equal to 1 and this choice is unique. Then p is defined as the unique irreducible monic polynomial in $k[X]$ having θ as a root. We say that p is the **minimal polynomial defining** θ.

If θ is not algebraic over k, then it is called **transcendental over** k. An extension K of k is called an **algebraic extension** if all elements of K are algebraic over k.

The following two theorems describe the structure of the fields generated by algebraic and transcendental elements over k.

Theorem 4.1.4 *Let k be a field, K an extension of it, and $\theta \in K$ an algebraic element over k. Then, the field generated by θ over k is*

$$k(\theta) = \{p(\theta) : p \in k[X], \deg(p) < n\},$$

where n is the degree of the minimal polynomial defining θ.

4.1 Field Extensions

$k(\theta)$ is a finite algebraic extension, and a basis for this extension is $\{1, \theta, \theta^2, \cdots, \theta^{n-1}\}$.

Proof Let p_0 be the minimal polynomial defining θ and let $n = \deg(p_0)$. If $\theta \in k$, then $k(\theta) = k$, and the conclusion of the theorem is evident.

Let us suppose that $\theta \notin k$; obviously, $n \geq 2$.

The polynomial ring $k[X]$ is an Euclidean domain so that, for every $p \in k[X]$ there exist $q, r \in k[X]$ such that $p = p_0 \cdot q + r$ and $\deg(r) < \deg(p_0) = n$. Then $p(\theta) = p_0(\theta) \cdot q(\theta) + r(\theta) = r(\theta)$. Thus, we have demonstrated that

$$\{p(\theta) : p \in k[X], \deg(p) < n\} = \{p(\theta) : p \in k[X]\} \equiv C. \tag{4.1}$$

Since C contains constant polynomials as well as the identity polynomial, $k \subseteq C$, and $\theta \in C$. It is easy to show that C forms a commutative ring with unity under the usual operations of addition and multiplication.

Let's demonstrate that the non-zero elements of C have multiplicative inverses. For every $\eta \in C$, $\eta \neq 0$, there exists $p \in k[X]$ such that $\eta = p(\theta)$ and $\deg(p) < n$. Since p_0 is irreducible, p_0 and p are coprime, which means there exist u and v in $k[X]$ such that $p_0 \cdot u + p \cdot v = 1$. Therefore, we have $p(\theta) \cdot v(\theta) = 1$, or $\eta \cdot v(\theta) = 1$. This implies that η is invertible.

Therefore C is a field, $k \subseteq C, \theta \in C$, so that $k(\theta) \subseteq C$. On the other hand, any subfield of K that contains both k and θ also contains the field C, hence $C = k(\theta)$.

According to (4.1), for every $\eta \in k(\theta)$ there exists $p = a_0 + a_1 X + \cdots + a_{n-1} X^{n-1} \in k[X]$ such that $\eta = a_0 + a_1 \theta + \cdots + a_{n-1} \theta^{n-1}$. Then the set $\{1, \theta, \theta^2, \cdots, \theta^{n-1}\}$ ia a basis of $k(\theta)$.

Let's demonstrate that $k(\theta)$ is an algebraic extension.

As we have observed from the above, for any $\eta \in k(\theta)$, there exists an $p \in k[X]$ with $\deg(p) < n$ such that $\eta = p(\theta)$. Therefore, for every $\eta \in k(\theta), \eta \neq 0$,

$$\begin{cases} \eta = a_0^1 \cdot 1 + a_1^1 \cdot \theta + a_2^1 \cdot \theta^2 + \cdots + a_{n-1}^1 \cdot \theta^{n-1} \\ \eta^2 = a_0^2 \cdot 1 + a_1^2 \cdot \theta + a_2^2 \cdot \theta^2 + \cdots + a_{n-1}^2 \cdot \theta^{n-1} \\ \cdots \\ \eta^{n+1} = a_0^{n+1} \cdot 1 + a_1^{n+1} \cdot \theta + a_2^{n+1} \cdot \theta^2 + \cdots + a_{n-1}^{n+1} \cdot \theta^{n-1} \end{cases}, \tag{4.2}$$

where $a_i^j \in k$, for every $i \in \{0, \cdots, n-1\}$, and every $j \in \{1, \cdots, n+1\}$.

We substitute $1, \theta, \theta^2, \ldots, \theta^{n-1}$ from the first n relationships in (4.2) into the last relationship and determine a polynomial $g \in k(X)$ such that $g(\eta) = 0$. This will show that η is algebraic. Here's the procedure:

For $j = 1$ there exists $i_1 \in \{0, \ldots, n-1\}$ such that $a_{i_1}^1 \neq 0$ (we assumed that $\eta \neq 0$). We can thus determine from the first equation of (4.2) θ^{i_1} as a function of η

and θ^i, where $i \in \{0, \cdots, n-1\} \setminus \{i_1\}$. We replace the expression for θ^{i_1} determined in this way into the other relationships in (4.2) and obtain

$$\begin{cases} \eta^2 - b_1 \eta = \sum_{i \neq i_1} b_i^2 \cdot \theta^i \\ \eta^3 - b_2 \eta = \sum_{i \neq i_1} b_i^3 \cdot \theta^i \\ \cdots \\ \eta^{n+1} - b_n \eta = \sum_{i \neq i_1} b_i^{n+1} \cdot \theta^i \end{cases}, \qquad (4.3)$$

where $b_l = a_{i_1}^{l+1} \cdot (a_{i_1}^1)^{-1} \in k$ and $b_i^j \in k$, for every $l \in \{1, 2, \cdots, n\}, i \in \{0, \cdots, n-1\} \setminus \{i_1\}$, and every $j \in \{2, \cdots, n+1\}$.

If in the first relationship of (4.3) all coefficients b_i^2 are zero, then $\eta^2 - 1 \cdot \eta = 0$, and therefore η is algebraic over k.

If there exists $i_2 \in \{0, \ldots, n-1\} \setminus \{i_1\}$ such that $b_{i_2}^2 \neq 0$, then, from the first relationship in (4.3), we express θ^{i_2} as a function of θ^i where $i \in \{0, \cdots, n-1\} \setminus \{i_1, i_2\}$, and also in terms of η^2 and η. Substituting this expression into the other relationships in (4.3), we obtain

$$\begin{cases} \eta^3 - c_{2,2} \cdot \eta^2 - c_{2,1} \cdot \eta = \sum_{i \neq i_1, i_2} c_i^3 \cdot \theta^i \\ \cdots \\ \eta^{n+1} - c_{n,2} \cdot \eta^2 - c_{n,1} \cdot \eta = \sum_{i \neq i_1, i_2} c_i^{n+1} \cdot \theta^i \end{cases}, \qquad (4.4)$$

where $c_{lm} \in k$ and $c_i^j \in k$, for every $l \in \{2, \cdots, n\}, m = 1, 2, i \in \{0, \cdots, n-1\} \setminus \{i_1, i_2\}$, and every $j \in \{3, \cdots, n+1\}$, and so on.

After at most n such substitutions, we obtain a relationship of the form

$$\eta^{n+1} - e_1 \cdot \eta^n - e_2 \cdot \eta^{n-1} - \cdots - e_{n+1} = 0,$$

where $e_i \in k$, for every $i \in \{1, \cdots, n+1\}$, and this demonstrates that η is algebraic over k. □

Theorem 4.1.5 *Let k be a field, K an extension of it, and $\theta \in K$ a transcendental over k. Then, the field generated by θ over k is*

$$k(\theta) = \left\{ \frac{p(\theta)}{q(\theta)} : p, q \in k[X], q \neq \underline{0} \right\} \subseteq K.$$

Proof We note that for $q \neq 0$, $q(\theta) \neq 0$ (θ is not algebraic over k); therefore, $k(\theta)$ is correctly constructed as a subset of K.

It is evident that $k(\theta)$ is a field, and $\theta \in k(\theta)$ with $k \subset k(\theta)$. Furthermore, any extension of k that contains θ will also clearly contain $k(\theta)$. □

4.1 Field Extensions

Remark 4.1.6 An extension with respect to a transcendental element cannot be finite. Indeed, let's assume that, even though θ is transcendental over k, there exists a finite number of elements $x_1, \cdots, x_n \in k(\theta)$ such that, for every $x \in k(\theta)$, $x = a_1 x_1 + \cdots + a_n x_n$, with $a_1, \cdots a_n \in k$. Then

$$\begin{cases} \theta = a_1^1 \cdot x_1 + \cdots + a_n^1 \cdot x_n \\ \theta^2 = a_1^2 \cdot x_1 + \cdots + a_n^2 \cdot x_n \\ \cdots \\ \theta^{n+1} = a_1^{n+1} \cdot x_1 + \cdots + a_n^{n+1} \cdot x_n \end{cases},$$

where $a_i^j \in k$, for every $i \in \{1, \cdots, n\}$, and every $j \in \{1, \cdots, n+1\}$. By a procedure analogous to the one used in the proof of Theorem 4.1.4, we can eliminate $x_1, \cdots, x_n$ from these $n+1$ equations, and the result will be an equation in θ with coefficients from k. This contradicts the hypothesis that θ is transcendental.

Examples 4.1.7

(1) Let $\mathbb{Q}$ be the field of rational numbers. $\sqrt{2}$ is an algebraic element over $\mathbb{Q}$. It follows that

$$\mathbb{Q}(\sqrt{2}) = \{p(\sqrt{2}) : p \in \mathbb{Q}[X]\} = \{a + b \cdot \sqrt{2} : a, b \in \mathbb{Q}\}.$$

(2) π is transcendental over $\mathbb{Q}$; therefore

$$\mathbb{Q}(\pi) = \left\{ \frac{p(\pi)}{q(\pi)} : p, q \in \mathbb{Q}[X] \right\}.$$

In the following, we will focus on a special class of fields—Liouville fields.

4.1.1 Liouville Fields

In this subsection, we will focus on a special category of objects: fields of holomorphic complex functions on open sets in $\mathbb{C}$.

We will recall some definitions and results related to complex functions of a complex variable.

Let $\mathbb{C}$ be the set of complex numbers with its usual structure of a complete normed field, and let $D \subset \mathbb{C}$ be an open set. A function $f : D \to \mathbb{C}$ is **differentiable at** $z_0 \in D$ if there exists the limit $\lim_{z \to z_0} \frac{f(z) - f(z_0)}{z - z_0} \in \mathbb{C}$; this limit is denoted by $f'(z_0)$, and it is called the **derivative of the function** f **at** z_0.

Let $z_0 = x_0+iy_0$ and $f(z) = u(x, y)+iv(x, y)$, for every $z = x+iy \in D$. Then, f is differentiable at z_0 if and only if the real functions of two real variables u and v are differentiable at $(x_0, y_0) \in \mathbb{R} \times \mathbb{R}$ and satisfy the Cauchy-Riemann conditions:

$$\begin{cases} \dfrac{\partial u}{\partial x}(x_0, y_0) = \dfrac{\partial v}{\partial y}(x_0, y_0) \\ \dfrac{\partial u}{\partial y}(x_0, y_0) = -\dfrac{\partial v}{\partial x}(x_0, y_0) \end{cases}.$$

In this case $f'(z_0) = \dfrac{\partial u}{\partial x}(x_0, y_0) + i \dfrac{\partial v}{\partial x}(x_0, y_0) = \dfrac{\partial v}{\partial y}(x_0, y_0) - i \dfrac{\partial u}{\partial y}(x_0, y_0)$.

A function f is **holomorphic on** D if it is differentiable at all points in D. A function f defined on a disk $D \subset \mathbb{C}$ is holomorphic on D if and only if f is analytic on D (i.e., it can be represented as a Taylor series expansion around any point in D).

A point $a \in D$ is an **isolated singular point** for $f : D \setminus \{a\} \to \mathbb{C}$ if f is holomorphic on $D \setminus \{a\}$. Isolated singular points come in three types: removable singularities, poles, or essential singularities.

The point a is a **removable singularity** of f if there exists a holomorphic function on D, g, such that $f(z) = g(z)$, for every $z \in D \setminus \{a\}$. The singular point a of the function f is removable if and only if $\lim_{z \to a} f(z) \in \mathbb{C}$. For example, for the function $f : \mathbb{C} \setminus \{0\} \to \mathbb{C}$, $f(z) = \frac{\sin z}{z}$, the point $a = 0$ is a removable singularity.

The point a is a **pole** or **non-essential singularity** of f if there exists a holomorphic function on D, g, with $g(a) \neq 0$, and there is $n \in \mathbb{N}^*$ such that $f(z) = \dfrac{g(z)}{(z-a)^n}$, for every $z \in D \setminus \{a\}$ (n is the order of the pole a). A point z is a pole of the function f if and only if the Laurent series of f around the point a ($f(z) = \sum_{k=-n}^{+\infty} a_k \cdot (z-a)^k$) has a finite principal part. For example, the function $f : \mathbb{C} \setminus \{0, i, -i\} \to \mathbb{C}$, $f(z) = \dfrac{1}{z^2(z^2+1)}$, has a double pole (of order 2) at 0, and simple poles at i and $-i$.

The point a is an **essential singularity** of f if the principal part of the Laurent series of f around the point a ($\sum_{k=-\infty}^{-1} a_k \cdot (z-a)^k$) is infinite. For example, the function $f : \mathbb{C} \setminus \{0\} \to \mathbb{C}$, $f(z) = \sin \dfrac{1}{z}$, has an essential singularity at $a = 0$.

We can now state the definition of Liouville fields.

Definition 4.1.8 Let $D \subset \mathbb{C}$ be an open set, and let $L(D)$ be a set of functions $f : D \setminus I_f \to \mathbb{C}$, where each function f has, separately, a at most countable set of isolated singularities I_f. $L(D)$ is called a **Liouville field** of functions on D if the following conditions are satisfied:

(a) $(L(D), +, \cdot)$ forms a field with respect to the usual addition and multiplication operations, and $\mathbb{C} \subset L(D)$ (i.e., $L(D)$ contains constant functions).

4.1 Field Extensions

(b) For any $f \in L(D)$, f is holomorphic on $D \setminus I_f$.
(c) For any $f \in L(D)$, f' (the derivative of f) also belongs to $L(D)$.

When the set D is understood, we will denote it as L instead of $L(D)$.

Example 4.1.9 The simplest example of a Liouville field is the field of complex rational functions:

$$R(\mathbb{C}) = \left\{ \frac{P}{Q} : P, Q \in \mathbb{C}[X], Q \neq \underline{0} \right\} \quad (\underline{0} \text{ is the neutral element in } (\mathbb{C}[X]), +)).$$

It is evident that $R(\mathbb{C})$ is a field containing constant functions. For any $P, Q \in \mathbb{C}[X]$ (P and Q are polynomials with complex coefficients), $f = \frac{P}{Q}$ is a uniformly continuous and holomorphic function on $\mathbb{C} \setminus I_f$, where I_f is a finite set of poles (f is a meromorphic function). Additionally,

$$\left(\frac{P}{Q} \right)' = \frac{P'Q - PQ'}{Q^2} \in R(\mathbb{C}).$$

Liouville fields allow extensions by algebraic or transcendental elements. We are interested in under what conditions such extensions are themselves Liouville fields.

Definition 4.1.10 Let L be a Liouville field on the open set $D \subseteq \mathbb{C}$, and let $\theta : D \setminus I_\theta \to \mathbb{C}$ be a holomorphic function on D except for a countable set of isolated singularities I_θ. If $L(\theta)$, the field generated by θ over L, is a Liouville field than $L(\theta)$ is the **Liouville extension** of L generated by θ.

In general, the extension $L(\theta)$ of a Liouville field L is not a Liouville field.

Theorem 4.1.11 *Let L be a Liouville field on the open set $D \subseteq \mathbb{C}$ and let $\theta : D \setminus I_\theta \to \mathbb{C}$ be a holomorphic function on D except for a countable set of isolated singularities I_θ. If θ is an algebraic element over L, then the Liouville extension of L generated by θ is $L(\theta) = \{P(\theta) : P \in L[X], \deg(P) \leq n\}$, where $n + 1$ is the degree of the minimal polynomial defining θ.*

Proof Theorem 4.1.4 implies that

$$L(\theta) = \{P(\theta) : P \in L[X], \deg(P) \leq n\} = \{P(\theta) : P \in L[X]\}$$

is the field generated by θ.

For every $P \in L[X]$, $P(\theta)$ is holomorphic on $D \setminus I_\theta$.

Let $P_0 = a_0 + a_1 X + \cdots + a_n X^n + X^{n+1} \in L[X]$ be the minimal polynomial defining θ. Therefore $a_0, \cdots, a_n \in L$, $P_0(\theta) = 0$, and $P(\theta) \neq 0$, for every $P \in L[X]$ with $\deg(P) \leq n$.

Then $a_0(z) + a_1(z)\theta(z) + \cdots + a_n(z)\theta^n(z) + \theta^{n+1}(z) = 0$. If we differentiate this identity, we obtain

$$\left[a_0'(z) + a_1'(z) \cdot \theta(z) + \cdots + a_n'(z) \cdot \theta^n(z) \right] +$$
$$+ \left[a_1(z) + 2a_2(z) \cdot \theta(z) + \cdots + na_n(z) \cdot \theta^{n-1}(z) + (n+1)\theta^n(z) \right] \cdot \theta'(z) = 0.$$

$P = a'_0 + a'_1 X + \cdots + a'_n X^n$, $Q(X) = a_1 + 2a_2 X + \cdots + n a_n X^{n-1} + (n+1)X^n \in L[X]$ and $\deg(Q) \leqslant n$. It follows that $Q(\theta) \neq 0$. Then $\theta' = -\dfrac{P(\theta)}{Q(\theta)}$.

Since $L(\theta)$ is a field, there exists $Q_1 \in L[X]$ such that $Q_1(\theta) = [Q(\theta)]^{-1}$ and then $\theta' = -P(\theta) \cdot Q_1(\theta) \in L(\theta)$.

Now, for every $P = b_0 + b_1 X + \cdots + b_n X^n \in L[X]$, $[P(\theta)]' = [b'_0 + \cdots + b'_n \theta^n] + [b_1 + \cdots + n b_n \theta^{n-1}] \cdot \theta' \in L(\theta)$, so that $L(\theta)$ is a Liouville field. □

Theorem 4.1.12 *Let L be a Liouville field of functions on the open set $D \subset \mathbb{C}$, and let $\theta : D \setminus I_\theta \to \mathbb{C}$ be a holomorphic function on D except for a countable set of isolated singularities I_θ (θ can be either algebraic or transcendental over L). We denote*

$$L(\theta) = \left\{ \frac{P(\theta)}{Q(\theta)} : P, Q \in L[X], Q(\theta) \neq 0 \right\}.$$

The necessary and sufficient condition for $L(\theta)$ to be a Liouville field on D is that $\theta' \in L(\theta)$. In this case, $L(\theta)$ is the Liouville extension of L generated by θ.

Proof It is evident that $L(\theta)$ is a field under the usual operations of addition and multiplication, and that $L(\theta)$ is an extension of L containing θ. Furthermore, it can be easily observed that any function f in $L(\theta)$ is holomorphic on $D \setminus I_f$, where I_f is a set at most countable of isolated singularities of f. It follows that $L(\theta)$ will be a Liouville field on D if and only if, for any $f \in L(\theta)$, f' belongs to $L(\theta)$.

If $L(\theta)$ is a Liouville field, then it is evident that $\theta' \in L(\theta)$.

Conversely, assuming that $\theta' \in L(\theta)$, let $f \in L(\theta)$ be an arbitrary function. Then, there exist two polynomials $P, Q \in L[X]$ with $Q(\theta) \neq 0$ such that $f(z) = \dfrac{P(\theta(z))}{Q(\theta(z))}$, for every $z \in D \setminus I_f$. In this case

$$f'(z) = \frac{P'(\theta(z)) \cdot Q(\theta(z)) - P(\theta(z)) \cdot Q'(\theta(z))}{Q^2(\theta(z))} \cdot \theta'(z).$$

Since $\theta' \in L(\theta)$, there exist $P_1, Q_1 \in L[X]$ with $Q_1(\theta) \neq 0$ such that $\theta'(z) = \dfrac{P_1(\theta(z))}{Q_1(\theta(z))}$, for every $z \in D \setminus I_\theta$; then

$$f' = \frac{[P'(\theta) \cdot Q(\theta) - P(\theta) \cdot Q'(\theta)] \cdot P_1(\theta)}{Q^2(\theta) \cdot Q_1(\theta)} = \frac{P_2(\theta)}{Q_2(\theta)}$$

with $P_2, Q_2 \in L[X]$ and $Q_2(\theta) \neq 0$.

It follows that $f' \in L(\theta)$ and so $L(\theta)$ is a Liouville field.

It is evident that any Liouville field extension of L that contains θ also contains $L(\theta)$. □

4.1 Field Extensions

If θ is algebraic over L, then for any $Q \in L[X]$ with $Q(\theta) \neq 0$, there exists $Q_1 \in L[X]$ such that $[Q(\theta)]^{-1} = Q_1(\theta)$ (see Theorem 4.1.4). Thus, for any $P \in L[X]$, $\dfrac{P(\theta)}{Q(\theta)} = P(\theta) \cdot Q_1(\theta) = P_1(\theta)$, where $P_1 \in L[X]$, and therefore $L(\theta) = \{P(\theta) : P \in L[X]\}$. Taking into account Theorem 4.1.11, $\theta' \in L(\theta)$; thus, the previous theorem reduces to Theorem 4.1.11.

If θ is transcendental over L, then, in general, $L(\theta)$ is not a Liouville field. We can illustrate this situation by considering $L = R(\mathbb{C})$ and $\theta(z) = \arcsin z$. If we replace the field $R(\mathbb{C})$ with the field $L_1 = R(\mathbb{C})(\sqrt{1-z^2})$, then $L_1(\theta)$ is a Liouville field.

However, there are some remarkable transcendental elements θ for which $L(\theta)$ is a Liouville field. Before presenting them, let's recall the definitions of some of the elementary complex functions.

4.1.2 Exponential and Logarithmic Functions

The exponential and logarithmic functions play a special role among the other elementary functions. In the introduction of this chapter, we noted that among the elementary functions, we must consider trigonometric and hyperbolic functions as well as their inverses. All of these functions are defined using the exponential and logarithmic functions. Then, as we will observe, any elementary function belongs to algebraic extensions, exponential extensions, logarithmic extensions of the Liouville field of rational functions, or combinations of such extensions.

The function $f : \mathbb{C} \to \mathbb{C}$, defined by

$$f(z) \equiv e^z = e^x \cdot (\cos y + i \sin y), \text{ for every } z = x + iy \in \mathbb{C},$$

is called the **exponential function**. The real part, u, and the imaginary part, v, are defined by the following expressions: $\begin{cases} u(x, y) = e^x \cdot \cos y \\ v(x, y) = e^x \cdot \sin y \end{cases}$.

We observe that u and v are differentiable functions on $\mathbb{R}^2$ and satisfy the Cauchy-Riemann conditions at every point $z = x + iy \in \mathbb{C}$. This implies that f is holomorphic on $\mathbb{C}$, i.e. an entire function; its derivative is $f'(z) = e^z = f(z)$, for every $z \in \mathbb{C}$.

We recall the following two useful formulas in complex analysis:

$$e^z = \lim_{n \to \infty} \left(1 + \frac{z}{n}\right)^n, \text{ for every } z \in \mathbb{C},$$
$$\varepsilon^z = 1 + \frac{z}{1!} + \frac{z^2}{2!} + \cdots + \frac{z^n}{n!} + \cdots, \text{ for every } z \in \mathbb{C}.$$

The exponential function has several properties that immediately follow from its definition:

$$e^{z_1+z_2} = e^{z_1} \cdot e^{z_2}, \text{ for every } z_1, z_2 \in \mathbb{C},$$
$$e^{z+2\pi i} = e^z, \text{ for every } z \in \mathbb{C},$$
$$e^{\pm iy} = \cos y \pm i \sin y, \text{ for every } y \in \mathbb{R},$$
$$e^0 = 1, e^z \neq 0, \text{ for every } z \in \mathbb{C}.$$

With the help of this function, we can introduce other four entire functions (holomorphic on $\mathbb{C}$): "sine", "cosine", "hyperbolic sine", and "hyperbolic cosine":

$$\sin z = \frac{e^{iz} - e^{-iz}}{2i} = z - \frac{z^3}{3!} + \frac{z^5}{5!} + \cdots + (-1)^n \frac{z^{2n+1}}{(2n+1)!} + \cdots$$
$$\cos z = \frac{e^{iz} + e^{-iz}}{2} = 1 - \frac{z^2}{2!} + \frac{z^4}{4!} + \cdots + (-1)^n \frac{z^{2n}}{(2n)!} + \cdots$$
$$\sinh z = -i \sin iz = \frac{e^z - e^{-z}}{2} = z + \frac{z^3}{3!} + \frac{z^5}{5!} + \cdots + \frac{z^{2n+1}}{(2n+1)!} + \cdots$$
$$\cosh z = \cos iz = \frac{e^z + e^{-z}}{2} = 1 + \frac{z^2}{2!} + \frac{z^4}{4!} + \cdots + \frac{z^{2n}}{(2n)!} + \cdots$$

With the help of the exponential function, we can also define the tangent function: $\tan : \mathbb{C} \setminus \{\frac{\pi}{2} + k\pi : k \in \mathbb{Z}\} \to \mathbb{C}$, by

$$\tan z = \frac{\sin z}{\cos z} = -i \frac{e^{2iz} - 1}{e^{2iz} + 1}.$$

It is a meromorphic functions with simple poles at points $z = \frac{\pi}{2} + k\pi, k \in \mathbb{Z}$.

Let now the multi-valued function $\text{Log} : \mathbb{C}^* \to 2^{\mathbb{C}}$ be defined by

$$\text{Log}(z) = \{w \in \mathbb{C} : e^w = z\}, \text{ for any } z \in \mathbb{C}^* = \mathbb{C} \setminus \{0\}.$$

If $z = |z|(\cos(\arg z) + i \sin(\arg z))$ is the trigonometric form of the number $z \in \mathbb{C}^*$, then $|z| > 0$, and

$$\text{Log}(z) = \{\log |z| + i(\arg z + 2k\pi) : k \in \mathbb{Z}\}.$$

For each $k \in \mathbb{Z}$, the multi-valued function defined by

$$\text{Log}_k(z) = \log |z| + i(\arg z + 2k\pi), \text{ for every } z \in \mathbb{C}^*,$$

is called a branch of Log. The branches of Log are not uniformly defined on $\mathbb{C}^*$; they become uniform if we consider their restrictions to $D = \mathbb{C} \setminus \{z \in \mathbb{C} : \text{Re}(z) \leqslant 0, \text{Im}(z) = 0\}$. Thus, for any $k \in \mathbb{Z}$, the function

$$\log_k : D \to \mathbb{C}, \log_k(z) = \log |z| + i(\arg z + 2k\pi),$$

is the uniform branch of the multi-valued function Log; it is also holomorphic on D, and $(\log_k(z))' = \frac{1}{z}$, for every $z \in D$.

$\log_0$ is called the principal branch of Log and is denoted as $\log : D = \mathbb{C} \setminus \{z \in \mathbb{C} : \operatorname{Re}(z) \leqslant 0, \operatorname{Im}(z) = 0\} \to \mathbb{C}$,

$$\log(z) = \log|z| + i \arg z, \text{ for every } z \in D.$$

This is the complex **logarithm function**.

The logarithm function allows expressions for many other elementary functions. Let $\alpha \in \mathbb{C}$ and let the multi-valued function $F : \mathbb{C}^* \to 2^{\mathbb{C}}$ be defined by $F(z) = e^{\alpha \operatorname{Log} z} = \{e^{\log|z|+i(\arg z + 2k\pi)} : k \in \mathbb{Z}\}$, for every $z \in \mathbb{C}^*$. The restriction to the set $D = \mathbb{C} \setminus \{z \in \mathbb{C} : \operatorname{Re}(z) \leqslant 0, \operatorname{Im}(z) = 0\}$ of the principal branch ($k = 0$) of the multi-valued function F is called the **power function**. Therefore, the power function is a function $f : D \to \mathbb{C}$ defined by

$$f(z) = z^{\alpha} = e^{\alpha \log z} = e^{\alpha(\log|z|+i \arg z)}, \text{ for every } z \in D.$$

In the particular case where $n \in \mathbb{N}^*$ and $\alpha = \frac{1}{n}$, we obtain the **nth root function**, $f : D \to \mathbb{C}$, defined by

$$f(z) \equiv \sqrt[n]{z} = \sqrt[n]{|z|} \cdot \left(\cos \frac{\arg z}{n} + i \sin \frac{\arg z}{n}\right).$$

Let the multi-valued function $\operatorname{Arcsin} : \mathbb{C} \to 2^{\mathbb{C}}$ be defined by $\operatorname{Arcsin}(z) = \{w \in \mathbb{C} : \sin w = \frac{e^{iw}-e^{-iw}}{2i} = z\} = -i \operatorname{Log}(iz \pm \sqrt{1-z^2})$. This function has a double infinity of branches. Considering the principal branches of the multi-valued functions radical and logarithm, we obtain the function

$$\arcsin z = -i \log\left(iz + \sqrt{1-z^2}\right).$$

Proceeding in a similar manner, we obtain the functions:

$$\arccos z = -i \log(z + \sqrt{z^2-1}),$$
$$\arctan z = \frac{1}{2i} \log \frac{i-z}{i+z},$$
$$\operatorname{arsinh} z = \log(z + \sqrt{z^2+1}),$$
$$\operatorname{arcosh} z = \log(z + \sqrt{z^2-1}).$$

4.1.3 Elementary Functions

In Example 4.1.9, we noticed that the field of rational functions over $\mathbb{C}$, denoted as $R(\mathbb{C})$, is a Liouville field. Let's observe that θ, where $\theta(z) = e^z$, is a transcendental element over $R(\mathbb{C})$. Furthermore, if $D = \mathbb{C} \setminus \{z \in \mathbb{C} : \operatorname{Re}(z) \leqslant 0, \operatorname{Im}(z) = 0\}$

and $\theta(z) = \log z$, then the set of restrictions of rational functions to D, denoted as $R(D)$, is a Liouville field, and θ is a transcendental element over $R(D)$.

In these two cases, we are interested in determining the Liouville extensions of the fields $R(\mathbb{C})$ and $R(D)$. In fact, we will consider a more general case.

Definition 4.1.13 Let L be a Liouville field on the open set $D \subset \mathbb{C}$, let $u \in L$ be a non-identically zero function and let I_u be the countable set of isolated singularities of u.

(i) If the element θ, defined as $\theta(z) = \log(u(z))$ is transcendental over L, then we say that θ is **logarithmic** over L. We will implicitly assume that we restrict the functions of L to an open subset $D_1 \subseteq D$ on which the principal branches of the logarithmic function involved is uniform (in this case $D_1 = D \setminus \{z \in \mathbb{C} : \text{Re}(u(z)) \leq 0, \text{Im}(u(z)) = 0\}$).
We remark that $\theta'(z) = \dfrac{u'(z)}{u(z)}$ such that $\theta' \in L \subseteq L(\theta) = \left\{\dfrac{P(\theta)}{Q(\theta)} : P, Q \in L[X]\right\}$. According to Theorem 4.1.12, $L(\theta)$ is the Liouville extension of L generated by θ. Therefore

$$L(\theta) = \left\{\frac{a_0 \log^n(u) + \cdots + a_n}{b_0 \log^m(u) + \cdots + b_m} : a_k, b_l \in L\right\}.$$

$L(\theta)$ is called a **Liouville logarithmic extension** of L.

(ii) If the element $\theta : D \setminus I_u \to \mathbb{C}$, defined by $\theta(z) = e^{u(z)}$ is transcendental over L, then we say that θ is **exponential** over L. Since $\theta'(z) = u'(z) \cdot \theta(z)$, it follows that $\theta' \in L(\theta)$. According to Theorem 4.1.12,

$$L(\theta) = \left\{\frac{a_0 e^{nu(z)} + \cdots + a_n}{b_0 e^{mu(z)} + \cdots + b_m} : a_k, b_l \in L\right\}$$

is the Liouville extension of L generated by θ. $L(\theta)$ is called a **Liouville exponential extension** of L.

Definition 4.1.14 Let L be a Liouville field of functions on the open set $D \subset \mathbb{C}$, let $\theta : D \setminus I_\theta \to \mathbb{C}$ be a uniform and holomorphic function on D except for a countable set of isolated singularities I_θ and let $L(\theta)$ be the Liouville extension of L generated by θ. $L(\theta)$ is called an **elementary Liouville extension** if θ is algebraic, logarithmic or exponential over L. Therefore, an elementary Liouville extension of L, $L(\theta)$, is the smallest Liouville field that contains L and an algebraic, logarithmic, or exponential element θ over L. As previously noted in Theorem 4.1.11 and in (i) and (ii) of Definition 4.1.13, an elementary Liouville extension of L is of the form $L(\theta) = \{P(\theta) : P \in L[X], \deg(P) \leq n\}$ if θ is algebraic and $n+1$ is the degree of the minimal polynomial defining θ and $L(\theta) = \left\{\dfrac{P(\theta)}{Q(\theta)} : P, Q \in L[X]\right\}$ if θ is transcendental logarithmic or exponential.

A Liouville field G is called an **elementary extension** of L if it is obtained through a finite number of successive elementary Liouville extensions of L. This means that $G = L(\theta_1, \theta_2, \cdots, \theta_n)$ where, for every $k = 1, \cdots, n$, θ_k is algebraic, logarithmic, or exponential over $L_{k-1} = L(\theta_1, \cdots, \theta_{k-1})$ and $L_0 = L$. **We will implicitly assume that G is a Liouville field formed by restricting functions to an open subset $D_1 \subseteq D$ on which the principal branches of the logarithmic functions involved are uniform.** A function f is called **elementary with respect to the Liouville field L** if it belongs to an elementary extension of L.

A function f is an **elementary function** if it is elementary with respect to the Liouville field of rational functions (see Example 4.1.9).

Based on the definitions given in Sect. 4.1.2, the functions sin, cos, tan, sinh, cosh, arcsin, arccos, arctan, arsinh, and arcosh are elementary functions.

4.2 Liouville's Principle

The problem of integration in finite terms can now be formulated as follows: Let f be an elementary function; we will first determine an elementary extension L of R such that $f \in L$. Then, we need to find an elementary extension G of L such that $g = \int f \in G$, or show that such an extension does not exist. Finally, in the case where such an extension G exists, we must explicitly determine the function g. The solution to this problem is essentially based on Liouville's principle. However, before stating and proving it, we need a few auxiliary results. The following propositions represent adaptations of the theorems 12.2 and 12.3, from [1].

Proposition 4.2.1 *Let L be a Liouville field, let $u \in L$ be a non-identically zero function and let θ, $\theta(z) = \log u(z)$, be logarithmic over L.*

For every polynomial $P \in L[X]$, $P = a_0 + a_1 X + a_2 X^2 + \cdots + a_n X^n$, with $n > 0$, and $a_n \neq 0$, there exists $Q \in L[X]$ such that $\deg(Q) = \begin{cases} n-1, a_n \in \mathbb{C}, \\ n, a_n \in L \setminus \mathbb{C} \end{cases}$ and $[P(\theta)]' = Q(\theta)$.

Proof

$$P(\theta(z)) = a_0(z) + a_1(z)\theta(z) + a_2(z)\theta^2(z) + \cdots + a_n(z)\theta^n(z), \text{ for every } z \in D_1.$$

Differentiating the above relationship, we obtain

$$(P(\theta))'(z) = (a_0'(z) + a_1'(z)\theta(z) + a_2'(z)\theta^2(z) + \cdots + a_n'(z)\theta^n(z)) +$$

$$+ (a_1(z) + 2a_2(z)\theta(z) + \cdots + na_n(z)\theta^{n-1}(z))\theta'(z).$$

Taking into account that $\theta'(z) = \frac{u'(z)}{u(z)}$, we obtain

$$(P(\theta))'(z) = \left[a_0'(z) + a_1(z) \cdot \frac{u'(z)}{u(z)}\right] + \left[a_1'(z) + 2a_2(z) \cdot \frac{u'(z)}{u(z)}\right]\theta(z) + \cdots +$$

$$+ \left[a_{n-1}'(z) + na_n(z) \cdot \frac{u'(z)}{u(z)}\right]\theta^{n-1}(z) + a_n'(z)\theta^n(z).$$

So $Q = \left[a_0' + a_1 \cdot \frac{u'}{u}\right] + \left[a_1' + 2a_2 \cdot \frac{u'}{u}\right]X + \cdots + \left[a_{n-1}' + na_n \cdot \frac{u'}{u}\right]X^{n-1} + a_n'X^n \in L[X]$ and $[P(\theta)]' = Q(\theta)$.

If a_n is a constant function ($a_n \in \mathbb{C}$), then $a_n' = 0$ and $\deg(Q) \leq n - 1$. If we suppose that $a_{n-1}' + na_n \cdot \frac{u'}{u} = \underline{0}$, then $(na_n\theta + a_{n-1})' = \underline{0}$ and so $na_n\theta + a_{n-1} \in \mathbb{C}$, which is absurd, since θ is transcendental over L. Therefore $a_n' = 0$ and $a_{n-1}' + na_n \cdot \frac{u'}{u} \neq \underline{0}$ which means that $\deg(Q) = n - 1$.

If a_n is not constant ($a_n \in L \setminus \mathbb{C}$), then $a_n' \neq 0$ and so $\deg(Q) = n$. □

Proposition 4.2.2 *Let L be a Liouville field on the open set $D \subset \mathbb{C}$ and let $u \in L$ be a non-identically zero function. We suppose that the function $\theta : D \setminus I_u \to \mathbb{C}$, $\theta(z) = e^{u(z)}$ is exponential over L, where I_u is the countable set of isolated singularities of u (see the (ii) of Definition 4.1.13).*

For every polynomial $P \in L[X]$, $P = a_0 + a_1X + a_2X^2 + \cdots + a_nX^n$, with $n > 0$, and $a_n \neq 0$, there exists $Q \in L[X]$ such that $\deg(Q) = \deg(P)$ and $(P(\theta))' = Q(\theta)$.

In addition, $P(\theta)$ divides $Q(\theta)$ if and only if P is monomial (i.e. $a_0 = a_1 = \cdots = a_{n-1} = 0$).

Proof

$$P(\theta(z)) = a_0(z) + a_1(z)\theta(z) + a_2(z)\theta^2(z) + \cdots + a_n(z)\theta^n(z), \text{ for every } z \in D.$$

Differentiating the above relationship, we obtain

$$(P(\theta))'(z) = (a_0'(z) + a_1'(z)\theta(z) + a_2'(z)\theta^2(z) + \cdots + a_n'(z)\theta^n(z)) +$$

$$+ (a_1(z) + 2a_2(z)\theta(z) + \cdots + na_n(z)\theta^{n-1}(z))\theta'(z).$$

Taking into account that $\theta'(z) = u'(z) \cdot \theta(z)$, we obtain $(P(\theta))'(z) =$

$$= a_0'(z) + \left[a_1'(z) + a_1(z)u'(z)\right]\theta(z) + \cdots + \left[a_n'(z) + na_n(z)u'(z)\right]\theta^n(z).$$

So $Q = a_0' + \left[a_1' + a_1u'\right]X + \cdots + \left[a_n' + na_nu'\right]X^n \in L[X]$ and $(P(\theta))' = Q(\theta)$.

We remark that $a_n' + na_nu' = \theta^{-n}(a_n\theta^n)'$; if we suppose that $a_n' + na_nu' = \underline{0}$, then $a_n\theta^n \in \mathbb{C}$, which is absurd, since θ is transcendental over L. Therefore $a_n' + na_nu' \neq \underline{0}$, which means that $\deg(Q) = n = \deg(P)$.

4.2 Liouville's Principle

If $a_0 = a_1 = \cdots = a_{n-1} = 0$, then $P = a_n X^n$ and $Q = [a_n' + na_n u']X^n$, and it is evident that $P(\theta)$ divides $Q(\theta)$.

Conversely, assuming that $P(\theta)$ divides $Q(\theta)$; then there exists $b \in L, b \neq \underline{0}$, such that $Q(\theta) = bP(\theta)$ or

$$(a_0' - ba_0) + (a_1' + a_1 u' - ba_1)\theta + \cdots + (a_n' + na_n u' - ba_n)\theta^n = \underline{0}.$$

Since θ is transcendental over L, it follows that all coefficients of the polynomial in θ above must be zero. Let's assume, by contradiction, that P is not a monomial; then there exists $k < n$ such that $a_k \neq 0$; since $a_k' + ka_k u' - ba_k = 0 = a_n' + na_n u' - ba_n$, it follows that

$$\frac{a_k'}{a_k} + ku' = \frac{a_n'}{a_n} + nu'. \qquad (*)$$

On the other hand

$$\left(\frac{a_n}{a_k} \cdot \theta^{n-k}\right)' = \left(\frac{a_n}{a_k}\right)' \theta^{n-k} + \frac{a_n}{a_k}(n-k)\theta^{n-k}u' =$$

$$= \theta^{n-k}\left[\frac{a_n'}{a_k} - \frac{a_n a_k'}{a_k^2} + (n-k)\frac{a_n}{a_k} \cdot u'\right] = \frac{a_n}{a_k} \cdot \theta^{n-k}\left[\frac{a_n'}{a_n} - \frac{a_k'}{a_k} + (n-k)u'\right].$$

Using the equality $(*)$ in the relationship above, we obtain $\left(\frac{a_n}{a_k} \cdot \theta^{n-k}\right)' = \underline{0}$, which means that $\left(\frac{a_n}{a_k} \cdot \theta^{n-k}\right) \in \mathbb{C}$, which is absurd, since θ is transcendental over L. Therefore, if $P(\theta)$ divides $Q(\theta)$, then $a_0 = a_1 = \cdots = a_{n-1} = 0$. □

Let L be a Liouville field and $f \in L$; we are interested in determining under what conditions f admits an antiderivative elementary with respect to L. In other words, under what conditions there exists a function g belonging to an elementary extension of G of L such that $g' = f$. The answer to this question is provided by Liouville's principle, which we present below.

Theorem 4.2.3 (Liouville's Principle) *Let L be a Liouville field and let $f \in L$. The function f admits an antiderivative elementary with respect to L if and only if there exist $m \in \mathbb{N}^*, u_0, u_1, \cdots, u_m \in L$ and $c_1, \cdots, c_m \in \mathbb{C}$ such that*

$$f = u_0' + \sum_{k=1}^{m} c_k \cdot \frac{u_k'}{u_k}.$$

In this case, $g = \int f = u_0 + \sum_{k=1}^{m} c_k \log u_k$ is an antiderivative of f.

Proof We assume that there exists a function g, elementary with respect to L, such that $g' = f$. This means that there is G, an elementary extension of L, such that $g \in G$. It follows that there exists a natural number n such that G is obtained after n successive elementary Liouville extensions of L.

We will prove the theorem by induction on n.

If $n = 0$, then $G = L$ and then $g \in L$. We then choose $u_0 = g$, $c_1 = \cdots = c_m = 0$, and non-zero arbitrary elements $u_1, \cdots, u_m \in L$. It follows that

$$f = g' = u_0' + \sum_{k=1}^{m} c_k \cdot \frac{u_k'}{u_k}.$$

We now assume that the theorem is proven for any Liouville field L and for any elementary extension of it with at most $n - 1$ elements.

Now we assume that the antiderivative of f, g, belongs to an elementary extension of L, denoted as G, with n elements. Therefore, there exist $\theta_1, \cdots, \theta_n$ such that $G = L(\theta_1, \cdots, \theta_n)$, and, for every $k = 1, \cdots, n$, θ_k is algebraic, logarithmic or exponential over $L(\theta_1, \cdots, \theta_{k-1})$. It can be easily shown that $L(\theta_1, \cdots, \theta_n) = L(\theta_1)(\theta_2, \cdots, \theta_n)$. Then $f \in L \subseteq L(\theta_1)$ and its antiderivative, $g \in L(\theta_1)(\theta_2, \cdots, \theta_n)$ (an elementary extension with $n - 1$ elements of $L(\theta_1)$). The inductive hypothesis assures us that there exist $v_0, v_1, \cdots, v_m \in L(\theta_1)$ and $c_1, \cdots, c_m \in \mathbb{C}$ such that

$$f = v_0' + \sum_{k=1}^{m} c_k \cdot \frac{v_k'}{v_k}. \tag{4.5}$$

For the sake of convenience in writing, we will further denote θ_1 as θ. Therefore in (4.5), $v_0, v_1, \cdots, v_m \in L(\theta)$ and $c_1, \cdots, c_m \in \mathbb{C}$. The demonstration concludes if we show that the relationship (4.5) can be rewritten by replacing $v_i \in L(\theta)$ with $u_i \in L$.

We will treat separately the cases where θ is algebraic, logarithmic, or exponential over L.

(1) θ **is algebraic over** L. Let $a_0 + a_1 X + \cdots + a_n X^n + X^{n+1} \in L[X]$ be the minimal polynomial defining θ, and let $\theta = \theta_0, \theta_1, \cdots, \theta_n$ be its $n + 1$ roots. Theorem 4.1.11 assures us that the Liouville extension of L is $L(\theta) = \{P(\theta) : P \in L[X], \deg(P) \leq n\}$. Then there exist $P_0, P_1, \cdots, P_m \in L[X]$ such that $v_k = P_k(\theta)$ and $\deg(P_k) \leq n$, for every $k = 0, 1, \cdots, m$. It follows that

$$f = [P_0(\theta)]' + \sum_{k=1}^{m} c_k \cdot \frac{[P_k(\theta)]'}{[P_k(\theta)]}. \tag{4.6}$$

4.2 Liouville's Principle

In Eq. (4.6), the left-hand side does not depend on $\theta = \theta_0$. Therefore, the equation can be rewritten by replacing θ with any of the other roots $\theta_1, \cdots, \theta_n$; thus, we obtain:

$$f = [P_0(\theta_j)]' + \sum_{k=1}^{m} c_k \cdot \frac{[P_k(\theta_j)]'}{[P_k(\theta_j)]}, \, j = 0, 1, \cdots, n. \quad (4.7)$$

If we add Eqs. (4.7), we find:

$$g' = f = \frac{1}{n+1} \cdot \sum_{j=0}^{n} \left\{ [P_0(\theta_j)]' + \sum_{k=1}^{m} c_k \cdot \frac{[P_k(\theta_j)]'}{[P_k(\theta_j)]} \right\},$$

from where

$$g = \frac{1}{n+1} \cdot \sum_{j=0}^{n} P_0(\theta_j) + \sum_{k=1}^{m} c_k \cdot \left(\sum_{j=0}^{n} \log P_k(\theta_j) \right) = \quad (4.8)$$

$$= \frac{1}{n+1} \cdot \sum_{j=0}^{n} P_0(\theta_j) + \sum_{k=1}^{m} c_k \cdot \left[\log \left(\prod_{j=0}^{n} P_k(\theta_j) \right) \right].$$

We remark that, for every $k = 1, \cdots, m$, $Q_k(\theta_0, \theta_1, \cdots, \theta_n) = \prod_{j=0}^{n} P_k(\theta_j)$ and $Q_0(\theta_0, \theta_1, \cdots, \theta_n) = \sum_{j=0}^{n} P_0(\theta_j)$ are symmetric polynomials in $n+1$ variables $\theta_0, \theta_1, \cdots, \theta_n$. But every symmetric polynomial in the variables $\theta_0, \theta_1 \cdots, \theta_n$ admits a unique expression in terms of the elementary variables $\varepsilon_0 = \theta_0 + \theta_1 + \cdots + \theta_n$, $\varepsilon_1 = \theta_0\theta_1 + \cdots + \theta_r\theta_s + \theta_{n-1}\theta_n, \cdots, \varepsilon_n = \theta_0\theta_1 \cdots \theta_n$. According to Vieta's relationships, ε_l is calculated using the coefficients of the above minimal polynomial as follows:

$$\varepsilon_j = \sum_{0 \leqslant j_0 < j_1 < \cdots < j_l \leqslant n} \theta_{j_0} \cdot \theta_{j_1} \cdots \theta_{j_l} = (-1)^{j+1} a_{n-j} \in L, \text{ for every } j = 0, \cdots n.$$

It follows that $u_0 = \frac{1}{n+1} \cdot \sum_{j=0}^{n} P_0(\theta_j) \in L$ and, for every $k = 1, \cdots, m$, $u_k = \prod_{j=0}^{n} P_k(\theta_j) \in L$ and therefore $g = u_0 + \sum_{k=1}^{m} c_k \cdot \log u_k$ or

$$f = u_0' + \sum_{k=1}^{m} c_k \cdot \frac{u_k'}{u_k}, \text{ where } u_0, u_1, \cdots, u_m \in L.$$

(2) **θ is logarithmic over L.** Let $u \in L$ such that $\theta = \log u$ is transcendental over L. According to Theorem 4.1.12 and Definition 4.1.13, the Liouville extension of L generated by θ is $L(\theta) = \left\{ \frac{P(\theta)}{Q(\theta)} : P, Q \in L[X] \right\}$. It follows that, for

every $k = 0, 1, \cdots, m$, the functions v_k in formula (4.5) are of the form $v_k = \dfrac{P_k(\theta)}{Q_k(\theta)}$, where $P_k, Q_k \in L[X]$. A simple calculation shows us that, for every $k = 1, \cdots, m$:

$$\frac{v_k'}{v_k} = \frac{[P_k(\theta)]' Q_k(\theta) - P_k(\theta)[Q_k(\theta)]'}{Q_k^2(\theta)} \cdot \frac{Q_k(\theta)}{P_k(\theta)} = \frac{[P_k(\theta)]'}{P_k(\theta)} - \frac{[Q_k(\theta)]'}{Q_k(\theta)}.$$

So we can rewrite (4.5) as:

$$f = \left[\frac{P_0(\theta)}{Q_0(\theta)}\right]' + \sum_{k=1}^{p} c_k \cdot \frac{[P_k(\theta)]'}{P_k(\theta)}, \tag{4.9}$$

where, $P_0, Q_0 \in L[X]$ are relatively prime, Q_0 is monic (the leading coefficient is 1), and for every $k = 1, \cdots, p$, $P_k \in L$ or $P_k \in L[X]$ are irreducible monic polynomials with $\deg(P_k) > 0$, P_k are all different and $c_k \neq 0$.

By factoring the polynomial Q_0, we obtain $Q_0(\theta) = \prod_{j=1}^{s}(Q_j(\theta))^{r_j}$, where Q_j are distinct irreducible monic polynomials and $r_j \in \mathbb{N}^*$. Now, the unique partial fraction decomposition of $\dfrac{P_0(\theta)}{Q_0(\theta)}$ leads us to

$$\frac{P_0(\theta)}{Q_0(\theta)} = P(\theta) + \sum_{j=1}^{s} \sum_{l=1}^{r_j} \frac{P_{jl}(\theta)}{[Q_j(\theta)]^l}, \tag{4.10}$$

where $P, P_{jl}, Q_j \in L[X]$ and $0 \leqslant \deg(P_{jl}) < \deg(Q_j)$, for every $j = 1, \cdots, s$ and $l = 1, \cdots, r_j$.

With this decomposition, relationship (4.9) becomes

$$f = [P(\theta)]' + \sum_{j=1}^{s} \sum_{l=1}^{r_j} \left(\frac{[P_{jl}(\theta)]'}{[Q_j(\theta)]^l} - \frac{l P_{jl}(\theta)[Q_j(\theta)]'}{[Q_j(\theta)]^{l+1}}\right) + \sum_{k=1}^{p} c_k \cdot \frac{[P_k(\theta)]'}{P_k(\theta)}. \tag{4.11}$$

The left-hand side of Eq. (4.11) does not depend on θ. It follows that the terms depending on θ in the right-hand side that do not cancel each other out must equal zero.

We recall that θ is logarithmic over L. Then, Proposition 4.2.1 tells us that for any monic irreducible polynomial $R \in L[X]$ with $\deg(R) > 0$, there exists another polynomial $S \in L[X]$ such that $\deg(S) = \deg(R) - 1$ and $[R(\theta)]' = S(\theta)$. From this, it follows that $R(\theta)$ cannot divide $[R(\theta)]'$.

We will apply this observation in (4.11), successively for $R = Q_j$ and $R = P_k$.

Let $j \in \{1, \cdots, s\}$ be arbitrary, and let $R = Q_j$. Since $Q_j(\theta)$ does not divide $[Q_j(\theta)]'$, the right-hand side of Eq. (4.11) contains only one term with

4.2 Liouville's Principle

the denominator $[Q_j(\theta)]^{r_j+1}$ (we observe that $r_j + 1 \geq 2$, and, $P_k(\theta)$ being irreducible, $[Q_j(\theta)]^{r_j+1} \neq P_k(\theta)$, for any $k = 1, \cdots, p$). Since this term does not appear on the left-hand side, it should not appear on the right-hand side of Eq. (4.11) either; because j was chosen arbitrarily, it means that the double sum should not appear on the right-hand side of relationship (4.11).

Let $k \in \{1, \cdots, p\}$ be arbitrary, so that $\deg(P_k) > 0$, and let $R = P_k$. Since the polynomials P_k are distinct from each other, and as $P_k(\theta)$ does not divide $[P_k(\theta)]'$, the right-hand side of Eq. (4.11) contains only one term with the denominator $P_k(\theta)$. Since this term does not appear on the left-hand side, it should not appear on the right-hand side of Eq. (4.11) either. As a result, from the second sum that appears in (4.11), only the terms for which $\deg(P_k) = 0$ (i.e. $P_k \in L$) remain. If we denote these P_k with $u_k \in L$ for $k = 1, \cdots, q$, then (4.11) can be written as:

$$f = [P(\theta)]' + \sum_{k=1}^{q} c_k \cdot \frac{u'_k}{u_k}. \qquad (4.12)$$

Now, if we suppose that $\deg(P) > 0$, then, according again to Proposition 4.2.1, there exists $Q \in L[X]$ such that $[P(\theta)]' = Q(\theta)$. Since the left-hand side of Eq. (4.12) does not depend on θ, it must be the case that $\deg(Q) = 0$. If the coefficient of the leading term in P is in $\mathbb{C}$, then $\deg(P) = 1$. Thus $P(\theta) = a+b\cdot\theta$, where $a \in L$ and $b \in \mathbb{C}$ and $[P(\theta)]' = a'+b\cdot\theta' = a'+b\cdot\frac{u'}{u}$. In the case where the coefficient of the leading term in P is in $L\setminus\mathbb{C}$, $0 = \deg(Q) = \deg(P)$ and so $P(\theta) = a \in L$. In both cases, Eq. (4.12) can be written in the form:

$$f = a' + b \cdot \frac{u'}{u} + \sum_{k=1}^{q} c_k \cdot \frac{u'_k}{u_k}, \quad a, u, u_k \in L, b, c_k \in \mathbb{C}.$$

(3) θ **is exponential over** L. Let $u \in L$ such that $\theta = e^u$ is transcendental over L.

Similar to the logarithmic case, we can write Eq. (4.5) in the form of (4.11), where we recall that $P, P_{jl}, Q_j, P_k \in L[X]$, P_k, Q_j are distinct, irreducible, monic polynomials, $r_j \in \mathbb{N}^*$, $0 \leq \deg(P_{jl}) < \deg(Q_j)$ and $\deg(P_k) \geq 0$, for every $k = 1, \cdots p$, $j = 1, \cdots s$ and $l = 1, \cdots r_j$.

According to Proposition 4.2.2, for every polynomial $R \in L[X]$ with $\deg(R) > 0$ there exists $S \in L[X]$ such that $\deg(S) = \deg(R)$ and $[R(\theta)]' = S(\theta)$. Moreover, $R(\theta)$ divides $S(\theta)$ if and only if R is a monomial (i.e. $R = a \cdot X^n$ for some $a \in L$ and $n \in \mathbb{N}$).

We will apply this observation in (4.11), successively for $R = Q_j$ and $R = P_k$.

Let $j \in \{1, \cdots, s\}$ be arbitrary, and let $R = Q_j$. If Q_j is not monomial, then $Q_j(\theta)$ does not divide $[Q_j(\theta)]'$. Through a similar reasoning as in point (2), it follows that the term j in the double sum on the right-hand side of (4.11) should not appear. Hence, it follows that in the double sum, only the terms corresponding to those j for which Q_j is a monomial can remain. Similarly, in the second sum in (4.11), only the terms corresponding to those k for which P_k is a monomial or $P_k \in L$ can appear; let's assume that, for every $k = 1, \cdots, q$, $P_k \in L$ and that, for every $k = q + 1, \cdots p$, P_k is monomial. Furthermore, since Q_j and P_k are monic irreducible and monomial polynomials, they can only be of the form $Q_j = X = P_k$, for every $j = 1, \cdots, s, k = q + 1, \cdots, p$. Finally, because the polynomials Q_j are distinct from each other, and likewise, the polynomials P_k are distinct, it follows that there is only one $j \in \{1, \cdots, s\}$ and one $k \in \{q + 1, \cdots, p\}$ for which $Q_j = X = P_k$. Let us suppose $Q_1 = X = P_{q+1}$.

Taking all of this into account, the exponential case Eq. (4.11) is rewritten as:

$$f = [P(\theta)]' + \sum_{l=1}^{r_1} \left(\frac{[P_{1l}(\theta)]}{\theta^l}\right)' + \sum_{k=1}^{q} c_k \cdot \frac{u_k'}{u_k i} + c_{q+1} \cdot \frac{\theta'}{\theta}, \qquad (4.13)$$

where $u_k = P_k \in L$, for every $k = 1, \cdots, q$. Since $0 \leqslant \deg(P_{1l}) < \deg(Q_1) = 1$, $\deg(P_{1l}) = 0$ so that $P_{1l} \in L$. Considering additionally the fact that $\theta' = u' \cdot \theta$, (4.13) takes the form

$$f = \left[\sum_{l=-r}^{n} a_l \cdot \theta^l\right]' + \sum_{k=1}^{q} c_k \cdot \frac{u_k'}{u_k} = \sum_{l=-r}^{n} \left[a_l \cdot \theta^l\right]' + \sum_{k=1}^{q} c_k \cdot \frac{u_k'}{u_k} = \qquad (4.14)$$

$$= \sum_{l=-r}^{n} [a_l' \theta^l + l a_l u' \theta^l] + \sum_{k=1}^{q} c_k \cdot \frac{u_k'}{u_k} = \sum_{l=-r}^{n} b_l \cdot \theta^l + \sum_{k=1}^{q} c_k \cdot \frac{u_k'}{u_k},$$

where $a_l \in L$, $b_l = a_l' + l a_l \cdot u' \in L$, for every $l = -r, \cdots, n$, and $u_k \in L, c_k \in \mathbb{C}$, for every $k = 1, \cdots, q$. Since the left-hand side of Eq. (4.13) does not depend on θ, only b_0 remains in the first sum on the right-hand side, so that

$$f = a_0' + \sum_{k=1}^{q} c_k \cdot \frac{u_k'}{u_k}.$$

Conversely, we suppose that there exist $m \in \mathbb{N}^*$, $u_0, u_1, \cdots, u_m \in L$ and $c_1, \cdots, c_m \in \mathbb{C}$ such that

$$f = u_0' + \sum_{k=1}^{m} c_k \cdot \frac{u_k'}{u_k}.$$

Then $g = u_0 + \sum_{k=1}^{m} c_k \cdot \log u_k$ is an antiderivative of f. For every $k = 1, \cdots, m$ we denote by $\theta_k = \log u_k$; if we remark that $u_k \in L \subseteq L_{k-1} = L(\theta_1, \cdots, \theta_{k-1})$, then θ_k is logarithmic over L_{k-1}. Therefore $L_k = L(\theta_1, \cdots, \theta_k)$ is an elementary Liouville extension of L_{k-1} and then $L_m = L(\theta_1, \cdots, \theta_m)$ is an elementary extension of L. Since $g \in L_m$, g is an antiderivative of f elementary with respect to L. □

4.2 Liouville's Principle

An immediate consequence of Liouville's theorem is a well-known result in high school mathematics.

Corollary 4.2.4 *Every rational function admits an elementary antiderivative.*

Proof Let $f = \dfrac{P}{Q} \in R(\mathbb{C})$ where $P, Q \in \mathbb{C}[X]$ and $Q \neq \underline{0}$ (see Example 4.1.9).

By factoring the polynomial Q, we obtain $Q = c \cdot \prod_{j=1}^{s}(X - a_j)^{r_j}$, where $c \in \mathbb{C}$ and, for every $j = 1, \cdots, s$, $a_j \in \mathbb{C}$ is the multiple root of order $r_j \in \mathbb{N}^*$ of the polynomial Q ($\sum_{j=1}^{s} r_j = \deg(Q)$). Let now the partial fraction decomposition of f,

$$f = \frac{P}{Q} = P_0 + \sum_{j=1}^{s} \sum_{l=1}^{r_j} \frac{c_{jl}}{(X - a_j)^l}, \qquad (4.15)$$

where, for every $j = 1, \cdots, s$ and $l = 1, \cdots, r_j$, $P_0 \in \mathbb{C}[X]$ and $c_{jl} \in \mathbb{C}$.

If $P_0 = \sum_{k=0}^{n} a_k X^k$, then $P_1 = \displaystyle\sum_{k=0}^{n} \frac{a_k}{k+1} X^{k+1} \in \mathbb{C}[X]$ and $P_1' = P_0$.

With this notation, Eq. (4.15) can be rewritten as:

$$f(z) = P_1'(z) + \sum_{\substack{j \in \{1,\cdots,s\} \\ r_j = 1}} \frac{c_{j1}}{z - a_j} + \sum_{\substack{j \in \{1,\cdots,s\} \\ r_j \geq 2}} \left(\frac{c_{j1}}{z - a_j} + \sum_{l=2}^{r_j} \frac{c_{jl}}{(z - a_j)^l} \right) =$$

$$= P_1'(z) + \sum_{j=1}^{s} c_{j1} \cdot \frac{(z - a_j)'}{z - a_j} - \sum_{\substack{j \in \{1,\cdots,s\} \\ r_j \geq 2}} \sum_{l=2}^{r_j} \frac{c_{jl}}{l-1} \cdot \left[\frac{1}{(z - a_j)^{l-1}} \right]' = \qquad (4.16)$$

$$= \left[P_1(z) - \sum_{\substack{j \in \{1,\cdots,s\} \\ r_j \geq 2}} \sum_{l=2}^{r_j} \frac{c_{jl}}{l-1} \cdot \frac{1}{(z - a_j)^{l-1}} \right]' + \sum_{j=1}^{s} c_{j1} \cdot \frac{(z - a_j)'}{z - a_j}.$$

In (4.16) we denote by $u_0(z) = P_1(z) - \displaystyle\sum_{\substack{j \in \{1,\cdots,s\} \\ r_j \geq 2}} \sum_{l=2}^{r_j} \frac{c_{jl}}{l-1} \cdot \frac{1}{(z - a_j)^{l-1}}$, and by $u_j(z) = z - a_j$; then $c_j = c_{j1} \in \mathbb{C}$, $u_j \in R(\mathbb{C})$, for $j = 0, \cdots, s$, and

$$f = u_0' + \sum_{j=1}^{s} c_j \cdot \frac{u_j'}{u_j}.$$

Therefore f admits as an elementary antiderivative the function

$$g = u_0 + \sum_{j=1}^{s} \log u_j.$$

□

Examples 4.2.5

(1) Let's show that $\int e^{z^2} dz$ is not an elementary function. Let $\theta : \mathbb{C} \to \mathbb{C}$ be defined by $\theta(z) = e^{z^2}$; θ is exponential over $R(\mathbb{C})$. Let $L = R(\mathbb{C})(\theta)$ be the Liouville extension of the field $R(\mathbb{C})$ of rational functions and let $f = \theta \in L$; we remark that any function in L is of the form $\frac{P(\theta)}{Q(\theta)}$, where $P, Q \in R(\mathbb{C})[X]$. If we suppose that the antiderivative of f, $g = \int f$, is elementary with respect to L, then, there exist $m \in \mathbb{N}^*$, $u_0, u_1, \cdots, u_m \in L$ and $c_1, \cdots, c_m \in \mathbb{C}$ such that $f = \theta = u_0' + \sum_{k=1}^{m} c_k \cdot \frac{u_k'}{u_k}$ and $g = \int f = u_0 + \sum_{k=1}^{m} c_k \log u_k$ is an antiderivative of f. Taking into account the properties of logarithms, it is easily observed that in the previous relation we can choose $u_0 = \frac{P(\theta)}{Q(\theta)}$ and, for every $k = 1, \cdots, m$, $u_k = P_k(\theta)$, where $P, Q, P_k \in R(\mathbb{C})[X]$.

The left-hand side of the relation $\theta = u_0' + \sum_{k=1}^{m} c_k \cdot \frac{u_k'}{u_k}$ has no poles. Consequently, u_0 must be a polynomial in the variable θ, and for any $k = 1, \cdots, m$, u_k is constant. Therefore, the previous relation is rewritten as

$$\theta = u_0'. \tag{4.17}$$

Let $u_0 = \sum_{j=0}^{n} a_j \theta^j$, where, for every $j = 0, \cdots n$, $a_j \in R(\mathbb{C})$. Since $\theta'(z) = 2z\theta(z)$, the relation (4.17) becomes

$$\theta(z) = u_0'(z) = a_0'(z) + \sum_{j=1}^{n} \left[a_j'(z) + 2jza_j(z) \right] \cdot \theta^j(z). \tag{4.18}$$

θ is transcendental over $R(\mathbb{C})$; so that the relationship (4.18) is an identity in the variable θ. Therefore $n = 1$, $a_0'(z) = 0$ and

$$a_1'(z) + 2za_1(z) = 1. \tag{4.19}$$

The right-hand side of relation (4.19) has no poles; thus, it follows that $a_1 = P \in \mathbb{R}(\mathbb{C})[X]$. However, in this situation, $\deg[P'(z) + 2zP(z)] \geq 1 > 0 = \deg(1)$, and this shows that Eq. (4.19) has no solution. Consequently, $g = \int f$ is not elementary with respect to L.

4.2 Liouville's Principle

(2) If, in the previous example, we consider the function to be $f(z) = z \cdot \theta(z)$, then the relation (4.18) from the previous example becomes

$$z\theta(z) = u_0'(z) = a_0'(z) + \sum_{j=1}^{n}\left[a_j'(z) + 2jza_j(z)\right] \cdot \theta^j(z).$$

Also, this relation is an identity and then $n = 1$, $a_0'(z) = 0$ and

$$a_1'(z) + 2za_1(z) = z.$$

These conditions are satisfied if $a_0(z) = c \in \mathbb{C}$ and $a_1(z) = \frac{1}{2}$. Therefore we arrive at the obvious antiderivative of f, $\int f(z)dz = c + \frac{1}{2}e^{z^2}$.

(3) Let $\theta(z) = \log z$ and $L = R(\mathbb{C})(\theta)$, where $R(\mathbb{C})$ is the Liouville field of rational functions. Then $L = \left\{\dfrac{P(\theta)}{Q(\theta)} : P, Q \in R(\mathbb{C})[X]\right\}$.

Let $f \in L$ be the function defined by $f(z) = \dfrac{2\theta(z) - 2}{(\theta(z) - z)^2}$. In order for f to have an elementary antiderivative with respect to L, we need to find $m \in \mathbb{N}^*$, $u_0, u_1, \cdots, u_m \in L$ and $c_1, \cdots, c_m \in \mathbb{C}$ such that $f = u_0' + \sum_{k=1}^{m} c_k \cdot \dfrac{u_k'}{u_k}$.

There exist $P_k, Q_k \in R[X]$ such that $u_k = \dfrac{P_k(\theta)}{Q_k(\theta)}$, for every $k = 0, \cdots, m$. Since $\dfrac{u_k'}{u_k} = \dfrac{[P_k(\theta)]'}{P_k(\theta)} - \dfrac{[Q_k(\theta)]'}{Q_k(\theta)}$, the previous relationship can be rewritten in the form $f = \dfrac{[P_0(\theta)]' \cdot Q_0(\theta) - P_0(\theta) \cdot [Q_0(\theta)]'}{[Q_0(\theta)]^2} + \sum_{k=1}^{p} c_k \cdot \dfrac{[P_k(\theta)]'}{P_k(\theta)}$, where $P_0, Q_0 \in R(\mathbb{C})[X]$ are relatively prime, and $P_k \in R(\mathbb{C})[X]$ are irreducible monic polynomials, for every $k = 1, \cdots, p$. Since $f(z) = \dfrac{2\theta(z) - 2}{(\theta(z) - z)^2}$ it follows that $Q_0(\theta)(z) = \theta(z) - z$ and $c_k = 0$, for every $k = 1, \cdots, p$. Then $[Q_0(\theta)]'(z) = \dfrac{1-z}{z}$ and

$$2\theta(z) - 2 = [P_0(\theta)]'(z) \cdot \theta(z) - z \cdot [P_0(\theta)]'(z) - P_0(\theta)(z) \cdot \dfrac{1-z}{z}.$$

Since this equality is an identity, $P_0(\theta) = g \in R(\mathbb{C})$ and then

$$2\theta(z) - 2 = g'(z) \cdot \theta(z) - z \cdot g'(z) - g(z) \cdot \dfrac{1-z}{z}.$$

Therefore $g(z) = 2z$ and so

$$f(z) = \left[\frac{2z}{\theta(z) - z}\right]'.$$

In conclusion, the function g, defined by $g(z) = \dfrac{2z}{\log z - z}$, is an antiderivative of f, elementary with respect to $L = R(\mathbb{C})(\theta)$.

4.3 Particular Cases

In this paragraph, we aim to analyze several classes of functions for which, using Liouville's principle, we decide on the conditions under which they admit antiderivatives expressible by elementary functions and calculate these antiderivatives.

We will specify the conditions under which we are working. Let L be a Liouville field on $D \subset \mathbb{C}$, and let $p \in L$ be such that its antiderivative θ ($\theta' = p$) is transcendental over L (either logarithmic or exponential).

Since $\theta' = p \in L \subset L(\theta)$, from Theorem 4.1.12, it follows that $L(\theta) = \left\{\dfrac{P(\theta)}{Q(\theta)} : P, Q \in L[X]\right\}$ is a Liouville field on D ($L(\theta)$ is the smallest Liouville extension of L that contains the transcendental element θ).

In the following, we will present the conditions under which integrals of the form $\displaystyle\int f(\theta(z), z) dz$ are elementary over $L(\theta)$, where $f : \mathbb{C} \times D \to \mathbb{C}$ has the property that $f(\theta, z) = A_0(z) \cdot \theta + A_1(z)$ or, more generally, $f(\theta, z) = \displaystyle\sum_{k=0}^{n} \binom{n}{k} A_k(z) \cdot \theta^{n-k}$.

4.3.1 Integrating Expressions of the Form $A_0\theta + A_1$

Theorem 4.3.1 *Let L be a Liouville field on the open set $D \subseteq \mathbb{C}$ and let $p, A_0, A_1 \in L$ such that $\theta(z) = \int p(z) dz$ is elementary with respect to L and transcendental over L, i.e. θ is a logarithmic or exponential over L.*

$\displaystyle\int [A_0(z)\theta(z) + A_1(z)] dz$ *is elementary with respect to $L(\theta)$ if and only if there exist $c \in \mathbb{C}, a \in L$ and A elementary with respect to L such that:*

$$\begin{cases} A_0(z) = c \cdot p(z) + a'(z), \\ A_1(z) = a(z) \cdot p(z) + A'(z). \end{cases}$$

4.3 Particular Cases

In this case

$$\int [A_0(z) \cdot \theta(z) + A_1(z)] \, dz = \frac{c}{2} \cdot \theta^2(z) + a(z) \cdot \theta(z) + A(z).$$

The above integral defines a function in $L(\theta)$ if and only if c, a, A can be chosen such that $c \in \mathbb{C}, a, A \in L$.

Proof As mentioned above, the smallest Liouville field containing L and θ is
$$L(\theta) = \left\{ \frac{P(\theta)}{Q(\theta)} : P, Q \in L[X] \right\}.$$

Necessity Let's assume that $F(\theta, z) = \int [A_0(z)\theta(z) + A_1(z)] \, dz$ defines an elementary function with respect to $L(\theta)$; according to Liouville's principle, there exist $m \in \mathbb{N}^*$, there exist $c_1, \cdots, c_m \in \mathbb{C}$, and there exist $u_0, u_1, \cdots, u_m \in L(\theta)$ such that

$$F(\theta, z) = \sum_{k=1}^{m} c_k \cdot \log(u_k(\theta, z)) + u_0(\theta, z).$$

Taking the derivative of the above relationship with respect to the variable θ, we obtain:

$$\frac{\partial F}{\partial \theta} = \sum_{k=1}^{m} c_k \cdot \frac{1}{u_k} \cdot \frac{\partial u_k}{\partial \theta} + \frac{\partial u_0}{\partial \theta}.$$

Because u_k are rational functions in θ, $\dfrac{\partial u_k}{\partial \theta}$ are also rational functions in θ, and thus, $\dfrac{\partial F}{\partial \theta} \in L(\theta)$. On the other hand

$$\frac{d}{dz} F(\theta(z), z) = A_0(z) \cdot \theta(z) + A_1(z) = \frac{\partial F}{\partial \theta}(\theta(z), z) \cdot \theta'(z) + \frac{\partial F}{\partial z}(\theta(z), z) =$$

$$= \frac{\partial F}{\partial \theta}(\theta(z), z) \cdot p(z) + \frac{\partial F}{\partial z}(\theta(z), z),$$

from where it follows that $\dfrac{\partial F}{\partial z} \in L(\theta)$. Therefore, it follows that

$$\frac{\partial F}{\partial \theta}, \frac{\partial F}{\partial z} \in L(\theta).$$

From the previous relationship we obtain that

$$A_0 \cdot \theta + A_1 = \frac{\partial F}{\partial \theta} \cdot p + \frac{\partial F}{\partial z}. \tag{4.20}$$

Since θ is transcendental over L, Eq. (4.20) is an identity in θ (otherwise, (4.20) would lead to a polynomial equation in θ with coefficients from L, which would contradict the transcendence of θ). Therefore

$$A_0(z) \cdot \theta(z) + A_1(z) \equiv_\theta \frac{d}{dz} F(\theta(z), z). \tag{4.21}$$

From (4.21) it follows that, for every $\gamma \in \mathbb{C}$,

$$A_0(z)[\theta(z) + \gamma] + A_1(z) = \frac{d}{dz} F(\theta(z) + \gamma, z),$$

or, by integrating with respect to z from $c_0 \in \mathbb{C}$ to $z \in \mathbb{C}$,

$$\int_{c_0}^{z} \{A_0(z)[\theta(z) + \gamma] + A_1(z)\}\, dz = F(\theta(z) + \gamma, z) - F(\theta(c_0) + \gamma, c_0).$$

We denote by $-F(w(c_0) + \gamma, c_0) = c(\gamma)$, and then

$$\int_{c_0}^{z} \{A_0(z)[\theta(z) + \gamma] + A_1(z)\}\, dz = F(w(z) + \gamma, z) + c(\gamma). \tag{4.22}$$

We differentiate Eq. (4.22) with respect to γ twice, and we obtain successively

$$\int_{c_0}^{z} A_0(z)\, dz = \frac{\partial F}{\partial \theta}(\theta(z) + \gamma, z) + c'(\gamma), \tag{4.23}$$

$$0 = \frac{\partial^2 F}{\partial \theta^2}(\theta(z) + \gamma, z) + c''(\gamma). \tag{4.24}$$

If we set $\gamma = 0$ in (4.24) and denote $c = -c''(0) \in \mathbb{C}$, we obtain

$$\frac{\partial^2 F}{\partial \theta^2}(\theta, z) = c. \tag{4.25}$$

From (4.25) it follows that there exists $a_1 : D \to \mathbb{C}$ such that

$$\frac{\partial F}{\partial \theta}(\theta, z) = c \cdot \theta + a_1(z), \tag{4.26}$$

4.3 Particular Cases

and since $\dfrac{\partial F}{\partial \theta} \in L(\theta)$ and θ is transcendental over L, relationship (4.26) is an identity in θ. It follows that if we assign a constant value θ_0 to θ in (4.26), $a_1(z) = \dfrac{\partial F}{\partial \theta}(\theta_0, z) - c \cdot \theta_0 \in L$.

We observe that the first term in Eq. (4.23) does not depend on γ, so if we set $\gamma = 0$, we obtain

$$\int_{c_0}^{z} A_0(z)dz = \frac{\partial F}{\partial \theta}(\theta, z) + c'(0). \tag{4.23'}$$

We substitute (4.26) into (4.23') and obtain:

$$\int_{c_0}^{z} A_0(z)dz = c \cdot \theta(z) + a_1(z) + c'(0),$$

or, if we denote by $a(z) = a_1(z) + c'(0)$, then $a \in L$ and

$$\int_{c_0}^{z} A_0(z)dz = c \cdot \theta(z) + a(z). \tag{4.27}$$

Differentiating Eq. (4.27) with respect to z yields

$$A_0(z) = c \cdot p(z) + a'(z), \text{ where } c \in \mathbb{C} \text{ and } a \in L.$$

Then it follows that

$$F(\theta, z) = \int \left[c \cdot p(z) \cdot \theta(z) + a'(z) \cdot \theta(z) + A_1(z) \right] dz =$$

$$= \frac{c}{2} \cdot \theta^2(z) + a(z) \cdot \theta(z) + \int [A_1(z) - a(z) \cdot p(z)] dz.$$

Let us denote

$$H(\theta, z) = \int [A_1(z) - a(z) \cdot p(z)] dz = F(\theta, z) - \frac{c}{2} \cdot \theta^2(z) - a(z) \cdot \theta(z).$$

Then H elementary with respect to $L(\theta)$ and applying the Liouville's principle to this function, there exist $n \in \mathbb{N}^*$, $d_1, \cdots, d_n \in \mathbb{C}$ and $v_0, v_1, \cdots, v_n \in L(\theta)$ such that

$$H(\theta, z) = \sum_{k=1}^{n} d_k \cdot \log(v_k(\theta, z)) + v_0(\theta, z). \tag{4.28}$$

Similarly to how we demonstrated for the function F, we obtain:

$$\frac{\partial H}{\partial \theta}, \frac{\partial H}{\partial z} \in L(\theta),$$

and since θ is transcendental with respect to L, the following relationship (obtained by differentiating H with respect to z) is an identity in θ:

$$A_1(z) - a(z) \cdot p(z) = \frac{\partial H}{\partial \theta}(\theta, z) \cdot p(z) + \frac{\partial H}{\partial z}(\theta, z).$$

Then, for every $\gamma \in \mathbb{C}$,

$$A_1(z) - a(z) \cdot p(z) = \frac{d}{dz} H(\theta(z) + \gamma, z).$$

Integrating this relationship with respect to z from $c_0 \in \mathbb{C}$ to $z \in \mathbb{C}$, we obtain

$$\int_{c_0}^{z} [A_1(z) - a(z) \cdot p(z)] dz = H(\theta(z) + \gamma, z) - H(\theta(c_0) + \gamma, c_0),$$

or, if we denote $-H(\theta(c_0) + \gamma, c_0) = c_1(\gamma)$,

$$\int_{c_0}^{z} [A_1(z) - a(z) \cdot p(z)] dz = H(\theta(z) + \gamma, z) + c_1(\gamma). \tag{4.29}$$

We differentiate (4.29) with respect to γ and obtain

$$0 = \frac{\partial H}{\partial \theta}(\theta + \gamma, z) + c_1'(\gamma). \tag{4.30}$$

Let now $c_1 = -c_1'(0) \in \mathbb{C}$; it follows from (4.30)

$$\frac{\partial H}{\partial \theta}(\theta, z) = c_1.$$

Then there exists a function $A : D \to \mathbb{C}$ such that

$$H(\theta, z) = c_1 \cdot \theta + A(z). \tag{4.31}$$

The function $H(\theta, z) - c_1 \cdot \theta = A(z)$ does not depend on θ, so that $A(z) = H(\theta_0, z) - c_1 \cdot \theta_0$, where θ_0 is fixed in $\mathbb{C}$. From relationships (4.28) and (4.31), we deduce that

$$A(z) = \sum_{k=1}^{n} d_k \cdot \log(v_k(\theta_0, z)) + v_0(\theta_0, z) - c_1 \cdot \theta_0.$$

4.3 Particular Cases

Since the functions $v_0(\theta_0, \cdot), \cdots, v_n(\theta_0, \cdot) \in L$, it follows that A is an elementary function with respect to L.

We notice that the first term in relationship (4.29) does not depend on γ and therefore

$$\int [A_1(z) - a(z) \cdot p(z)] dz = H(\theta, z) + c_1(0).$$

Differentiating the last relationship with respect to z yields

$$A_1(z) - a(z) \cdot p(z) = c_1 \cdot p(z) + A'(z), \text{ or}$$

$$A_1(z) = (a(z) + c_1) \cdot p(z) + A'(z).$$

It is now evident that we can rename the function as $a + c_1$, and once again obtain a function in L, such that

$$\begin{cases} A_0(z) = c \cdot p(z) + a'(z), \\ A_1(z) = a(z) \cdot p(z) + A'(z), \end{cases}$$

where A is elementary with respect to L.

Sufficiency Suppose there exist $c \in \mathbb{C}, a \in L$, and an elementary function A with respect to L such that the conditions of the theorem are satisfied. Then

$$\int [A_0(z) \cdot \theta(z) + A_1(z)] dz = \int [c \cdot p(z) \cdot \theta(z) + a'(z) \cdot \theta(z) + a(z) \cdot p(z) + A'(z)] dz$$

$$= \frac{c}{2} \cdot \theta^2(z) + a(z) \cdot \theta(z) + A(z)$$

and then the antiderivative of $A_0(z) \cdot \theta(z) + A_1(z)$ is elementary with respect to $L(\theta)$.

We can immediately notice that $\int [A_0(z) \cdot \theta(z) + A_1(z)] dz \in L(\theta)$ if and only if A can be chosen in such a way that $A \in L$. □

Examples 4.3.2

(1) It is considered the integral:

$$F(\theta, z) = \int \left[\left(\frac{1}{z} + \frac{2}{(z+1)^2} \right) \cdot \theta(z) + \left(\frac{z-1}{z(z+1)} + \frac{2}{z} + z \right) \right] dz,$$

where

(a) $\theta(z) = \log z$, or
(b) $\theta(z) = \arctan z$.

It is observed that, if we denote $A_0(z) = \dfrac{1}{z} + \dfrac{2}{(z+1)^2}$ and $A_1(z) = \dfrac{z-1}{z(z+1)} + \dfrac{2}{z} + z$, then $A_0, A_1 \in R(\mathbb{C})$, where $R(\mathbb{C})$ is the Liouville field of rational functions (see Example 4.1.9). In both cases, (a) and (b), θ is transcendental with respect to $R(\mathbb{C})$ and $\theta' = p \in R(\mathbb{C})$ (in the case a) $\theta'(z) = p(z) = \dfrac{1}{z}$ and, in the case (b), $\theta'(z) = p(z) = \dfrac{1}{z^2+1}$). We also observe that the second condition from the previous theorem, $A_1 = a \cdot p + A'$, leads to $A' = A_1 - a \cdot p \in R(\mathbb{C})$ ($a \in R(\mathbb{C})$ from the first condition). Since A' is a rational function, A will be an elementary function with respect to $R(\mathbb{C})$ (according to Corollary 4.2.4, the antiderivative of any rational function is an elementary function with respect to $R(\mathbb{C})$).

It follows that the integral will be elementary with respect to $R(\mathbb{C})(\theta)$ if and only if we can find a complex number c and a function $a \in R(\mathbb{C})$ such that $A_0(z) = c \cdot p(z) + a'(z)$.

In case (a), the last relationship amounts to

$$\frac{1}{z} + \frac{2}{(z+1)^2} = \frac{c}{z} + a'(z).$$

It is observed that, if we choose $c = 1$ in the above relationship (for $c \neq 1$, $a(z) = (1-c) \cdot \log z - \dfrac{2}{z+1}$, and therefore, $a \notin R(\mathbb{C})$), we obtain $a(z) = -\dfrac{2}{z+1} + c_1$ and then $a \in R(\mathbb{C})$. We can then determine A from the relationship

$$A'(z) = \frac{z-1}{z(z+1)} + \frac{2}{z} + z + \left(\frac{2}{z+1} - c_1\right) \cdot \frac{1}{z} = \frac{3-c_1}{z} + z.$$

If we choose $c_1 = 3$, then $a(z) = \dfrac{3z+1}{z+1}$ and $A(z) = \dfrac{z^2}{2}$ and so $a, A \in R(\mathbb{C})$. It follows that the antiderivative F is

$$F(\theta(z), z) = \frac{1}{2} \cdot \theta^2(z) + \frac{3z+1}{z+1} \cdot \theta(z) + \frac{z^2}{2} = \frac{1}{2} \cdot \log^2 z + \frac{3z+1}{z+1} \cdot \log z + \frac{z^2}{2}.$$

We remark that $F(\theta, \cdot) \in R(\mathbb{C})(\theta)$.

In the case (b) the first relationship of Theorem 4.3.1 is

$$\frac{1}{z} + \frac{2}{(z+1)^2} = \frac{c}{z^2+1} + a'(z),$$

from where

$$a(z) = \log z - \frac{2}{z+1} - c \cdot \arctan z.$$

4.3 Particular Cases

We observe that, regardless of the value we assign to $c \in \mathbb{C}$, a does not belong to $R(\mathbb{C})$. This implies that the integral is not elementary with respect to $R(\mathbb{C})(\arctan)$.

(2) Let us consider the integral

$$F(\theta(z), z) = \int \left[\left(\frac{1}{z} + z \cdot e^{z^2} \right) \cdot \theta(z) + \frac{1}{2z} \cdot e^{z^2} + 1 \right] dz,$$

where $\theta(z) = \log z$.

If we denote by $A_0(z) = \frac{1}{z} + z \cdot e^{z^2}$ and by $A_1(z) = \frac{1}{2z} \cdot e^{z^2} + 1$, then we remark that $A_0, A_1 \in L = R(\mathbb{C})(\eta)$, where $R(\mathbb{C})$ is the Liouville field of rational functions, $\eta(z) = e^{z^2}$, and L is the elementary Liouville extension of $R(\mathbb{C})$ with the exponential η. Moreover, $\theta'(z) = p(z) = \frac{1}{z}$, and then $p \in L$.

We want to verify if $F(\theta, \cdot)$ is elementary with respect to $L(\theta)$. The conditions from the previous theorem can be written as follows:

$$\frac{1}{z} + z \cdot e^{z^2} = \frac{c}{z} + a'(z)$$
$$\frac{1}{2z} \cdot e^{z^2} + 1 = \frac{a(z)}{z} + A'(z).$$

From the first relationship, considering $c = 1$, we obtain $a(z) = \frac{1}{2} \cdot e^{z^2}$. Substituting this into the second relationship for a, we get $A(z) = z$. Therefore $a, A \in L = R(\mathbb{C})(\eta)$, and then the integral $F(\theta, \cdot) \in L(\theta) = R(\mathbb{C})(\eta, \theta)$:

$$F(\theta(z), z) = \frac{1}{2} \cdot \log^2 z + \frac{1}{2} \cdot e^{z^2} \cdot \log z + z.$$

4.3.2 Integrating a Polynomial in θ

Theorem 4.3.3 *Let L be a Liouville field on the open set $D \subseteq \mathbb{C}$, let $p \in L$ be such that $\theta(z) = \int p(z) dz$ is elementary with respect to L, and let $A_0, A_1, \cdots, A_n \in L$.*

$\int \sum_{k=0}^{n} \binom{n}{k} \cdot A_k(z) \cdot \theta^{n-k}(z) dz$ *is elementary with respect to $L(\theta)$ if and only if there exist $B_0 = c \in \mathbb{C}$, $B_1, \cdots, B_n \in L$ and B_{n+1} elementary with respect to L such that:*

$$\begin{cases} A_0(z) = c \cdot p(z) + B_1'(z) \\ A_1(z) = B_1(z) \cdot p(z) + \frac{1}{2} \cdot B_2'(z) \\ \cdots \cdots \\ A_n(z) = B_n(z) \cdot p(z) + \frac{1}{n+1} \cdot B_{n+1}'(z). \end{cases}$$

In this case

$$\int \sum_{k=0}^{n} \binom{n}{k} \cdot A_k(z) \cdot \theta^{n-k}(z) dz = \frac{1}{n+1} \cdot \sum_{k=0}^{n+1} \binom{n+1}{k} \cdot B_k(z) \cdot \theta^{n+1-k}(z).$$

The above integral defines a function in $L(\theta)$ if and only if B_{n+1} can be chosen such that $B_{n+1} \in L$.

Proof We recall that, since $\theta' = p \in L \subseteq L(\theta)$, the smallest Liouville field containing L and θ is $L(\theta) = \left\{ \dfrac{P(\theta)}{Q(\theta)} : P, Q \in L[X] \right\}$ (Theorem 4.1.12).

Necessity We suppose that

$$F(\theta, z) = \int \sum_{k=0}^{n} \binom{n}{k} \cdot A_k(z) \cdot \theta^{n-k}(z) dz$$

is an elementary function with respect to $L(\theta)$. The Liouville's principle assures us of the existence of a natural number $m \in \mathbb{N}^*$, m complex numbers $c_1, \cdots, c_m \in \mathbb{C}$ and $m+1$ functions $u_0, u_1, \cdots, u_m \in L(\theta)$ such that

$$F(\theta, z) = \sum_{k=1}^{m} c_k \cdot \log(u_k(\theta, z)) + u_0(\theta, z).$$

It follows then that

$$\frac{\partial F}{\partial \theta} = \sum_{k=1}^{m} c_k \cdot \frac{1}{u_k} \cdot \frac{\partial u_k}{\partial \theta} + \frac{\partial u_0}{\partial \theta}.$$

Since the functions u_k are rational in the variable θ, $\dfrac{\partial u_k}{\partial \theta} \in L(\theta)$, from where $\dfrac{\partial F}{\partial \theta} \in L(\theta)$. On the other hand

$$\sum_{k=0}^{n} \binom{n}{k} \cdot A_k(z) \cdot \theta^{n-k}(z) = \frac{d}{dz} F(\theta(z), z) = \frac{\partial F}{\partial \theta} \cdot p + \frac{\partial F}{\partial z}$$

from where it follows that $\dfrac{\partial F}{\partial z} \in L(\theta)$.

Since θ is transcendental over L, the following relationship is an identity in θ:

$$\sum_{k=0}^{n} \binom{n}{k} \cdot A_k(z) \cdot \theta^{n-k} = \frac{\partial F}{\partial \theta}(\theta, z) \cdot p(z) + \frac{\partial F}{\partial z}(\theta, z). \tag{4.32}$$

4.3 Particular Cases

Therefore

$$\sum_{k=0}^{n} \binom{n}{k} \cdot A_k(z) \cdot \theta^{n-k}(z) \equiv_\theta \frac{d}{dz} F(\theta(z), z). \quad (4.33)$$

In (4.33), we replace θ with $\theta + \gamma$, and we obtain

$$\sum_{k=0}^{n} \binom{n}{k} \cdot A_k(z) \cdot (\theta + \gamma)^{n-k} = \frac{d}{dz} F(\theta + \gamma, z), \text{ for every } \gamma \in \mathbb{C}. \quad (4.34)$$

We integrate in (4.34) from $c_0 \in \mathbb{C}$ to $z \in \mathbb{C}$ and find

$$\int_{c_0}^{z} \sum_{k=0}^{n} \binom{n}{k} \cdot A_k(z) \cdot [\theta(z)+\gamma]^{n-k} dz = F(\theta(z)+\gamma, z) - F(\theta(c_0)+\gamma, c_0). \quad (4.35)$$

We denote $-F(\theta(c_0) + \gamma, c_0) = c(\gamma)$ and obtain

$$\int_{c_0}^{z} \sum_{k=0}^{n} \binom{n}{k} \cdot A_k(z) \cdot [\theta(z) + \gamma]^{n-k} dz = F(\theta(z) + \gamma, z) + c(\gamma). \quad (4.36)$$

By differentiating Eq. (4.36) $(n-1)$-times with respect to γ, we obtain:

$$\int_{c_0}^{z} n![A_0(z)(\theta(z) + \gamma) + A_1(z)]dz = F_{\theta^{n-1}}^{(n-1)}(\theta + \gamma, z) + c^{(n-1)}(\gamma). \quad (4.37)$$

In (4.37), we set $\gamma = 0$ and denote $c^{(n-1)}(0) = c \in \mathbb{C}$.

$$\int_{c_0}^{z} n![A_0(z) \cdot \theta(z) + A_1(z)]dz = F_{\theta^{n-1}}^{(n-1)}(\theta, z) + c. \quad (4.38)$$

Since $F_{\theta^{n-1}}^{(n-1)} \in L(\theta)$, it follows that $\int_{c_0}^{z} n![A_0(z) \cdot \theta(z) + A_1(z)]dz \in L(\theta)$. Therefore, using the result established in Theorem 4.3.1, there exist $c \in \mathbb{C}$, and $b_1, b_2 \in L$ such that

$$A_0(z) = \frac{1}{n!} \cdot \left(c \cdot p(z) + b_1'(z) \right)$$

$$A_1(z) = \frac{1}{n!} \cdot \left(b_1(z) \cdot p(z) + b_2'(z) \right) \text{ and}$$

$$\int_{c_0}^{z} n![A_0(z) \cdot \theta(z) + A_1(z)]dz = \frac{c}{2} \cdot \theta^2(z) + b_1(z) \cdot \theta(z) + b_2(z).$$

Hence

$$F^{(n-1)}_{\theta^{n-1}}(\theta, z) = \frac{c}{2} \cdot \theta^2(z) + b_1(z) \cdot \theta(z) + b_2(z). \qquad (*)$$

If we integrate the last relationship with respect to θ, we obtain

$$F^{(n-2)}_{\theta^{n-2}}(\theta, z) = \frac{c}{3!} \cdot \theta^3(z) + \frac{b_1(z)}{2} \cdot \theta^2(z) + b_2(z) \cdot \theta(z) + b_3(z).$$

Since $F^{(n-2)}_{\theta^{n-2}} \in L(\theta)$ and θ is transcendental over L, the above relationship is an identity. Thus, giving θ a constant value $\theta_0 \in \mathbb{C}$, we obtain $b_3 \in L$, and so on.

After $n - 2$ successive integrations of the relationship $(*)$, we obtain:

$$F'_\theta(\theta, z) = \frac{c}{n!} \cdot \theta^n + \frac{b_1(z)}{(n-1)!} \cdot \theta^{n-1} + \cdots + \frac{b_{n-1}(z)}{1!} \cdot \theta + b_n(z)$$

and, with the same argument as above, $b_k \in R$, for every $k = 1, \cdots, n$.

For every $k = 0, \cdots, n$, we denote now $B_k = \frac{k!}{n!} \cdot b_k$, where $b_0 = c \in \mathbb{C}$; then $B_0, B_1, \cdots, B_n \in L$ and

$$F'_\theta(\theta, z) = \sum_{k=0}^{n} \binom{n}{k} \cdot B_k(z) \cdot \theta^{n-k}(z). \qquad (4.39)$$

If we integrate Eq. (4.39) once again with respect to θ, we obtain

$$F(\theta, z) = \sum_{k=0}^{n} \binom{n}{k} \cdot B_k(z) \cdot \frac{1}{n-k+1} \cdot \theta^{n-k+1}(z) + \frac{1}{n+1} \cdot B_{n+1}(z) = \qquad (4.40)$$

$$= \frac{1}{n+1} \cdot \sum_{k=0}^{n} \binom{n+1}{k} \cdot B_k(z) \cdot \theta^{n+1-k}(z) + \frac{1}{n+1} \cdot B_{n+1}(z).$$

From (4.40), $B_{n+1} = \Phi(\theta, \cdot)$, where

$$\Phi(\theta, \cdot) = (n+1) \cdot F(\theta, \cdot) - \sum_{k=0}^{n} \binom{n+1}{k} \cdot B_k \cdot \theta^{n+1-k}.$$

From (4.39), we remark that $\dfrac{\partial \Phi}{\partial \theta} = 0$; it follows that there is $\theta_0 \in \mathbb{C}$ such that $\Phi(\theta, \cdot) = \Phi(\theta_0, \cdot)$. On the other hand

$$\Phi(\theta, z) = (n+1) \cdot \left(\sum_{k=1}^{m} c_k \cdot \log(u_k(\theta, z)) + u_0(\theta, z) \right) -$$

$$- \sum_{k=0}^{n} \binom{n+1}{k} \cdot B_k(z) \cdot \theta^{n+1-k}(z),$$

4.3 Particular Cases

so that

$$B_{n+1}(z) = (n+1) \cdot \sum_{k=1}^{m} c_k \cdot \log(u_k(\theta_0, z)) + u_0(\theta_0, z) - \sum_{k=0}^{n} \binom{n+1}{k} \cdot B_k(z) \cdot \theta_0^{n+1-k},$$

which shows that B_{n+1} is elementary with respect to L.

We can rewrite (4.40) as

$$\int \sum_{k=0}^{n} \binom{n}{k} \cdot A_k(z) \cdot \theta^{n-k} dz = \frac{1}{n+1} \cdot \sum_{k=0}^{n+1} \binom{n+1}{k} \cdot B_k(z) \cdot \theta^{n+1-k}.$$

We differentiate the above relationship with respect to z, and we obtain

$$\sum_{k=0}^{n} \binom{n}{k} \cdot A_k \cdot \theta^{n-k} = \frac{1}{n+1} \cdot \sum_{k=0}^{n+1} \binom{n+1}{k} \cdot B'_k \cdot \theta^{n+1-k} + \qquad (4.41)$$

$$+ \frac{1}{n+1} \cdot \sum_{k=0}^{n} \binom{n+1}{k} \cdot B_k \cdot (n+1-k) \cdot \theta^{n-k} \cdot p =$$

$$= \frac{1}{n+1} \cdot \sum_{k=0}^{n} \left[\binom{n+1}{k+1} B'_{k+1} + \binom{n+1}{k} B_k (n+1-k) p \right] \cdot \theta^{n-k}.$$

The relationship (4.41) is an identity in θ, and therefore, we can equate the coefficients of like powers of θ in both sides. This results in:

$$\begin{cases} B'_0 = 0, \\ \binom{n}{0} A_0 = \frac{1}{n+1} \left[\binom{n+1}{1} B'_1 + \binom{n+1}{0} B_0 \cdot (n+1) \cdot p \right] \\ \quad \ldots \ldots \\ \binom{n}{k} \cdot A_k = \frac{1}{n+1} \left[\binom{n+1}{k+1} B'_{k+1} + \binom{n+1}{k} B_k (n+1-k) \cdot p \right] \\ \quad \ldots \ldots \\ \binom{n}{n-1} A_{n-1} = \frac{1}{n+1} \left[\binom{n+1}{n} B'_n + \binom{n+1}{n-1} B_{n-1} \cdot 2 \cdot p \right] \\ \binom{n}{n} A_n = \frac{1}{n+1} \left[\binom{n+1}{n+1} B'_{n+1} + \binom{n+1}{n} B_n \cdot p \right] \end{cases}.$$

The conditions required in the theorem immediately follow from the above relationships.

It can be easily observed from the above proof that $F(\theta, \cdot) \in L(\theta)$ if and only if $B_{n+1} \in L$.

Sufficiency Suppose there exist $c = B_0 \in \mathbb{C}$, $B_1, \cdots, B_n \in L$, and B_{n+1} elementary with respect to L such that, for any $k = 0, \cdots, n$,

$$A_k = B_k \cdot p + \frac{1}{k+1} \cdot B'_{k+1}.$$

Then

$$F(\theta, z) = \int \sum_{k=0}^{n} \binom{n}{k} \cdot A_k(z) \cdot \theta^{n-k} dz =$$

$$= \sum_{k=0}^{n} \binom{n}{k} \int \left(B_k(z) \cdot p(z) + \frac{1}{k+1} \cdot B'_{k+1}(z) \right) \cdot \theta^{n-k} dz =$$

$$= \frac{c}{n+1} \cdot \theta^{n+1} + \sum_{k=1}^{n} \binom{n}{k} \cdot \int B_k(z) \cdot p(z) \cdot \theta^{n-k} dz +$$

$$+ \sum_{k=1}^{n+1} \binom{n}{k-1} \cdot \frac{1}{k} \cdot \int B'_k(z) \cdot \theta^{n+1-k} dz = \frac{c}{n+1} \cdot \theta^{n+1} +$$

$$+ \sum_{k=1}^{n} \left(\binom{n}{k} \cdot \int B_k(z) \cdot p(z) \cdot \theta^{n-k} dz + \frac{1}{n+1} \cdot \binom{n+1}{k} \cdot \int B'_k(z) \cdot \theta^{n+1-k} dz \right)$$

$$+ \frac{1}{n+1} \cdot B_{n+1}(z) = \frac{c}{n+1} \cdot \theta^{n+1} + \sum_{k=1}^{n} \left(\binom{n}{k} \cdot \int B_k(z) \cdot p(z) \cdot \theta^{n-k} dz +\right.$$

$$+ \frac{1}{n+1} \cdot \binom{n+1}{k} \cdot B_k(z) \cdot \theta^{n+1-k} -$$

$$\left. - \frac{n+1-k}{n+1} \cdot \binom{n+1}{k} \cdot \int B_k(z) \cdot p(z) \cdot \theta^{n-k} dz \right) + \frac{1}{n+1} \cdot B_{n+1}(z) =$$

$$= \frac{1}{n+1} \cdot \sum_{k=0}^{n+1} \binom{n+1}{k} \cdot B_k(z) \cdot \theta^{n+1-k}.$$

It can be observed that $F(\theta, \cdot)$ is elementary with respect to $L(\theta)$.

□

4.3 Particular Cases

Examples 4.3.4

(1) Let us consider the integral

$$\int \left(z^2 \cdot \theta^2(z) + \frac{z}{z+1} \right) dz.$$

If we denote with $A_0(z) = z^2$, $A_1(z) = 0$ and $A_2(z) = \frac{z}{z+1}$, then the functions A_0, A_1, and A_2 belong to the field $R(\mathbb{C})$ of rational functions. Let's assume that $\theta(z) = \log z$; θ is transcendental over $R(\mathbb{C})$. The relationships from the previous theorem are written in our case:

$$z^2 = \frac{c}{z} + B_1'(z)$$

$$0 = \frac{B_1(z)}{z} + \frac{1}{2} \cdot B_2'(z)$$

$$\frac{z}{z+1} = \frac{B_2(z)}{z} + \frac{1}{3} \cdot B_3'(z).$$

We deduce from the above relationships:

$$c = 0$$
$$B_1(z) = \frac{z^3}{3}$$
$$B_2(z) = -\frac{2z^3}{9}$$
$$B_3(z) = 3z + \frac{2}{9}z^3 - 3\log(z+1).$$

It can be observed that $B_0, B_1, B_2 \in R(\mathbb{C})$, and B_3 is elementary with respect to $R(\mathbb{C})$. This implies that the integral is elementary with respect to $R(\mathbb{C})(\theta)$ and that

$$\int \left(z^2 \cdot \theta^2(z) + \frac{z}{z+1} \right) dz = \frac{1}{3} \cdot z^3 \cdot \log^2 z - \frac{2}{9} \cdot z^3 \cdot \log z - \log(z+1) + z + \frac{2}{27} \cdot z^3.$$

We note that the integral does not belong to $R(\mathbb{C})(\theta)$.

(2) Let's calculate the integral of $\int \theta^n(z) dz$, where $\theta(z) = \log z$. We remark that, in this case, $A_0(z) = 1$ and $A_1(z) = \cdots = A_n(z) = 0$. Then $A_0, A_1, \cdots, A_n \in R(\mathbb{C})$. The relationships from the previous theorem are:

$$1 = \frac{c}{z} + B_1'(z)$$
$$0 = \frac{B_1(z)}{z} + \frac{1}{2} \cdot B_2'(z)$$
$$\cdots\cdots$$
$$0 = \frac{B_n(z)}{z} + \frac{1}{n+1} \cdot B_{n+1}'(z).$$

It follows that $c = 0$, and $B_k(z) = (-1)^{k-1} k! z$, for every $k = 1, \cdots, n+1$. Then $B_{n+1} \in R(\mathbb{C})$ and so the integral belongs to $R(\mathbb{C})(\theta)$:

$$\int \log^n z \, dz = \frac{1}{n+1} \sum_{k=1}^{n+1} (-1)^{k-1} k! \binom{n+1}{k} z \log^{n-k+1} z.$$

4.4 Exercises

(1) Let's show that $\int \dfrac{e^z}{z^2} dz$ is not an elementary function.

   Indication: Let $\theta : \mathbb{C} \to \mathbb{C}$ be defined by $\theta(z) = e^z$; θ is exponential over $R(\mathbb{C})$. Let $L = R(\mathbb{C})(\theta)$ be the Liouville extension of the field $R(\mathbb{C})$ of rational functions and let $f(z) = \frac{\theta(z)}{z^2} \in L$. If we suppose that the antiderivative of f, $g = \int f$, is elementary with respect to L, then, there exist $m \in \mathbb{N}^*$, $u_0, u_1, \cdots, u_m \in L$ and $c_1, \cdots, c_m \in \mathbb{C}$ such that $f = u_0' + \sum_{k=1}^{m} c_k \cdot \frac{u_k'}{u_k}$ and $g = \int f = u_0 + \sum_{k=1}^{m} c_k \log u_k$ is an antiderivative of f. Taking into account the properties of logarithms, it is easily observed that in the previous relation we can choose $u_0 = \frac{P(\theta)}{Q(\theta)}$ and, for every $k = 1, \cdots, m$, $u_k = P_k(\theta)$, where $P, Q, P_k \in R(\mathbb{C})[X]$.

   The left-hand side of the relation $\frac{\theta(z)}{z^2} = \frac{[P(\theta(z))]' Q(\theta(z)) - P(\theta(z))[Q(\theta(z))]'}{Q(\theta(z))^2} + \sum_{k=1}^{n} c_k \frac{[P_k(\theta(z))]'}{P_k(\theta(z))}$ has a double pole at 0; it follows that $Q(\theta(z)) = z$ and that the polynomials P_k are constants, for every $k = 1, \cdots, n$. Consequently, u_0 must be a polynomial in the variable θ, and for any $k = 1, \cdots, m$, u_k is constant. Therefore, the previous relation is rewritten as

$$\theta = u_0'. \tag{4.42}$$

Let $u_0 = \sum_{j=0}^{n} a_j \theta^j$, where, for every $j = 0, \cdots n$, $a_j \in R(\mathbb{C})$. Since $\theta'(z) = 2z\theta(z)$, the relation (4.42) becomes

$$\theta(z) = u_0'(z) = a_0'(z) + \sum_{j=1}^{n} \left[a_j'(z) + 2jz a_j(z) \right] \cdot \theta^j(z). \tag{4.43}$$

θ is transcendental over $R(\mathbb{C})$; so that the relationship (4.43) is an identity in the variable θ. Therefore $n = 1$, $a_0'(z) = 0$ and

$$a_1'(z) + 2z a_1(z) = 1. \tag{4.44}$$

The right-hand side of relation (4.44) has no poles; thus, it follows that $a_1 = P \in R(\mathbb{C})[X]$. However, in this situation, $\deg[P'(z) + 2z P(z)] \geq 1 > 0 = \deg(1)$, and this shows that Eq. (4.44) has no solution. Consequently, $g = \int f$ is not elementary with respect to L.

(2) Let $R(\mathbb{C})$ be the Liouville field of rational functions (see Example 4.1.9) and let $\eta(z) = \log z$. Determine the form of the Liouville field extension $L = R(\mathbb{C})(\eta)$. It is considered the integral

$$F(\theta(z), z) = \int \left\{ \left[\frac{2}{z^2+1} + \frac{1}{z(\eta(z)+1)^2} \right] \theta(z) + \frac{\eta(z)}{(z^2+1)(\eta(z)+1)} \right\} dz,$$

where $\theta(z) = \arctan z$. Check if $F(\theta, \cdot)$ is elementary with respect to $L(\theta)$.

4.4 Exercises

Indication: According to Theorem 4.1.12, $L = R(\mathbb{C})(\eta) = \left\{ \frac{P(\eta)}{Q(\eta)} : P, Q \in R(\mathbb{C})[X] \right\}$. If $A_0(z) = \frac{2}{z^2+1} + \frac{1}{z(\eta(z)+1)^2}$ and $A_1(z) = \frac{\eta(z)}{(z^2+1)(\eta(z)+1)}$, then $p, A_0, A_1 \in L$, where $p(z) = \frac{1}{z^2+1} = \theta'(z)$. Additionally $F(\theta(z), z) = \int [A_0(z)\theta(z) + A_1(z)]dz$. (a) If $\eta(z) = \log z$, then there exist $c = 2 \in \mathbb{C}$, $a(z) = -\frac{1}{\log z+1}$, $A(z) = \arctan z$ such that $a \in L$, $A \in L(\theta)$ and $A_0 = c \cdot p + a'$, $A_1 = a \cdot p + A'$. So that $F(\theta, \cdot)$ is elementary with respect to $L(\theta)$ and $F(\theta(z), z) = \theta^2(z) + a(z)\theta(z) + A(z)$.

(3) Let $L = R(\mathbb{C})$ be the Liouville field of rational functions (see Example 4.1.9) and let $\eta(z) = \arctan z$; then $p = \theta' \in L$. Show that, if $A_0, A_1 \in L$, then $\int [A_0(z)\theta(z) + A_1(z)]dz$ is elementary with respect to $L(\theta)$ if and only if there exist $c \in \mathbb{C}$ and $a \in L$ such that $A_0(z) = \frac{c}{z^2+1} + a'(z)$. Verify if the integrals

$$\int \left(\frac{z^2+2}{z^2+1} \arctan z + \frac{1}{z^2+1} \right) dz, \quad \int \left(\frac{z^2+1}{z^2+2} \arctan z + \frac{1}{z^2+1} \right) dz \text{ are elementary relative to } L.$$

Indications: Under the assumptions of the exercise, the second condition of Theorem 4.3.1 is satisfied. Indeed, since $a \in L$, $A_1 - ap \in L = R(\mathbb{C})$, it therefore admits an elementary antiderivative (see Corollary 4.2.4). Thus, there exists an elementary A with respect to L such that $A_1 = ap + A'$.

Let us now apply this first part of the exercise to the two integrals. In the first case, $A_0(z) = \frac{z^2+2}{z^2+1}$, $A_1(z) = \frac{1}{z^2+1}$. The first relation from Theorem 4.3.1 requires the existence of $c \in \mathbb{C}$ and $a \in L$ such that $\frac{z^2+2}{z^2+1} = \frac{c}{z^2+1} + a'(z)$. It is observed that the relation is satisfied by $c = 1$, $a(z) = z$. $\int [A_0(z)\theta(z) + A_1(z)]dz = \frac{1}{2}\arctan^2 z + (z+1)\arctan z - \frac{1}{2}\log(z^2+1)$. For the second integral $A_0(z) = \frac{z^2+1}{z^2+2}$, $A_1(z) = \frac{1}{z^2+1}$. The relation $\frac{z^2+1}{z^2+2} = \frac{c}{z^2+1} + a'(z)$ is not verified for any $c \in \mathbb{C}$ and $a \in L$. Therefore, the second integral is not elementary with respect to $L(\theta)$.

(4) Let $L = R(\mathbb{C})(\sqrt{1-z^2})$ be the Liouville extension of the field of rational functions, $R(\mathbb{C})$, generated by $\sqrt{1-z^2}$.

  (a) Show that $L = \{\alpha(z) + \beta(z)\sqrt{1-z^2} : \alpha, \beta \in R(\mathbb{C})\}$.

  (b) Show that $\int \left(\frac{z}{\sqrt{1-z^2}} \arcsin z + \sqrt{1-z^2} \right) dz$ is elementary with respect to $L(\theta)$, where $\theta(z) = \arcsin z = -i \log(iz + \sqrt{1-z^2})$ is a logarithmic over L.

Indications: (a) $\eta(z) = \sqrt{1-z^2}$ is an algebraic element over $R(\mathbb{C})$ and the minimal polynomial defining η is $X^2 + (z^2 - 1)$. Theorem 4.1.11 ensures that the Liouville extension of $R(\mathbb{C})$ generated by η is $L = \{P(\eta) : P \in R(\mathbb{C})[X], \deg(P) \leq 1\} = \{\alpha + \beta\eta : \alpha, \beta \in R(\mathbb{C})\}$.

(b) Let's check if the conditions of Theorem 4.3.1 are satisfied for the mentioned integral. $A_0(z) = \frac{z}{\sqrt{1-z^2}}$, $A_1(z) = \sqrt{1-z^2}$, $A_0, A_1 \in L$. $\theta'(z) = p(z) = \frac{1}{\sqrt{1-z^2}}$, $p \in L$. For the integral to be elementary with respect to L, there must exist $c \in \mathbb{C}$, $a \in L$, and A elementary with respect to L such that $\begin{cases} A_0(z) = c \cdot p(z) + a'(z) \\ A_1(z) = a(z) \cdot p(z) + A'(z) \end{cases}$. The first relation is satisfied if we choose $c = 0$ and $a(z) = -\sqrt{1-z^2}$ ($a \in L$). We observe that, if $a \in L$, the second relation is always satisfied. Indeed, in this case $A_1(z) - a(z)p(z) \in L$ and thus $A_1(z) - a(z)p(z) = \alpha(z) + \beta(z)\sqrt{1-z^2}$, with $\alpha, \beta \in R(\mathbb{C})$. It follows that $A(z) = \int (\alpha(z) + \beta(z)\sqrt{1-z^2})dz$ is always elementary with respect to L. In our case $A(z) = \int (1 + \sqrt{1-z^2})dz = z + \frac{1}{2}\arcsin z + \frac{1}{2}z\sqrt{1-z^2}$. It follows that $\int \left(\frac{z}{\sqrt{1-z^2}} \arcsin z + \sqrt{1-z^2} \right) dz = \left(\frac{1}{2} - \sqrt{1-z^2} \right) \arcsin z + z + \frac{1}{2}z\sqrt{1-z^2}$.

References

1. Geddes, K. O., Czapor, S. R., & Labahn, G. (1992). *Algorithms for computer algebra*. Boston: Kluwer Academic Publishers.
2. Ostrowski M. A. (1945–1946). Sur l'integrabilité élémentaire de quelques classes d'expressions. *Commentarii Mathematici Helvetici, 18*, 283–308.
3. Risch, R. H. (1969). The problem of integration in finite terms. *Transactions of the American Mathematical Society, 139*, 167–189.
4. Ritt, J. F. (1948). *Integration in finite terms*. New York: Columbia University Press.
5. Rosenlicht, M. A. (1968). Liouville's theorem on functions with elementary integrals. *Pacific Journal of Mathematics, 24*, 153–161.
6. Rosenlicht, M. A. (1972). Integration in finite terms. *The American Mathematical Monthly, 79*, 963–972.

Index

Symbols
(f, g), 103
$<, \leq$, 14
$R(\mathbb{C})$, 181
$\lfloor x \rfloor$, 5
$\mathbb{N}$, 5, 45
$\mathbb{N}^*$, 5, 46
$\mathbb{Q}$, 5, 58
$\mathbb{Q}_p$, 40
$\mathbb{R}$, 3, 13
$\mathbb{R}_+, \mathbb{R}_-, \mathbb{R}^*, \mathbb{R}_+^*$, 3
$\mathbb{Z}$, 5, 53
$\mathbb{Z}_p$, 36
$f(x) = O(g(x))(x \to x_0)$, 128
$f(x) = o(g(x))(x \to x_0)$, 129
$f(x) \approx \sum_{n=1}^{\infty} a_n f_n(x)(x \to x_0)$, 133
$f(x) \sim g(x)$, 130
$f^n \underset{\mathbb{R}}{\to} f, f^n \to f$, 16
k-contraction, 90

A
Absolute value
　on a field, 25
　p-adic, 28
　trivial, 27
　usual, 25
Archimedean
　absolute value, 27
　norm, 27
　property, 5
Asymptotic approximation, 133
Asymptotic expansion, 133

Asymptotic sequence, 131
Asymptotically equal, 130
Asymptotically equivalent, 130

B
Babylonian algorithm, 97
Binet's formula, 80

C
Casorati determinant, 70
Cassini's identity, 125
Characteristic equation, 72
Characteristic of a field, 8
Complete normed field, 30
Completion of a field, 31
Congruence modulo q, 35
Contraction, 90
Convergent sequence, 16, 25

D
Decimal fraction, 11
Decimal part, 15
Dedekind complete, 2
Difference operator, 65

E
Elementary extension
　of a Liouville field, 187
Elementary function, 187
Elementary Liouville extension, 186

Elliptic integral
  of the first kind, 106
  of the second kind, 114
Equivalent
  absolute values, 26
  norms, 26
Euler's function Γ, 142
Exponential element, 186

F
Fibonacci numbers, 79, 80
Field, 2
  completion, 31
  Dedekind complete, 2
Function
  exponential, 183
  logarithm, 185
  nth root, 185
  power, 185

G
Golden ratio, 79

H
Heron's algorithm, 97, 103
Holomorphic function, 180
Homogeneous recurrence, 88

I
Integer part, 5, 15, 60
Integers, 5, 53
  p-adic, 36
Integral domain, 5

L
Landen transformation, 111
Legendre's formula, 115
Lemniscate of Bernoulli, 119
Liouville
  exponential extension, 186
  extension, 181
  field, 180
  logarithmic extension, 186
  principle, 189
Logarithmic element, 186
Lower decimal approximations, 18

M
Mean, 98
  arithmetic, 101
  contraharmonic, 101
  generated by two means, 103
  geometric, 101
  harmonic, 101
  logarithmic, 101
  of Neuman—Sándor, 101
  of Schwab-Borchardt, 101, 104
  quadratic, 101
Method of
  Lagrange, 81
  variation of parameters, 81
Metric induced
  by a norm, 25

N
Natural numbers, 4, 45
Newton's algorithm, 92
Non-Archimedean
  absolute value, 27
  norm, 27
Norm
  on a field, 25
  p-adic, 28
  trivial, 27
  usual, 25

O
Order big O, 128
Ordered field, 2
Order little o, 129

P
Peano's axioms, 46
Poincaré's theorem, 163
Power means, 99
Prime field, 8
Product of two real numbers
  in the decimal model, 21

R
Rational numbers, 5, 58
Real numbers, 3
  in decimal model, 13
Riccati recurrence, 89

S
Sequence
  convergent, 6
  Cauchy, 6, 30
  of successive approximations, 91

Index

Singular point
    essential, 180
    isolated, 180
    pole, 180
    removable, 180
Stirling's formula, 159, 162
Sum the two real numbers
    in the decimal model, 19

T
Theorem
    contraction principle, 90
    of Gauss, 106
    monotone convergence, 17
    of Ostrowski, 32
    representation of
        p-adic integers, 39
        p-adic numbers, 41

U
Upper decimal approximations, 18

W
Watson's lemma, 152
Well-ordering principle, 5

SPRINGER NATURE

GPSR Compliance

The European Union's (EU) General Product Safety Regulation (GPSR) is a set of rules that requires consumer products to be safe and our obligations to ensure this.

If you have any concerns about our products, you can contact us on ProductSafety@springernature.com

In case Publisher is established outside the EU, the EU authorized representative is:

Springer Nature Customer Service Center GmbH
Europaplatz 3
69115 Heidelberg, Germany

The manufacturer's authorised representative in the EU is Springer Nature Customer Service Centre GmbH, Europaplatz 3, 69115 Heidelberg, Germany. If you have any concerns regarding our products, please contact ProductSafety@springernature.com

Printed and bound by CPI Group (UK) Ltd, Croydon, CR0 4YY

26/03/2026

02078942-0007